DEPARTMENT

THE

Please
below
Boc'

Biological Solar Energy Conversion

Biological Solar Energy Conversion

Edited by

AKIRA MITSUI
Rosenstiel School of Marine and Atmospheric Science
University of Miami, Miami, Florida

SHIGETOH MIYACHI
Institute of Applied Microbiology
University of Tokyo, Bunkyo-ku, Tokyo

ANTHONY SAN PIETRO
Department of Plant Sciences
Indiana University, Bloomington, Indiana

SABURO TAMURA
Department of Agricultural Chemistry
University of Tokyo, Bunkyo-ku, Tokyo

Academic Press NEW YORK SAN FRANCISCO LONDON 1977

A Subsidiary of Harcourt Brace Jovanovich, Publishers

ACADEMIC PRESS, INC.
111 Fifth Avenue, New York, New York 10003

United Kingdom Edition published by
ACADEMIC PRESS, INC. (LONDON) LTD.
24/28 Oval Road, London NW1

Library of Congress Cataloging in Publication Data

Main entry under title:

Biological solar energy conversion.

    Papers presented at a conference held at the
Rosenstiel School of Marine and Atmospheric Science,
University of Miami, Nov. 15-18, 1976, under the
sponsorship of the United States–Japan Cooperative
Science Program, U.S. National Science Foundation, and
the Japanese Society for Promotion of Science.
    1.   Primary productivity (Biology)–Congresses.
2.   Photosynthesis–Congresses.   I.   Mitsui, Akira.
II.   United States–Japan Cooperative Science Program.
III.   United States.   National Science Foundation.
IV.   Nippon Gakajutsu Shinkókai.
QH91.8.P7B56   581.1'9121   77-5373
ISBN 0-12-500650-0
PRINTED IN THE UNITED STATES OF AMERICA

Contents

Section I   Capture and Utilization of Solar Energy

Section II   Synthesis of Organic Compounds from Carbon Dioxide

Section III   Nitrogen Fixation and Production of Single Cell Protein

Section IV   Practical Applications

Seminar Participants

T. **Akazawa**, Research Institute for Biochemical Regulation, School of Agriculture, Nagoya University, Chikusa, Nagoya, 464 Japan

J. A. **Bassham**, Lawrence Berkeley Laboratory, University of California, Berkeley, California 94720

N. I. **Bishop**, Department of Botany & Plant Pathology, Oregon State University, Corvallis, Oregon 97331

R. H. **Burris**, Department of Biochemistry, University of Wisconsin, Madison, Wisconsin 53706

M. **Gibbs**, Institute for Photobiology of Cells and Organelles, Brandeis University, Waltham, Massachusetts 02154

J. **Goldman**, Woods Hole Oceanographic Institute, Woods Hole, Massachusetts 02543

E. **Greenbaum**, Department of Biophysics and Biochemistry, The Rockefeller University, New York City, New York 10021

R. W. F. **Hardy**, Central Experiment Station, E. I. DuPont de Nemours, Wilmington, Delaware 19898

F. **Hayashi**, Criteria and Evaluation Division, Office of Pesticide Programs, U.S. Environmental Protection Agency, Washington, D.C. 20460

U. **Horstmann**, Marine Research Station, University of San Carlos, Cebu City, Philippines, (University of Kiel, Kiel, West Germany)

Y. **Inoue**, Laboratory of Plant Physiology, The Institute of Physical and Chemical Research, Wako-shi, Saitama, 351 Japan

R. **Ishii**, Faculty of Agriculture, University of Tokyo, Bunkyo-ku, Tokyo, Japan

G. A. **Jackson**, Woods Hole Oceanographic Institute, Woods Hole, Massachusetts 02543

B. **Kok**, Research Institute for Advanced Studies, Baltimore, Maryland 21227

A. I. **Krasna**, Department of Biochemistry, Columbia University, College of Physicians and Surgeons, New York, New York 10032

S. Kurita, Research Institute, Ishikawajima-Harima Heavy Industries Company, Ltd., Tokyo 135-91, Japan

E. S. Lipinsky, Batelle Columbus Laboratories, Columbus, Ohio 43201

A. Mitsui, Rosenstiel School of Marine and Atmospheric Science, University of Miami, Miami, Florida 33149

S. Miyachi, Institute of Applied Microbiology, University of Tokyo, Bunkyo-ku, Tokyo, Japan

J. Myers, Department of Zoology, University of Texas, Austin, Texas 78712

Y. Nakamura, Radioisotope Center, Faculty of Science, University of Tokyo, Bunkyo-ku, Tokyo, Japan

M. Nishimura, Department of Biology, Faculty of Science, Kyushu University, Fukuoka 812, Japan

W. J. North, Environmental Engineering Science, California Institute of Technology, Pasadena, California 91125

J. Ooyama, Fermentation Research Institute, Agency of Industrial Science and Technology, Ministry of International Trade and Industry, Inage, Chiba-city, Japan

W. J. Oswald, Department of Civil Engineering, University of California, Berkeley, California 94720

A. San Pietro, Department of Plant Sciences, Indiana University, Bloomington, Indiana 47401

G. H. Schmid, Max Planck Institut für Zuchtungsforchung, 5, Koln, 30 West Germany

K. Shibata, Laboratory of Plant Physiology, The Institute of Physical and Chemical Research, Wako-shi, Saitama 351, Japan

S. Tamura, Department of Agricultural Chemistry, University of Tokyo, Bunkyo-ku, Tokyo, Japan

A. Tanaka, Faculty of Agriculture, Hokkaido University, Sapporo, Japan

H. Teas, Department of Biology, University of Miami, Coral Gables, Florida

R. K. Togasaki, Department of Plant Sciences, Indiana University, Bloomington, Indiana 47401

N. E. Tolbert, Department of Biochemistry, Michigan State University, East Lansing, Michigan 48824

R. C. Valentine, Department of Agronomy and Range Science, University of California, Davis, California 95616

A. Watanabe, Seijo University, Setagaya-ku, Tokyo, Japan

D. L. Wise, Biochemical Engineering, Dynatech R/D Company, Cambridge, Massachusetts 02139

T. Yagi, Department of Chemistry, Shizuoka University, Shizuoka 422, Japan

O. Zaborsky, Division of Advanced Energy and Resources Research and Technology, National Science Foundation, Washington, D.C. 20550

I. Zelitch, Department of Biochemistry, Connecticut Agricultural Experiment Station, New Haven, Connecticut 06504

Preface

A conference on Biological Solar Energy Conversion was held at the Rosenstiel School of Marine and Atmospheric Science, University of Miami, on November 15-18, 1976, under the sponsorship of the United States-Japan Cooperative Science Program, U.S. National Science Foundation, and the Japanese Society for Promotion of Science. This volume consists of the formal papers presented during the conference.

Solar energy has now been duly recognized as our largest and ultimate nonfossil, nonnuclear energy resource. Mankind is totally dependent on this energy supply for his continued existence. Unfortunately, man has not yet learned to harness solar energy on a scale commensurate with his ever increasing energy requirements. On the other hand, the enormous magnitude of the solar radiation that reaches the land surfaces of the earth is so much greater than any of the foreseeable needs that it represents an inviting technical target. In planning this conference, we attempted to bring together a group of scientists who have made significant observations concerned with various aspects of Biological Solar Energy Conversion. The excellent exchange of information and ideas among the participants will surely further our understanding of these biological processes and lead hopefully to future practical application for the benefit of mankind.

We thank most sincerely Dr. William W. Hay, Dean, Rosenstiel School of Marine and Atmospheric Science, University of Miami, for his support and encouragement of this conference. It is a pleasure to acknowledge the kindness of Dr. Harris Stewart, Director, Atlantic Oceanographic and Meteorological Laboratories, NOAA, for providing space in his institute for the conference sessions. We express also our sincere gratitude to those participants who willingly served as Chairmen for the various sessions. The smooth running of the conference was due in great measure to the untiring efforts of the students and assistants in Dr. A. Mitsui's laboratory. Ms. Diana Rosner and Ms. JoAnn Radway of the University of Miami provided excellent editorial help. Lastly we acknowledge partial support for the conference from the U.S. Government Energy Research and Development Agency.

SECTION I. Capture and Utilization of Solar Energy

PHOTOHYDROGEN PRODUCTION IN GREEN ALGAE: WATER SERVES AS THE
PRIMARY SUBSTRATE FOR HYDROGEN AND OXYGEN PRODUCTION

by

Norman I. Bishop and Marianne Frick
Department of Botany and Plant Pathology
Oregon State University

and

Larry W. Jones
Department of Botany, University of Tennessee
Knoxville, Tennessee

I.   INTRODUCTION

The impact of the successful oil embargo as experienced by
the United States in 1973 was sufficiently severe to cause an increased awareness of and interest in possible alternate sources
of energy, including the bio-solar energy conversion device of
photosynthesis (NSF/NASA Report, 1972; NSF/RANN Report, 1973a;
NSF/RANN Report, 1973b; NSF/RANN Report, 1975). Nature's primary
process for the conversion of light energy to a biologically useful chemical form is photosynthesis. Obviously it is the method
which provided energy for the development of the earth's fossil
fuel supply and continues to provide the yearly organic energy
supply and reserves for the global population. Because of an
ever-increasing demand for all forms of energy throughout the
world it has become necessary to consider the possibilities of
(1) increasing the productivity of agriculture (the so-called
Green Revolution), (2) exploring the potential of increasing the
basic effeciency of the fundamental mechanism of photosynthesis
and (3) attempting to devise alternate mechanisms for the utilization of the primary stable photo-reductant generated in photosynthesis in a way more direct than that represented by the mechanism for carbon dioxide assimilation in green plant photosynthesis.
It is the primary purpose of this article to examine the last item
listed above in terms of the in vivo capacity of the green algae
to perform the light-dependent process of hydrogen production;
this reaction represents a natural process for utilization of the
energy provided by the generation of the primary photoreductant
of photosynthesis. If assumed that water serves as the electron
donor for this reaction then the photoevolution of hydrogen should
be comparable in magnitude to that of carbon dioxide assimilation
or oxygen evolution.

II.   HYDROGEN PRODUCTION BY GREEN ALGAE

During studies on the anaerobic metabolism of the green alga, Scenedesmus obliquus, Gaffron and Rubin (1942) noted that thoroughly adapted cells in the absence of both hydrogen and carbon dioxide produced hydrogen gas when illuminated. This phenomenon, since termed photohydrogen production, was separable from a much slower dark fermentative hydrogen metabolism through the action of dinitrophenol; this inhibitor caused complete suppresion of the dark hydrogen evolution and apparent stimulation of photohydrogen evolution (Gaffron and Rubin, 1942; Gaffron, 1944). They recognized that this form of metabolism was unique, that it was dependent upon the adaptable hydrogenase of Scenedesmus and that it was most likely representative of an anerobic photooxidation of some unknown intermediate formed in fermentation. Because of the esoteric nature of the problem, because of the apparent miniscule rates of hydrogen evolution and because of the inability of anerobically adapted cells to demonstrate a sustained production of hydrogen only limited additional studies were conducted during the intervening years and principally by subsequent student of Gaffron. The recent focusing of attention upon the potential of green algae to produce hydrogen photochemically has revived interest in this area as well as in the similar phenomenon, but not identical, in nitrogen fixing blue-green algae and photosynthetic bacteria. With the renewed interest it has been possible to ask additional questions about the precise mechanism for photohydrogen formation and to determine with some precision the role of the photosystems and the nature of the electron donor.

It is now apparent that the reactions catalyzed by anaerobically adapted algal cells which possess an adaptable hydrogenase are more numerous than originally suspected. These reactions include the following:

Light-Dependent Reactions:

1.   Photosynthesis:

$$CO_2 + 2H_2O + light \longrightarrow (CH_2O) + O_2 + H_2O$$

2.   Photoreduction:

$$CO_2 + 2H_2 + light \xrightarrow{H_2ase} (CH_2O) + H_2O$$

3. H_2 Photoproduction:

$$2H_2O + light \xrightarrow{H_2ase} 2H_2 + O_2$$

Dark Reactions:

4. Oxy-hydrogen Reactions:
$$2H_2 + O_2 \xrightarrow{H_2ase} 2H_2O$$

5. Dark CO_2 Fixation:
$$CO_2 + 2H_2 + energy \xrightarrow{H_2ase} (CH_2O) + H_2O$$

6. H_2 Production: $RH_2 \xrightarrow{H_2ase} R + H_2$

7. H_2 Uptake: $R + H_2 \xrightarrow{H_2ase} RH_2$

8. Respiration: $RH_2 + \frac{1}{2}O_2 \longrightarrow R + H_2O$

The increased complexity of the general "anaerobic" metabolism of algal cells resulting from the activation of the hydrogenase system, as summarized above, must be thoroughly evaluated and
understood in order to explain the many factors influencing photohydrogen production. The inhibition of reactions 1 and 3, for
example by dinitrophenol would markedly simplify the subsequent
reaction sequence. Similarly the presence of DCMU, a known inhibitor of PS-II activity and of oxygen evolution would allow only
reaction 2 to proceed in adapted cells.

Although the presence of an adaptable hydrogenase and the
ability of algal cells to perform photoreduction and the several
associated reactions listed above was originally analyzed for in
a restricted few species, notably Scenesdesmus obliquus, strain
D_3, it is now recognized that numerous additional species of algae
possess an adaptable hydrogenase and an anaerobic physiology comparable to Scenedesmus (Kessler, 1974). Recently, we have examined over 100 additional species from ten classes of algae for
their capacity for photoreduction and photohydrogen production.
Nearly all of the species possessing an adaptable hydrogenase were
members of the class Chlorophyceae and within either the order
Volvocales or Chlorococcales. The majority of species having the
potential for hydrogen photoproduction were found in the latter
order and included a number of species of Chlorella. However,
only two of these, Kirchneriella lunaris and Coelastrum probosci-
deum had a capacity for photohydrogen production greater than that
observed for Scenedesmus (Table 1). In Table I we have summarized
the observed rates of hydrogen and oxygen evolution noted in some
of the species studied. These values were obtained with the two
electrode system as described by Jones and Bishop, 1976; this
methodology allowed for an assessment of the immediate rates of
hydrogen metabolism and, consequently, only those values are presented in the data of Table I. It is important to stress that
these values do not represent sustained rates and, furthermore,
are not necessarily maximized values for hydrogen evolution.
Typical traces demonstrating the nearly simultaneous production

TABLE I
Rates of Hydrogen and Oxygen Evolution in Anaerobically Adapted
Cultures of Various Algal Species Containing Hydrogenase

Species	umoles H_2 ml cells/ hr	umoles H_2 mg Chl/ hr	umoles O_2 ml cells/ hr	umoles O_2 mg Chl/ hr
Scenedesmus obliquus (heterotrophic)	208.1	48.1	116.6	27.0
Scenedesmus obliquus (autotrophic)	128.8	20.2	134.1	15.9
Kirchneriella lunaris	204.5	32.1	89.0	14.0
Selenastrum sp.	75.2	11.8	112.5	17.6
Coelastrum proboscideum	156.4	24.6	145.8	22.9
Ankistrodesmus braunii	70.8	11.1	135.2	21.6
Ankistrodesmus falcatus	1.7	0.2	19.4	3.0
Chlorella vacuolata 211-8a	96.0	15.1	64.2	10.1
Chlorella vacuolata 211-8c	8.8	1.4	50.0	7.8
Chlorella protothecoides 211-8d	104.4	16.4	103.0	16.2
Chlorella vulgaris 211-11c	18.1	2.8	43.2	6.8
Chlorella sorokiniana 211-11d	92.4	14.5	184.2	28.9
Chlorella sorokiniana 211-11k	47.4	7.4	40.1	6.3
Chlorella sorokiniana 211-32	59.5	9.2	51.6	8.1
Chlorella sorokiniana 211-33	39.3	6.2	34.4	5.4
Chlorella sorokiniana 211-34	55.4	8.7	61.6	9.6
Chlorella fusca 343	14.2	2.2	60.6	9.5

All rates were determined during the initial one minute of illumination with corrections made for the subsequent dark uptake of either hydrogen or oxygen. All measurements made with 10 ul PCV resuspended in 1 ml of phosphate buffer, pH = 6.5. Temperature = 25 C. Adaptation time = 4 hours.

of hydrogen and oxygen in Scenedesmus and several Chlorella species are presented in Figures 1 and 2 respectively. From these traces it is obvious that the production and accumulation of oxygen in the reaction cell leads to a rapid inactivation of the hydrogen photoevolving capacity of all of the algae when the experiments are performed under either high light intensity or for prolonged periods of time.

A.   The Role of PS-I in Photohydrogen Production.

Since the early 1960's the basic mechanism of photosynthesis has been recognized to require two photosystems, photosystem I (PS-I) and photosystem II (PS-II), which are functionally connected by an intersystem electron transport system. This formulation in which water serves as the electron donor and NADP$^+$ as the terminal electron acceptor for these two photosystems represents the most generally accepted mechanism for photosysthesis of green plants. The experimental findings culminating in the elaboration of this formulation have had extensive reviewing (Boardman, N. K., 1970; Bishop, N.I., 1971a; Avron, M., 1967; Bishop, N.I., 1973; Brown, J.S., 1973; Trebst, A., 1974; Cheniae, G.M., 1970) and formulations representing this mechanism, i.e., the so-called Z-scheme of photosynthesis, occur in practically all general biology, botany and biochemistry textbooks.

The anaerobic, DCMU insensitive, assimilation of carbon dioxide and hydrogen gas by adapted cells of Scenedesmus, the process termed photoreduction (Gaffron, 1940), was shown to be an exclusive PS-I type reaction by Bishop and Gaffron (1962). In this reaction H$_2$ gas, through the intervention of the alga's hydrogenase system, serves as the source of electron for the photoreduction of carbon dioxide and is, in a qualified sense, similar to the process of carbon dioxide assimilation in bacterial photosynthesis. The initial attempts to determine if photohydrogen production were also a PS-I catalyzed reaction revealed that both photosystems were apparently involved (Bishop and Gaffron, 1963). Since both photoreduction and photohydrogen production require similar circumstances for their activation and function and, furthermore, appear to compete for the primary photoreductant generated by PS-I, the apparent participation of PS-II activity for hydrogen photoproduction seemed anamolous. However, if water serves as the electron donor for this reaction, as was originally suggested by Gaffron and Rubin, then the participation of both photosystems would be mandatory. The possibility of water serving as the substrate for hydrogen formation has been, and continues to be, a point of controversy. Additional lines of evidence confirming the requirement of at least PS-II, if not water photolysis, in hydrogen evolution will be discussed later.

The clearest lines of evidence demonstrating the absolute requirement for PS-I in photohydrogen evolution was obtained by mutational studies with Scenedesmus. Data summarizing the consequence of the loss of either P-700, cytochrome f or cytochrome

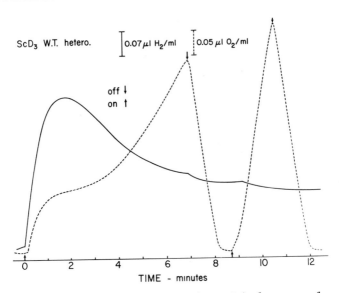

Figure 1. *Simultaneous production of hydrogen and oxygen by anaerobically adapted cultures of* Scenedesmus obliquus *measured in continuous light. Temperature = 25 C. Light intensity = 3 x 10^5 erg cm^{-2} sec^{-1}. Packed cell volume= 10ul.*

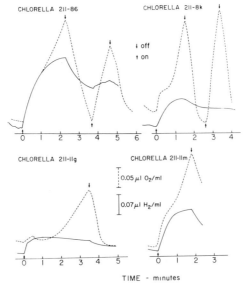

Figure 2. *Simultaneous production of hydrogen and oxygen by anaerobically adapted cultures of four Chlorella strains measured in continuous light. Experimental conditions as listed in Figure 1.*

TABLE II

WHOLE CELL AND CHLOROPLAST REACTIONS OF P-700 AND CYTOCHROME
DEFICIENT MUTATIONS OF SCENEDESMUS OBLIQUUS*

Mutant Strain	Photo-synthesis	P-700 Deficient Photo-reduction	Photo-hydrogen	DPC-DCPIP
PS-21	0	0	0	52.6
PS-24	5.6	4.5	2	75.7
PS-30	0	0	0	64.5
PS-46	5.7	2.8	3	63.2
PS-8	8.3	3.4	1	52.7
		Cytochrome Deficient		
PS-34	5.3	2.3	5.1	57.8
PS-50	6.8	3.5	1	71.2
PS-123	5.1	2.1	1	52.2
PS-134	1.4	tr	0	35.8
PS-141	10.4	3.5	8.9	92.8

* All values given are in percentages of the respective rates of comparable measurements performed on whole cells or chloroplasts of the wild-type strain. These values are: photosynthesis, 39.7 u moles O_2/hr - mg chlorophyll; photoreduction, 17.2 u moles CO_2/hr - mg chlorophyll; photohydrogen 50 u moles H_2/hr - mg chlorophyll; Diphenylcarbazide-dichlorophenolindolephenol, 90 u equivalents/hr - mg chlorophyll.

b-563 upon photohydrogen evolution and a variety of other partial reactions of whole cells and isolated chloroplast particles are shown in Table II.   Originally Bishop and Gaffron (1963) noted that mutant PS-8 of Scenedesmus was deficient in the capacity for photohydrogen evolution; concurrently, it was observed that this mutant lacked detectably P-700 (Weaver and Bishop, 1963; Butler and Bishop, 1963; Beinert and Kok, 1963). As indicated in Table II this mutant also showed decreased capacity for photohydrogen evolution. A number of additional mutants have been isolated which also lack P-700 (determined on the basis of light or chemically induced absorbancy changes at 697 nm and by SDS-PAGE); like mutant 8 these strains also do not evolve hydrogen in the light but, in general, show appreciable PS-II activity either in whole cell or chloroplast reactions (Table II).   The P-700 deficient mutants all possess an active ferredoxin and hydrogenase.

Also from the data of Table II it is clear that the genetic deletion of cytochrome f-553 (Bishop, 1971b) inhibits photosynthesis, photoreduction and photohydrogen evolution without strong inhibition of PS-II reactions. $HgCl_2$, which is a known inhibitor of the activity of plastocyanin (Katóh, 1972; Kimimura and Katoh, 1972, 1973), prevents hydrogen evolution (Bishop and Frick, unpublished results). DBMIB, a synthetic antagonist of the electron transport potential of plastoquinone, inhibits photohydrogen evolution in addition to its inhibitory effects upon PS-II type reactions (Ben-Amotz and Gibbs, 1975). Confirmatory evidence for the action of DBMIB upon the anaerobic hydrogen metabolism in Scenedesmus is shown in Figure 3; concentrations less than 5 x 10^{-6} M cause greater than 50% inhibition. The efficiency of this compound as an inhibitor is comparable to that of DCMU (See data of Figure 7).

From the results obtained with the various mutation types and inhibitors examined it is apparent that an intact and functional PS-I is an absolute prerequisite for the light-dependent hydrogen metabolism both in algae and in coupled chloroplast reactions leading to hydrogen formation. That P-700, in combination with an appropriate electron donor, upon its photoreduction drives the formation of the primary photoreductant (in this case either reduced ferredoxin or reduced NADP) which then interacts through the hydrogenase to form hydrogen gas appears to be a logical formulation. The requirement for an intact photosynthetic electron transport system to at least the level of plastoquinone indicates that the natural electron donor for hydrogen formation must enter into the in vivo mechanism at a point prior to plastoquinone.

B. Independence of Photohydrogen Evolution from an ATP-Dependent Reverse Electron Flow.

While it is clear that light mediates an electron flow from either an in situ organic hydrogen donor and/or water, it is also equally apparent that the ATP-dependent hydrogenase, which is involved in the hydrogen and nitrogen metabolism of the blue-

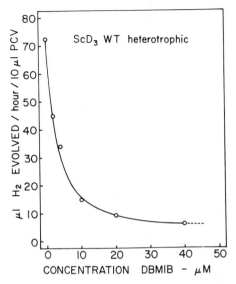

Figure 3.   Inhibition of photohydrogen evolution by the plastoquinone antagonist dibromo-methyl-isopropyl benzoquinone (DBMIB).   Temperature = 25 C.   Light intensity = 3 x 10^5 erg cm^{-2} sec^{-1}.

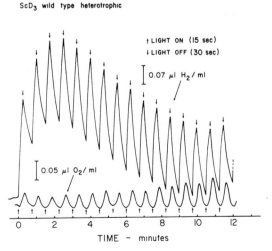

Figure 4.   The pattern of hydrogen and oxxgen production and consumption seen with the two electrode system and with modulated light and dark periods.   Experimental conditions as indicated in Figure 1.

green algae and photosynthetic bacteria, is not required for
photohydrogen evolution in green algae. The early observations
that dinitrophenol stimulates photohydrogen production but inhibits dark hydrogen fermentation reactions has been interpreted
by many as an additional classic example of uncoupling of electron
transport from phosphorylation. However, Gaffron and Rubin (1942)
offered a more appropriate interpretation in suggesting that the
observed augmentation of rates of hydrogen evolution by DNP was
due to the elimination of the two competitive reactions, photosynthesis and photoreduction. Subsequent studies on the action
of Cl-CCP on photosynthesis, photoreduction and photohydrogen
evolution (Bishop and Gaffron, 1963; Kaltwasser, Stuart and
Gaffron, 1969; Stuart and Kaltwasser, 1970; Stuart and Gaffron,
1971; and Stuart and Gaffron, 1972a and 1972c) revealed that the
alternate interpretation for the action of phosphorylation uncouplers was the more appropriate one. By inhibition of cyclic
photophosphorylation those reactions requiring ATP are prevented
and, consequently, more of the primary photoreductant is utilized
for hydrogen production (See Section II, reactions 1-8). Consideration of the various reactions which can be performed by anaerobically adapted cells of Scenedesmus reveals that the loss of
photosynthesis and photoreduction removes not only systems which
compete strongly for the primary photoreductant but two systems
which consume hydrogen gas (reactions 2 and 4) and also one which
produces oxygen gas (reaction 1), a classic inhibitor of hydrogenase activity.

The failure of uncouplers of phosphorylation to inhibit
photohydrogen evolution in the green algae underlines the differences in mechanisms for hydrogen evolution by this system and
that of the blue-green algae and photosynthetic bacteria. Earlier
Bishop and Gaffron (1963) stressed that the two then-known systems for producing hydrogen photochemically utilized different
mechanisms. At that time it was not apparent from the work of
Gest and his collegues (Gest et al, 1962; Ormerod and Gest, 1962)
that the ATP-dependent nature of hydrogen evolution in the photosynthetic bacteria was associated with their nitrogen fixing
capacity. Progress in this area now allows for a clearer dis-
tinction to be made between systems utilizing the ATP-dependent
and ATP-independent hydrogenase systems.

C.   Participation of PS-II in Photohydrogen Production.

In their original publication Gaffron and Rubin proposed that
the source of hydrogen was from either an organic hydrogen donor
or from water. Their predilection was toward the former interpretation because of their observations on the dark-hydrogen fermentation which appeared to result from a dehydrogenation of an organic donor system. In an elegant series of experiments Spruit
(1954, 1958) found that under the conditions required for photohydrogen evolution a simultaneous production of hydrogen and oxygen could be measured in a suspension of Chlorella cells. From
this observation, plus his measurements on the stoichiometry of

hydrogen and oxygen produced, Spruit favored the concept that the primary source of electrons for photohydrogen production was water. He recognized that because of the anaerobic circumstances of the algal cells that any oxygen produced by them would be rapidly scavenged by respiration and by the oxy-hydrogen reactions. He noted also the rapid dark uptake of hydrogen gas following a period of photohydrogen production (See Section II, reactions 4, 7 and 8). Combinations of these reactions would easily result in erroneous values for the stoichiometry and that only under ideal-ized conditions would it be possible to measure a value of two, i.e.,

$$2H_2O + light \xrightarrow{\text{PS-I and PS-II}}_{H_2ase} 2H_2 + O_2$$

We have repeated and extended the observations of Spruit with a two electrode system designed to provide simultaneous measurements on hydrogen and oxygen with a liquid phase system (Jones and Bishop, 1976).

In order to obtain a prolonged production of hydrogen gas and to establish conditions for a realistic evaluation of the stoichiometry of hydrogen and oxygen production it was necessary to perform experiments in periodic illumination where the length of the light period was sufficient to produce measurable amounts of both hydrogen and oxygen but where the accumulated oxygen con-centrations was below that which was inhibitory to the hydrogen-ase. A typical response of the algal system to 15 second illum-ination followed by 30 seconds of dark is illustrated in Figure 4. It can be seen that after the first light flash the amount of hydrogen and oxygen evolved per flash remains almost constant for about 20 flashes. Continuation of this regime eventually leads to inactivation of the system since the yield of oxygen per flash gradually increases. Modulation of either the light intensity of the flash or of the time between flashes allows for the appropri-ate circumstances for measuring hydrogen and oxygen stoichiometry. Data illustrating the dramatic influence of both of these para-meters are shown in Figures 5 and 6. The obtainment of constant rates of hydrogen and oxygen production per flash throughout at least 10 light flashes (Figure 4) provided data for a critical assessment of the stoichiometry over an anaerobic adaptation period of 8 hours (Table III). Under the experimental conditions employed for this experiment it is apparent that a value of 2 was approached only after about 6 hours of adaptation.

Under the experimental conditions employed for the ampero-metric determination of hydrogen and oxygen the normal kinetics of production and oxygen are quite different from those observed in the flashing light regime. When suspensions of cells are exposed to saturating white light a pattern like that shown pre-viously in Figure 1 is seen. Hydrogen gas appears immediately and after about a 7 sec delay oxygen begins to be detected. This delay is important and indicates, we believe, the internal con-sumption of oxygen by the reactions presented in Section II.

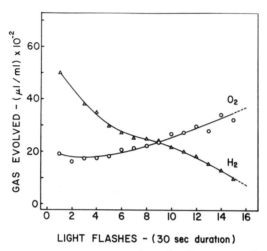

Figure 5. *Measurements on the simultaneous photoproduction of hydrogen and oxygen with 30 sec light and 30 sec dark periods with cultures of Scenedesmus adapted for 4 hours. White light intensity = 8.3 x 10⁴ erg cm⁻² sec⁻¹.*

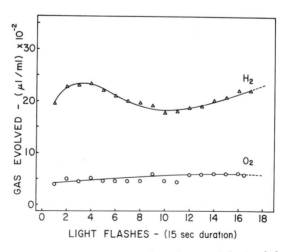

Figure 6. *Measurements as in Figure 5 but with a light-dark regime of 15 sec light - 30 sec dark.*

TABLE III

Changes in the Rates of Hydrogen and Oxygen Evolution and H_2/O_2
Stoichiometry During Anaerobic Adaptation of Scenedesmus

Time of Adaptation (hours)	umoles H_2 / hr	umoles O_2 / hr	$\frac{H_2}{O_2}$
1	1.81	1.97	0.92
2	2.94	2.29	1.28
4	3.23	2.42	1.33
6	3.89	2.17	1.79
8	4.23	2.18	1.94

*   *Values are calculated during the first one minute of
illumination with corrections made for the observed rate
of hydrogen and oxygen consumption observed in the following
dark period. All measurements made with 10 ul PCV in 2 ml
K_2H - KH_2PO_4 buffer. pH = 6.5.
Light intensity = 3 x 10^5 ergs/sec cm^2.*

The biphasic nature of the oxygen curve and the peaking of the
hydrogen curve represent both the activity of the oxy-hydrogen
reaction and the inactivation of the hydrogenase as a result of
the increased amount of oxygen. This inactivation is noted by
the failure of the cells to generate hydrogen during the second
period of illumination. The patterns shown in Figures 1 and 2
for the production and consumption of the two gases can be drastically changed by manipulation of parameters influencing the
availability of oxygen. For example, the addition of dithionite
to adapted algal cells prevents the appearance of oxygen and
stimulates the rate of hydrogen evolution by removing the competition introduced by the presence of oxygen. Dithionite does not
appear to act as an additional electron donor since only the rate
of hydrogen evolution is affected and not the final yield. Other
electron acceptors, such as nitrite and nitrate, inhibit hydrogen
production and stimulate oxygen production and, necessarily, the
inactivation of the hydrogenase (Bishop, et al, 1976).
     The obvious complexity of a system in which the primary
photoreductant can be utilized either to form hydrogen gas through
the mediation of the hydrogenase or to react directly with oxygen,
where both products of the light reaction, H_2 and O_2, can recom-
bine through the oxy-hydrogen reaction, where both gases can be
reabsorbed directly by the algae (reactions 7 and 8, Section II)
and where the essential enzyme, hydrogenase, is inactivated by

the oxygen produced by the system all serve to explain the variability of the stoichiometry. However, when these problems are recognized and the appropriate experiment is performed the predicted stoichiometry can be measured. But are such values fortuitious? Might they not arise from restraints placed upon the measurements and result from a simultaneous dehydrogenation of an organic hydrogen donor and normal photosynthetic generation of oxygen? These questions, plus many other unposed ones, are pertinent and need extra evaluation before an absolutely definitive answer can be provided.

However, it is relevant to ask whether the machinery of PS-II is required in photohydrogen production. We have indicated earlier that the light-dependent evolution of hydrogen appeared to have a requirement for both photosystems (Bishop and Gaffron, 1963). Although this observation has not been repeated or extended the findings remain pertinent since it has also been shown that inhibitors of PS-II activity, i.e., DCMU, hydroxylamine, and simazine, are equally effective against photohydrogen production. Although the ability of DCMU to inhibit this reaction has been a controversial subject (Healey, 1970; Stuart and Kaltwasser, 1970; Stuart and Gaffron, 1972a), it is now clear that its effectiveness can be variable. This variability is traceable, in part, to the algal species employed. whether the culture type utilized was obtained by photoautotrophic. heterotrophic or mixotrophic growth and whether inhibitory effects upon the initial rates of hydrogen production (the so-called fast phase) or upon the prolonged but slowed phase of hydrogen production (Stuart and Gaffron, 1971) are being described. A typical response of photohydrogen production by cells of <u>Scenedesmus</u> <u>obliquus</u>, obtained by either photoautotropic or heterotrophic growth, to DCMU concentration is shown in Figure 7. Clearly the DCMU exhibits greater inhibition with the heterotrophic cultures but the concentration range within which this compound is effective is approximately the same for both culture types. Nearly identical results have been obtained with comparable culture types of <u>Kirchneriella</u> <u>lunaris</u>, <u>Chlorella</u> <u>fusca</u>, <u>Selenastrum</u> sp. and numerous additional species of green algae (Bishop et al, 1976). With either photoautotrophic or mixotrophic cultures it is also of importance to evaluate the stage of the cell cycle at which cultures are selected for measurements. For example, synchronous cultures of Scenedesmus show a differential sensitivity of their photosynthesis to DCMU which is apparently dependent upon permeability characteristics of the different cell types formed during the algal cell cycle (Senger and Frickel-Faulstitch, 1975).

As part of the initial application of photosynthetic mutants of Scenedesmus to studies on the mechanism of photohydrogen evolution Bishop and Gaffron (1963) presented evidence that mutations blocked within PS-II, although having normal photoreductive capacities, were unable to sustain photohydrogen production. For reasons yet not clear, it was later reported that one of the PS-II

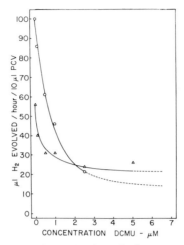

Figure 7. Inhibition of photohydrogen evolution by DCMU in photoautotrophic and heterotrophic cultures of Scene-desmus. Rates determined from inital slopes as shown in Figure 1. o —— o = heterotrophic; △——△ = autotrophic. Experimental conditions as indicated in Figure 1.

mutants employed in the earlier studies, mutant strain PS-11 had a normal capacity for photohydrogen evolution (Stuart and Kalt-wasser, 1970; Kaltwasser, Stuart and Gaffron, 1969). Consequently this aspect of the study has been reinvestigated utilizing the amperiometric technique for H_2 determination rather than the classic manometric technique. The results of our observations on ten independently isolated PS-II type mutants of Scenedesmus are summarized in Table IV. In no instance was a rate of photohydro-gen evolution detected which was greater than the rate of photo-synthesis in any of these mutants; according to one of the crit-eria utilized in selecting this mutant type all strains showed an unimpaired rate of photoreduction. It is now recognized that this mutant type represents the manifestations of a single gene mutation whose effect is pleiotropic. Numerous factors of PS-II including part of the plastoquinone pool, substance Q, C-550, cytochrome b-559 (H.P.) and the reaction center of PS-II, P-680, are affected by the mutation; stated otherwise these mutants have lost not only the primary reaction center of PS-II but also most of the known components of the reducing side of this photosystem (Bishop and Wong, 1971c; Bishop and Wong, 1974). If it is true that PS-II is essential for photohydrogen evolution then the only potential for this reaction in these mutations would be from an exclusive PS-I mediated hydrogen evolution; on occasion minute bursts of hydrogen have been observed. An alternate, and obvious, interpretation of the inability of the PS-II type mutants to

perform photohydrogen evolution is that they have lost preferentially the site for electron donation on the reducing side of PS-II
e.g., substance Q, which would be identical mechanistically to
DCMU inhibition. Hence the loss of oxygen evolving capacity in
these mutants would not be directly related to their inability to
produce hydrogen in the light. Evidence which contradicts this
interpretation has recently been obtained with a newly isolated
mutant of Scenedesmus (Bishop and Wong, unpublished data) which
lacks PS-II activity, has a normal low fluorescence which is
enchanced by DCMU, shows no major modifications of any of the
known components of PS-II and performs DCMU sensitive PS-II dependent electron transport with alternate electron donor-acceptor
systems. This low fluorescent mutant strain is apparently identical to the mutant strains of <u>Chlamydomonas</u> <u>reinhardtii</u> described
earlier (Epel and Levine, 1971; Epel, Butler and Levine, 1972;
and Epel and Butler, 1972). The Scenedesmus mutant, LF-1, has
neither photosynthesis nor photohydrogen production but retains
a normal photoreduction (Table IV). Although the low fluorescence
strains of Chlamydomonas and Scenedesmus show a decreased content
of cytochrome b-559 (H.P.), but a normal total cytochrome b-559
content, it is believed that this is not the underlying cause of
the loss of oxygen evolving capacity but rather is the indicator
of a fundamental change within the ultrastructure of the chloroplast membranes. Our preliminary results suggest that both the
reduction in content of the high potential form of cytochrome

TABLE IV

Whole Cell and Chloroplast Reactions of Mutations of Scenedesmus
    Obliquus Deficient in Photosystem II Activity*

Mutant Strain	Photo-synthesis	Photo-reduction	Photo-hydrogen	DPC-DCPIP
PS-4	0	104	0	0
PS-5	0	107	0	0
PS-11	0	98	2	2
PS-79	13	100	10	7
PS-84	12	100	7	5
PS-96	0	104	0	0
PS-102	0	101	0	0
PS-110	2	100	tr	1
PS-112	3	104	tr	2
LF-1	0	102	tr	87

* *All values given are in percentages of the respective rates
of comparable measurements performed on whole cells or chloroplasts of the wild-type strain. Actual values are as presented
in Table I.*

b-559 and the loss of oxygen evolving capacity are manifestations produced by modifications in the chloroplast lipids of LF-l. This alteration affects, consequently, only the water photolysis step, and by inference, photohydrogen evolution.

Additional lines of evidence from other mutant types support the interpretation drawn above. Specifically, mutations in which carotenoid biosynthesis is blocked develop PS-I but neither PS-II nor photohydrogen evolution activities. Light sensitive mutations which preferentially lose PS-II activity when exposed to high intensity irradiation also lose the potential for hydrogen evolution. Temperature sensitive mutations of Scenedesmus which lack PS-II activity when grown at a non-permissible temperature also lack photohydrogen evolution capacity (Bishop and Frick, unpublished data).

D. Nature of the Electron Donor in Photohydrogen Evolution.

As has been indicated earlier in this review, two schools of thought prevail concerning both the nature of the substance dehydrogenated during photohydrogen production and the photosystems utilized. In the original research of Gaffron and Rubin an organic hydrogen donor was favored as the substrate since the addition of glucose causes an increase in the amount of hydrogen evolved and because a period of active photosynthesis prior to adaptation stimulated hydrogen photoproduction. Kaltwasser, et al (1969) and Stuart and Gaffron (1971) attempted to provide additional specifics about the mechanism of glucose stimulation in Scenedesmus. Stuart and Gaffron showed that of thirteen potential substrates for glycolysis and intermediary metabolism of algae only glucose was preferentially utilized. Kaltwasser et al (1969) observed that the increased production of carbon dioxide, and supposedly also of hydrogen, resulted primarily from the breakdown of glucose by the Embden-Meyerhof pathway since the majority of the carbon dioxide evolved originiated from the 3 and 4 positions of labeled glucose. Although this deduction is logical as concerns the source of carbon dioxide evolved either in the dark or light anaerobic metabolism of adapted algal cells, it cannot predict any specifics about what portion of the glucose molecule is utilized for the production of hydrogen. Actual determination of the stoichiometry of hydrogen produced and glucose utilized showed that one mole of hydrogen gas was formed per two moles of glucose metabolized (Stuart and Gaffron, 1971). If photohydrogen production in anaerobic cells were to serve as an alternate mechanism for reoxidizing the reduced NAD formed in normal glycolysis (Kok, 1973) then a stoichiometry of 2, and not 0.5, would be expected. Since the green algae possess a mixed fermentative mechanism producing both lactic acid and alcohol not all of the reduced NAD would need to be reoxidized through the hydrogenase system. Perhaps because of these difficulties in evaluating mechanisms of photoanaerobic metabolism of algae Stuart and Gaffron (1971) concluded that "It is thus likely that H_2 photoproduction removes electrons from a "side path" of glucose (anaerobic)catabolism."

The glucose stimulation of photohydrogen production is restricted to the slow phase of photohydrogen production, is more apparent in starved or older cells, is maximally stimulated by Cl-CCP and is not inhibited by DCMU. Because measurements on this aspect of photohydrogen metabolism are performed over long periods of time, i.e., from 6-8 hours of manometric measurements, it is difficult to correlate observations on this aspect of the general phenomenon of the light-dependent evolution of hydrogen to that which occurs in periods of time less than 5-10 minutes of illumination. This latter phase, the rapid phase, requires the participation of both photosystems, is inhibited by DCMU, is not stimulated by glucose and appears to be correlated with a functional PS-II as we have discussed previously. Whether an organic hydrogen donor functions in both phases where electrons derived from this source can be donated either to PS-I or to both photosystems operating in a normal sequence must remain an open question at this time.

ACKNOWLEDGEMENTS

The authors wish to acknowledge the expert technical assistance of Ms. Marianne Frick and Mr. James Wong; their efforts in the reported studies on hydrogen metabolism of the green algae have been of paramount importance. Financial support for the research reported herein and for the preparation of the manuscript was provided through a grant from the National Science Foundation (BMS-7518023).

REFERENCES

Avron, M. (1967). In "Current Topics in Bioenergetics" (D.R. Sanadi, ed.) 2, 1.

Beinert, H. and Kok, B. (1963). In "Photosynthetic Mechanisms in Green Plants", (B. Kok and A. T. Jagendorf, eds), p. 131, NAS/NRC Publication 1145.

Ben-Amotz, A. and Gibbs, M. (1975). Biochem. Biophys. Res. Commun. 64, 355.

Bishop, N. I. and Gaffron, H. (1962). Biochem. Biophys. Res. Commun. 6, 471.

Bishop, N. I. and Gaffron, H. (1963). In "Photosynthetic Mechanisms of Green Plants" (B. Kok and A. T. Jagendorf, eds.) p. 441. NAS/NRC Publication 1145.

Bishop, N. I. (1966). Ann. Rev. Plant Physiol. 17, 185.

Bishop, N. I. (1971a). Ann. Rev. Biochem. 50, 197.

Bishop, N. I. (1971b). In "2nd International Congress on Photosynthesis Research", Stresa (G. Forti, M. Avron and A. Melandri, eds.) p. 459. Dr. W. Junk; The Hague.

Bishop, N. I. and Wong, J. (1971c). Biochim. Biophys. Acta 234, 433.

Bishop, N. I. (1973). In "Photophysiology" (A.C. Giese, ed.) VIII, 65.

Bishop, N. I. and Wong, J. (1974). Ber Deutsch. Bot. Ges. 87, 359.

Bishop, N. I. and Frick, M. (1976). Unpublished data.

Bishop, N. I. Frick, M. and Jones, L. W. (1976). Manuscript in preparation.

Boardman, N. K. (1970). Ann.Rev. Plant Physiol. 21, 115.

Brown, J. S. (1973). In "Photophysiology" (A.C. Giese, ed.) VIII, 97. Academic Press, New York.

Butler, W. L. and Bishop, N. I. (1963). In "Photosynthesetic Mechanisms in Green Plants" (B. Kok and A.T. Jagendorf, eds.) p. 91, NAS/NRC Publication 1145.

Cheniae, G. M. (1970). Ann. Rev. Plant Physiol. 21, 467.

Epel, B. L. and Levine, R. P. (1971). Biochem. Biophys. Acta 226, 154.

Epel, B. L., Butler, W. L. and Levine, R. P. (1972). Biochim. Biophys. Acta 275, 395.

Epel, B. L. and Butler, W. L. (1972). Biophysical J. 12, 922.

Gaffron, H. (1940). Am. J. Bot. 27, 273.

Gaffron, H. and Rubin, J. (1942). J. Gen. Physiol. 26, 209.

Gaffron, H. (1944). Bacteriol. Rev. 19, 1.

Gaffron, H. and Bishop, N. I. (1963). In "La Photosynthese", p. 645, No. 119 (Colloques Intern. Centre Natl. Recherche Sci.) Paris.

Gest, H., Ormerod, J. G and Ormerod, K. S. (1962). Arch. Biochem. Biophys. 97, 21.

Healey, F. P. (1970a). Plant Physiol. 45, 153.

Healey, F. P. (1970b). Planta 91, 220.

Jones, L. W. and Bishop, N. I. (1976). Plant Physiol. 57, 659.

Kaltwasser, H., Stuart, T. S. and Gaffron, H. (1969). Planta 89, 309.

Katoh, S. (1972). Plant and Cell Physiol. 13, 273.

Kessler, E. (1974). In "Algal Physiology and Biochemistry", (W. D. P. Steward, ed.) p. 456, Blackwell Scientific Pub. Ltd., London.

Kimimura, M. and Katoh, S. (1972). Biochim. Biophys. Acta 283, 279.

Kimimura, M. and Katoh, S. (1973). Biochim. Biophys. Acta 325, 167.

Kok, B. (1973). NSF/RANN Report. Proceedings of the Workshop on Bio-Solar Conversion, p. 22.

NSF/NASA Report, "Solar Energy as a National Energy Resource", Dec. 1972.

NSF/RANN Report (1972). "An Inquiry into Biological Energy Conversion". The University of Tennessee.

NSF/RANN Report (1973). "Proceedings of the Workshop on Bio-Solar Conversion". (A. San Pietro and S. Lien).

Ormerod, J. G. and Gest, H. (1962). Bacteriol. Rev. 26, 51.

Senger, H. and Frickel-Faulstitch, G. (1975). In "Proceedings of the 3rd International Congress on Photosynthesis," (M. Avron, ed.) Vol. I, p. 715.

Spruit, C. J. P. (1954). Proc. Intern. Photobiol. Congr. 1st, Amsterdam, p. 323.

Spruit, C. J. P. (1958). Mededel. Landbouwhogeschool, 58, 1.

Stuart, T. S. and Kaltwasser, H. (1970). Planta 91, 220.

Stuart, T. S. and Gaffron, H. (1971). Planta 100, 228.

Stuart, T. S. and Gaffron, H. (1972a). Planta 106, 91.

Stuart, T. S. and Gaffron, H. (1972b). Ibid p. 101.

Stuart, T. S. and Gaffron, H. (1972c). Plant Physiol. 50, 136.

Trebst, A. (1974). Ann. Rev. Plant Physiol. 25, 423.

Weaver, E. C. and Bishop, N. I. (1963). Science 140, 1095

HYDROGEN PRODUCTION BY MARINE PHOTOSYNTHETIC ORGANISMS AS A POTENTIAL ENERGY RESOURCE

Akira Mitsui and Shuzo Kumazawa

Rosenstiel School of Marine and Atmospheric Science
University of Miami
Miami, Florida 33149

I.   INTRODUCTION

In recent years most nations have clearly recognized the impending shortage of fossil fuels. In accordance with this crisis the development of alternative fuel sources has been greatly accelerated. Although many new concepts have been presented (e.g. 3, 16, 19, 63, 82, 105-107), solar energy conversion is one of the mechanisms which would put the least amount of strain on our natural resources. This article deals with one form of solar energy conversion, the biological and biochemical photoproduction of hydrogen.

The capability of many photosynthetic microorganisms to produce hydrogen gas has been recognized for many years (22, 29, 30). The process itself is directly or indirectly linked to the light dependent photosynthetic pathway. However, the solar energy conversion efficiency of the process has always been relatively low.   This observation, and the fact that hydrogen production is often sensitive to oxygen inhibition, have largely precluded the use of this theoretical pathway in applied research.

Our recent experimental results indicate that there may be hope for resolving some of these problems and thereby enhancing the future application of hydrogen photoproduction. Active research in this area may result in the development of innovative and valuable sources of fuel.

II.   ADVANTAGES OF BIOLOGICAL HYDROGEN PHOTOPRODUCTION SYSTEM (72)

There are numerous advantages to the biological approach to hydrogen photoproduction as opposed to thermo-chemical and physical methods:

1) The biological system could be operated at low physiological temperatures (i.e., $10-40°C$) as opposed to the normally high temperatures required for chemical or physical production of hydrogen (400 to $1000°K$).

2) The only major input into the system would be solar energy and a hydrogen donor, probably water (salt water).

3) The production of hydrogen would not involve the formation of pollutants, as in the case of fossil fuel refineries.

4) The fuel produced, H_2 gas, would be clean burning (i.e., yielding H_2O).

5) The system would readily lend itself to a potentially beneficial and profitable program of multiple utilization (68); including the production of food for human and animal consumption, fuel as methane and alcohol, and commercially usable chemical byproducts.

III. THE WORKING HYPOTHESES OF HYDROGEN PRODUCTION RESEARCH

The photoproduction of hydrogen by intact cells was first observed in algae by Gaffron and Rubin in 1942 (22) and in photosynthetic bacteria by Gest and Kamen in 1949 (29) (30). After these pioneering experiments, hydrogen photoproduction by intact cells was extensively studied by several laboratories (5, 6, 9, 11, 13, 14, 21, 22, 25-33, 36, 38, 41, 42, 44-48, 50-55, 58, 60, 64, 67, 69-74, 80, 81, 83, 84, 86, 91-102, 108, 109).

Through these and other experiments the mechanism of hydrogen evolution is starting to be elucidated, as diagrammatically represented in Figures 1, 2, and 3. Basically, solar radiation is absorbed by some form of pigment, for example, the light harvesting chlorophyll protein. This energy is transmitted to the electron through the reaction center chlorophyll

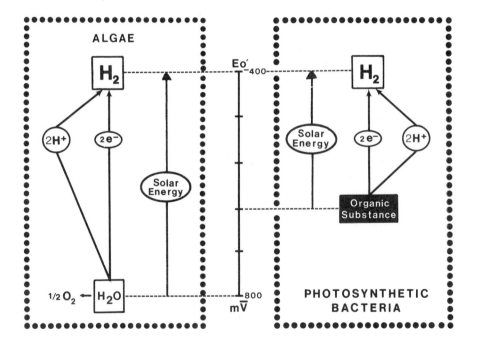

Figure 1   Evolution of H_2 by Photosynthetic Organisms.

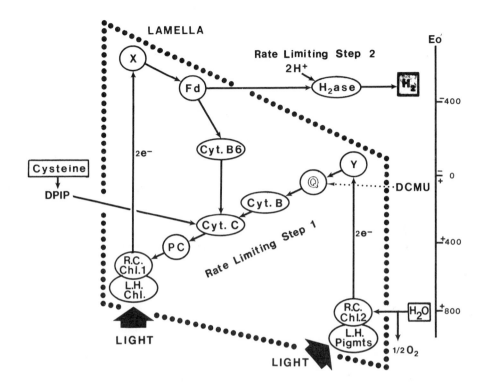

Figure 2   Electron Transport System and Rate Limiting Steps of Hydrogen
Photoproduction (after 66, 72).
Rate limiting reactions in hydrogen production are indicated in
the Figure.
L.H. Pig., light harvesting pigment proteins; R.C. Chl. I and
R.C. Chl. II, reaction center chlorophyll protein of photosystems I and II, respectively; X and Y, primary electron
acceptors of photosystems I and II; Q, plastoquinone; Cyt B,
cytochrome b; Cyt C, high redox potential algal cytochrome C;
PC, plastocyanine; Fd, ferredoxin; H₂ ase, hydrogenase; DPIP,
2, 6-dichlorophenol indophenol, an artificial electron carrier;
DCMU, dichlorophenyl dimethylurea, an inhibitor of photosystem II; MV, methyl viologen, an artificial electron carrier;
Eo' and mV, redox potential in millivolts at pH 7.0. The
components contained within the perimeter of the dashed lines
represent the core of the photosynthetic electron transport
system in the lamella.

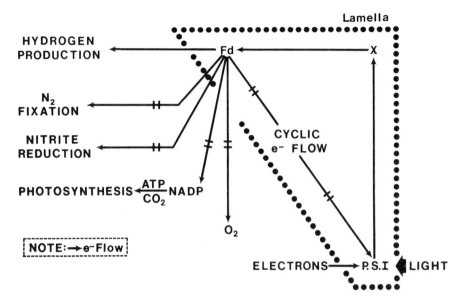

Figure 3   Alternate Pathways of Electron Flow in Living Cells

protein.   The source of the electron is some electron donor compound
(example, water).   This energized electron is the key to the process of
hydrogen gas evolution.   The electrons for the H_2 production step follow
essentially the same pathway as those leading to carbon fixation and other
associated functions (Figures 2 and 3).   The actual shunting of electrons
from the photosynthetic-carbon fixation pathway could occur just after
ferredoxin (or another primary electron acceptor), and the transfer of
electrons from ferredoxin (or another primary electron acceptor) to $2H^+$
could take place in three different ways (see Figure 4).   The transfer could
occur through: 1) an ATP independent hydrogenase catalyst; 2) an ATP
dependent nitrogenase system; or 3) NADP and hydrogenase mediated
system (or analogous systems).   Of course, all species of algae and
photosynthetic bacteria do not contain all three hydrogen evolution
systems.   Energetically very little is known about the relative efficiency of
each system.   The key element in the process of H_2 production is solar
energy input.   Solar radiation drives the photosynthetic system by supplying
the energy for the transfer of electrons from the donor compound to $2H^+$.
This donor substance varies from species to species.   Photosynthetic
bacteria require an electron donor such as an organic compound (30, 32) or
a sulfur compound (64).   However, algae may utilize water as their electron
donor (11, 21, 41, 42, 45, 46, 51, 96-102).   Figure 1 illustrates major
differences between algal and photosynthetic bacterial systems.
     However, it should be pointed out that the question of electron donors
of algae has been a very controversial issue.   Many hypotheses have been
presented along with experimental data.   For example, King and Gibbs (53),
and Jones and Bishop (44) recently have shown that hydrogen production in

some algal species may not be directly linked to the photolysis of water. Rather they suggest that some organic compound serves as the electron donor in such systems. Whether the electron donor of algal systems is water or organic substances, could depend on genetic groups, culture conditions, and physicochemical experimental conditions. Energetically there seem to be some basic differences between photosynthetic bacteria and algae. The minimum solar energy input for these two systems is quite different, as illustrated in Figure 1 (this figure is only an energy schematic and does not show the detailed structural differences between the algal and bacterial systems). It should also be pointed out that algal biophotolysis systems usually need two light reactions (see Figure 2) and bacterial hydrogen production systems may require only one.

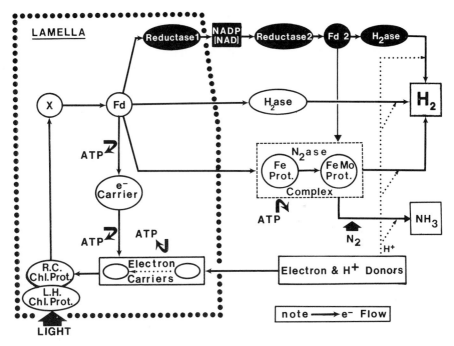

Figure 4   ATP Dependent and Independent Photoproduction of Hydrogen

In addition to intact cell experiments with hydrogen photoproduction, several research groups have initiated studies with cell free systems. During 1961-1963, my coworkers and I (2, 75, 78, see also 66) succeeded in constructing a working cell free hydrogen photoproduction system. This cell free system consisted of chloroplasts of plant leaves or lamellae of blue-green algae; methyl viologen or ferredoxin; and bacterial hydrogenase (bacterial hydrogenase was isolated from Desulfovibrio, Chromatium and Clostridium). Cysteine or ascorbate was the electron donor of the system.

This research was quickly followed by a demonstration that hydrogen could be produced from water (112). More recently Abeles (1964) (1) successfully promoted hydrogen photoproduction by a cell free green algal system. From the standpoint of energy production, earlier experiments were reexamined. In 1972-1973, Krampitz (56) demonstrated cell free photo-production of hydrogen using an NADP mediated system in the presence of chloroplasts, ferredoxin and bacterial hydrogenase. Ben-Amotz and Gibbs (7); Benemann et al. (8), and Rao et al. (89) reported cell free photoproduction of hydrogen using a system similar to that described above (chloroplast-ferredoxin-bacterial hydrogenase). Ben-Amotz and Gibbs used dithiothreitol as an electron donor. Rao et al. used chloroplasts which were stabilized by glutaraldehyde.

Numerous other studies of hydrogen production through chemical pathways have been proposed and researched. These studies have been reviewed in other literature sources (12, 15, 17, 82).

IV.   THE PROBLEM OF SOLAR ENERGY BIOCONVERSION EFFICIENCY

The problem which has plagued researchers in their attempt to practically apply the above mechanism of H_2 production is the question of efficiency.

Since the production of hydrogen gas is an offshoot of the photosynthetic pathway, the solar energy conversion efficiency is directly related to the metabolic demands of the plant cells being used in the system (see Figure 3).

This, in large part, is why the theoretical maximum of 30% photosynthetic solar energy conversion efficiency (54, 62) is never observed in nature, as can be seen in Figure 5. As such, the problem of increasing bioconversion efficiency involves a complex of variables associated with photosynthesis and plant metabolism in general, including the following important considerations: 1) Metabolic Potential--the metabolic potential of plants is associated with their photosynthetic capability, which in turn determines their H_2-producing potential. Different species exhibit different metabolic potentials. Even within the same species the attainment of this potential is dependent on the existence of proper environmental conditions (i.e., sunlight, temperature, salinity, nutrient availability); 2) Competing Pathways for Electrons--as Figure 3 illustrates, there are numerous pathways which the excited electron from photosystem I (PS I) can take. These include CO_2 fixation, O_2 reduction, cyclic electron flow, N_2 fixation, nitrite reduction or hydrogen production (55, 66, 67). Under natural conditions the percent of electrons flowing to each of these pathways will depend on the metabolic state of the cell; 3) Oxygen Inactivation of Hydrogenase and Nitrogenase--in the natural environment, O_2 produced during the photolysis of water will, at certain concentrations, inhibit the activity of some types of hydrogenase and nitrogenase enzymes which catalyze the H_2-producing reaction. This phenomenon, where it occurs, reduces the net yield of H_2 gas; 4) Rate Limiting Reactions-- certain steps in the transfer of electrons to H_2 production, especially electron flow between photosystem I and photosystem II, and hydrogenase and nitrogenase catalysis, are rate limiting (see Figure 2).

Figure 5

SOLAR ENERGY BIOCONVERSION EFFICIENCY

IN BIOLOGICAL SYSTEMS (AFTER 54, 62)

I. THEORETICAL MAXIMUM: PHOTOCHEMICAL REACTION (CELL FREE)

$$E = 1.3eV/2 \times h\nu = \textbf{30\%}$$

IN TERMS OF THE AVERAGE SOLAR ENERGY IN U.S.

$$E = (0.30) \quad (7.5 \times 10^8 \text{ KCAL ACRE}^{-1}\text{YR}^{-1})$$
$$= 2.25 \times 10^8 \text{ KCAL ACRE}^{-1}\text{YR}^{-1})$$

II. THEORETICAL MAXIMUM: INTACT CELLS:

$$E = \textbf{12\%}$$
$$= 0.90 \times 10^8 \text{ KCAL ACRE}^{-1}\text{YR}^{-1}$$

III. OBSERVED AGRICULTURAL EFFICIENCY:

$$E(\text{SUGAR CANE}) = \textbf{2.5\%}$$
$$= 0.19 \times 10^8 \text{ KCAL ACRE}^{-1}\text{YR}^{-1}$$

V. LIVING CELL AND CELL FREE APPROACHES IN ENHANCING HYDROGEN PHOTOPRODUCTION (67)

A) Living Cell H_2 Production

Figure 6 illustrates methods of increasing the hydrogen photoproduction capability in living cell system. These include 1) survey and selection of proper organisms; 2) the development of a method of efficient removal of oxidizing agents such as oxygen; 3) physical and chemical treatments of the cell; and 4) use of specific metabolic inhibitors. 5) In addition, some of the many isolated strains could be genetically altered in order to increase the efficiency of hydrogen production.

B) Cell Free H_2 Production. (66)

Another extension of intact cell H_2 photoproduction research would be the use of a cell-free approach. It is well known that natural, intact-cell biochemical systems operate under a network of checks and balances. This system helps to maintain organisms within a state of homeostasis (equilibrium). This precludes excessive buildup of any end product, and it also sets an upper limit to the rates of hydrogen production and nitrogen fixation (as well as other biochemical products). One of the most efficient

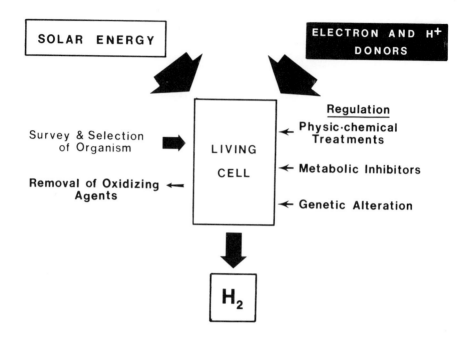

Figure 6   Microbial Process of Hydrogen Photoproduction: Living Cell
System.

ways of removing these restraints would be to isolate the hydrogen
producing system from its   cellular environment, creating a cell free
system.

The intact and cell-free approaches are bound together in two
respects: 1) by their mutual use of the algal and photosynthetic bacterial
photosynthetic production mechanisms, and 2) by the similarities in
experimental methodologies used to study both approaches. The general
format of a cell-free approach is outlined in Figure 7.

The initial step is to collect, isolate and culture different species
(genetic groups) of marine algae and photosynthetic bacteria, which will
serve as a source for the "pool of potential components" for the cell-free
system. The next step is to acquire information about the structure (e.g.,
X-ray crystallography) and function of these components so that "selection
processes" can be carried out.

The next steps in the process include "stabilization" (e.g., prevention
of chlorophyll protein photobleaching), immobilization, and the alteration
of the molecular structure of components.

Stabilization is one of the most important factors in successful cell-free system projects, primarily because components of cell-free H_2 photo-production systems are more unstable than their intact cell counterparts.

Finally, the selected components can be combined or arranged on a membrane or column to make the reaction system. At this stage tests must be made to determine the physicochemical conditions, solar energy intensities, and the electron and H^+ donors, which together yield the highest H_2 gas production rates.

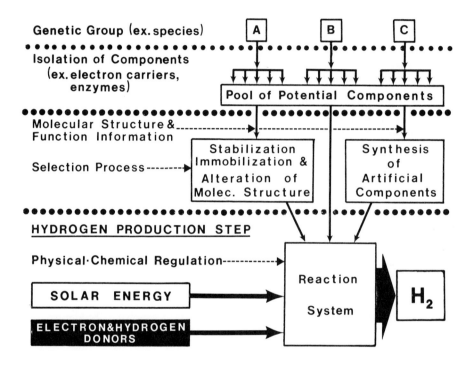

Figure 7   Biochemical Process of Hydrogen Photoproduction: Cell Free System

VI.   OUR APPROACH AND RESULTS OF HYDROGEN PHOTO-PRODUCTION BY MARINE ORGANISMS

Several new solutions to the bioconversion efficiency problem have been presented in recent years (4, 12, 15, 18, 23, 34, 37, 43, 54, 62, 66, 72). Our approach is based on the belief that there are undiscovered organisms which exhibit special characteristics suited for application. The methodology used to attack this problem and our results to date will be described below.

The first step was to survey the tropical marine environment for species with high hydrogen producing capabilities. Most of the early experiments with intact-cell systems utilized freshwater species of algae and photosynthetic bacteria. However, the marine environment contains many species of photosynthetic bacteria and algae which are capable of the same function.

The advantages of using marine photosynthetic microorganisms in mass cultures are many. The most apparent of these is the availability of salt water. Since many regions of the world suffer from lack of fresh water (79), the ability to use salt water becomes an important factor. Secondly, salt water, in its natural state, abounds with many nutrients (including CO_2, magnesium, sulfate, potassium) essential to the growth of photosynthetic organisms. In tropical and subtropical marine environments photosynthetic organisms (bacteria and algae) can be found in abundance year round. Since the tropical environment is well endowed with a wide diversity of marine algae and bacterial species it is an ideal location for the study of genetically different hydrogen photoproduction systems. We felt that a survey of hydrogen-producing tropical marine algae and photosynthetic bacteria would reveal species with exceptionally high hydrogen photoproducing capabilities. In addition, utilizing marine species would permit the use of salt water as a hydrogen donor. The impending shortage of fresh water in many areas of the world made this a key advantage. Other key elements included: 1) isolation and identification of individual species from mixed samples; 2) screening for H_2 production capabilities; 3) determination of culture conditions which maximize H_2 production and growth; and 4) the selection of a hydrogen donor which would be both efficient and inexpensive.

A)   Collection of Marine Photosynthetic Organisms

Water and sediment samples and sea grasses, mangrove leaves and seaweeds were collected throughout the Biscayne Bay area and in other selected regions of South Florida, the Florida Keys, the Bahamas and Colombia. In each area samples were obtained from several different levels in the water column, including surface, midwater and bottom samples. More than 300 samples were collected.

With the station data obtained, efforts were begun to find any relationships which might exist between microorganismal species and the geographical areas, macrophytes and environmental conditions from which they were collected. Much of this information was based on the use of the scanning electron microscope (SEM) (74, 87, 88).

The SEM pictures of both dead and living macrophyte surfaces showed the existence of a wide diversity of microorganisms. These included numerous types of blue-green algae and photosynthetic bacteria. It was also interesting to note that these epiphytic associations were specific to some extent (76, 87, 88).

B)   Isolation of Hydrogen Producing Organisms

Subsequent to the collection of algae and photosynthetic bacteria, mixed cultures were established in enriched seawater media. Two different methods were used to obtain single species cultures from these mixed cultures: 1) Agar shake culture method in both aerobic and anaerobic conditions in enriched seawater medium, and 2) Gas-Pack plate methods in different mixtures of various gases (cf., O_2, H_2, N_2, CO_2). So far, more than 60 species of blue-green algae and photosynthetic bacteria have been isolated with these two methods.

Isolated strains were morphologically characterized by the scanning electron microscope (87), and their absorption spectra were compared (70, 90).

Then their optimum growth conditions were determined in each species and strain (70, 90). These include optimum pH and salinity, and vitamin B_{12} and nutrient requirements.

C)   Finding of Highly Active and Stable Hydrogen Photoproducing Species

Using these isolated strains as a working base we have performed screening experiments for long term (3 to 7 days) hydrogen production capability. This approach diverges from the more common short term (e.g. 1-15 minute) hydrogen production experiments performed using hydrogen electrodes or manometry. We have chosen this "long term" method because it makes it possible to discuss the commercially important question of stability and quantity of H_2 production. Most previous experiments show that hydrogen photoproduction lasts only a very short time (a few minutes to 1 hour).

In the first set of experiments several strains of blue-green algae were shown to produce H_2 gas in the dark (see Table 1). Approximately one half of the tested species produced hydrogen in the dark (59).

The fact that so many strains of tropical marine blue-green algae produce H_2 in the dark coincides with the results of Ben-Amotz et al. (6) with temperate and cold water green and red marine algae. This leads to the observation that a wide diversity of marine species contain H_2 producing mechanisms, probably including some form of hydrogenase (or nitrogenase). However, the rates of dark hydrogen production are relatively low.

Light hydrogen production capabilities of marine blue-green algae were also tested.

Three different types of hydrogen production in the light were observed (59, 60).

1) Hydrogen produced in the dark and not in the light both with and without oxygen trapping agents (chromous chloride) as shown in Figure 8.

2) In some cases hydrogen production occurred in the dark at moderate rates and accelerated in the light in the presence of an oxygen trapping agent (see Figure 9). This acceleration depends on the given light intensities. Without an oxygen trapping agent, hydrogen produced in the dark was consumed photochemically.

SPECIES	RATE HYROGEN EVOLUTION		
	H_2 μl/mg Chl a/hr	H_2 μl/mg dry wt/hr	H_2 μl/mg protein/hr
MIAMI BG 2M	3.7	0.06	0.25
MIAMI BG 3S	TRACE	TRACE	TRACE
MIAMI BG 4S	47.0	0.76	1.80
MIAMI BG 4M	22.0	—	—
MIAMI BG 4B	TRACE	TRACE	TRACE
MIAMI BG 7	TRACE	TRACE	TRACE
MIAMI BG 9	3.5	0.04	0.08
MIAMI BG II4S*	42.0	0.30	0.50
MIAMI BG II5S	0.0	0.00	0.00
MIAMI BG 142	0.0	0.00	0.00
MIAMI BG LF 1	0.0	0.00	0.00
MIAMI BG FP 4	24.0	—	—
MIAMI BG FP 2	5.6	0.14	0.29

THESE RATES WERE MEASURED AFTER 20 HRS ANAEROBIC CONDITION

*THESE RATES WERE MEASURED AFTER 40 HRS ANAEROBIC CONDITION

Table 1   Rate of Dark Evolution of Hydrogen in Marine Blue-green Algae

3) In others, hydrogen was produced in the light with and without an oxygen trapping agent (as Figure 10 illustrates), but not in the dark.

Most of the strains produce hydrogen at rates of 0.0002-0.1 m moles H_2 per mg chlorophyll per day or 0.002-1.9 μmoles H_2 per mg dry weight per day (59). However, one marine blue-green algal strain (Miami BG7) has exhibited high light-dependent production rates. In subsequent experiments (77) the culture conditions of this strain were altered (i.e., minus combined nitrogen) yielding exceptionally high H_2 production rates (see Figures 11 and 12). Under the new conditions, Miami BG7 produced 230 μmoles of hydrogen per mg chlorophyll per hour, or 1 ml of H_2 gas per ml reaction mixture within a three-day illumination period. Furthermore, the H_2 gas was continually produced at approximately the same rate throughout the 5-7 day period (based on later tests). This experiment was repeated yielding the same results. During this period, the gas pressure within the reaction flasks increased to 1.3 atmospheres. 50% more hydrogen was produced when the pressure in the reaction vessels was kept constant at ambient pressure. Since many researchers have found H_2 production and nitrogen fixation in blue-green algae to be linked to heterocysts, at first it was hypothesized that Miami BG7 might have heterocysts, especially in nitrogen free media. However, observation by light microscopy and by SEM showed no heterocysts.

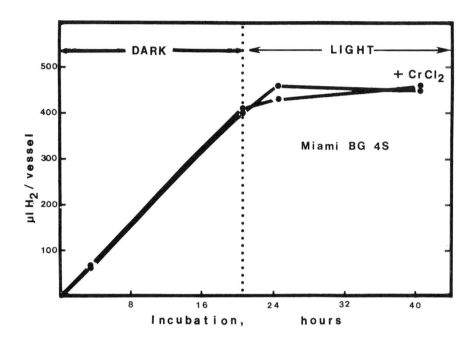

Figure 8   Hydrogen Production by a Marine Blue-green Alga Miami BG4S
            Alga was cultured in marine blue-green algal medium containing
            combined nitrogen (59). The harvested algae were suspended in
            combined nitrogen free algal culture medium at pH 7.0.
            Incubation was made at $25^\circ C$, under argon gas atmosphere.
            0.42 mg of chlorophyll a was contained in the 3 ml algal
            suspension (vessel).   The   light   intensity   was 4μ
            Einstein/m^2/sec.   Hydrogen   was   determined   by   gas
            chromatography.

In order to more closely examine the nature of this unique strain, a
number of experiments are presently in progress.

VII.   FUTURE PLANS OF HYDROGEN PHOTOPRODUCTION RESEARCH

The results of our experiments have enhanced the feasibility of
utilizing hydrogen photoproduction in algae as a source of fuel. Now that
we have discovered, isolated and successfully cultured several different
types of marine hydrogen producers, including a highly productive species
of blue-green algae, it is time to mount a more detailed investigation of
ways to increase and stabilize hydrogen production.

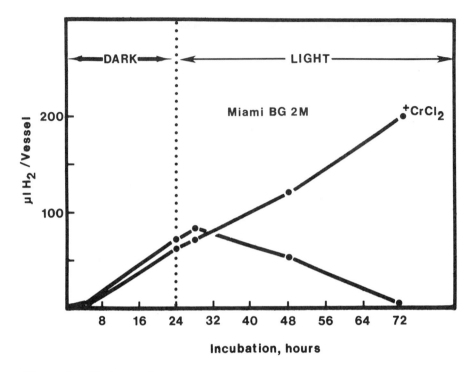

Figure 9   Hydrogen Production by a Marine Blue-green Alga Miami BG 2M
The experimental conditions were the same as in Figure 8.
0.82 mg chlorophyll a was contained in the 3 ml algal suspension
(vessel).

Four basic questions must be more thoroughly answered before a large scale system can be developed:

1)   Is it possible to increase and stabilize H_2 production rate and solar energy conversion efficiency through environmental manipulations?

2)   Is it possible to economically solve or at least circumvent the problem of oxygen inhibition of H_2 photoproduction exhibited by some of the strains?

3)   Are the hydrogen and electron donors in these strains water? and

4)   How does the metabolic state of the cultures affect hydrogen production?

In terms of gross hydrogen production it is possible to make an estimate of the amount of gas which could be made available for commercial use. Based on our experiments with Miami BG7 and the estimate made by California Institute of Technology (Jet Propulsion Laboratory) (49) for the nation's hydrogen needs in the year 2000

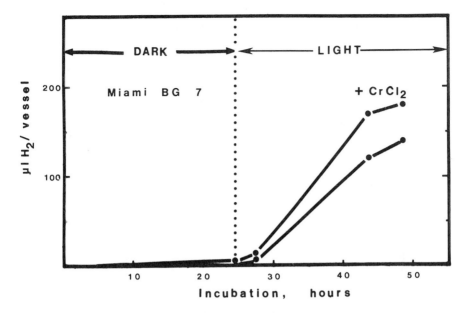

Figure 10   Hydrogen Production by a Marine Blue-green Alga Miami BG 7
The experimental conditions were the same as in Fig. 8.
0.024 mg chlorophyll a was contained in the 3 ml algal suspension (vessel).

$(28 \times 10^{12} m^3 H_2$: includes new technologies), it is possible to calculate the
size of culture required to meet these needs:
For a culture 1 m deep a total area of $(215 \ km)^2$ or $(134 \ miles)^2$
would be needed to supply all of the U.S.'s hydrogen energy
needs in the year 2000.
Looking at this from another point of view it is possible to calculate
what size culture would be needed to supply the electrical demands of an
average sized house (2 bedrooms) during the summer months (peak use
period) in southern Florida (i.e., 1000 kwhr/month, estimate by Florida
Power and Light). Using the same production rates as above, a culture 1 m
deep and having a total area of $(8 \ m)^2$ or $(26 \ feet)^2$ would be adequate to
meet these needs.
It is obvious that these figures are highly tentative and involve many
simplifying assumptions. Nevertheless they do provide some indication of
the great potential usefulness of this approach.
In addition to these utilization potentials there are many avenues for
the future development and advancement of the H_2 production concept and
for its application to practical problems.

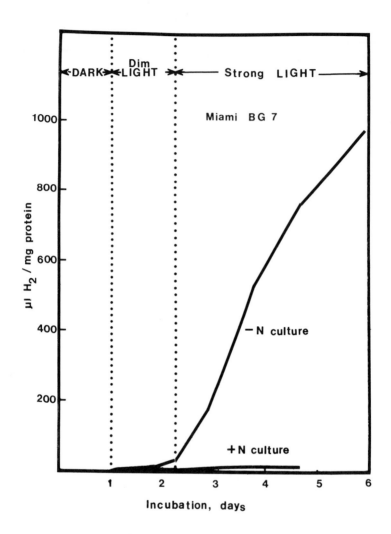

Figure 11   Comparison of Hydrogen Production by a Marine Blue-green
Alga Miami BG 7, Cultured With and Without Combined
Nitrogen.
The experimental conditions were the same as in Figure 8
except light intensities.   Dim light, 4.0 μEinstein/m^2/sec;
strong light, 63 μEinstein/m^2/sec.

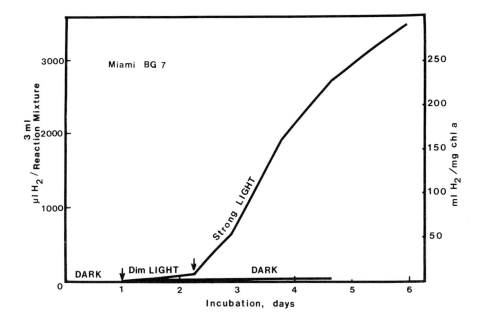

Figure 12   Light Dependent Hydrogen Production by a Marine Blue-green Alga; Miami BG 7
Alga was cultured in combined nitrogen free medium. The experimental conditions were the same as in Figure 11.

A.)   A Floating Hydrogen Production System.

Considering our orientation towards marine research, the first extension which comes to mind is the development of a floating hydrogen production system. The advantages afforded by such a system would include: 1) the conservation of land area, 2) the opportunity to utilize the natural kinetic energy resources of water masses (e.g. tides, currents, heat capacity) to help maintain and operate large culture systems, and 3) immediate availability of salt water for culture media. We envision that the first part of such experiments would involve the design of a relatively portable floating system electronically connected to a monitoring station and portable lab. Initial tests could be run on the feasibility of floating H_2 production cultures. In addition, experiments could begin for: 1) utilizing the natural heat capacity of large water masses in the maintenance of culture temperature equilibrium, 2) harnessing natural water movements (tides, currents) for culture stirring, 3) utilizing solar heat for sterilization of newly injected saltwater culture media, through the use of solar energy concentration mechanisms (e.g., furnell lense) and 4) studying basic problems in harvesting cultures and collecting H_2 produced in floating systems.

B.)   Hydrogen Production System for Land Use.

As discussed above, a culture tank of 1 m x 8 m x 8 m would be needed to supply the energy needs of an average household. Based on this estimate, it will be conceivable to examine the possibility of developing a hydrogen production system for household and commercial use. Both projects would demand the intimate cooperation of consultants and organizations including engineers, architects, chemists, and gas and electric company technical experts. Many technical difficulties are expected during the development of these systems.

VIII. MULTIPLE UTILIZATION POTENTIAL FOR HYDROGEN PRODUCTION CULTURE SYSTEM (68)

Figure 13 shows that there are many potential uses and benefits which could come from the development of a hydrogen photoproduction technology. The integration of these "offshoots" into the central H_2 production goal could increase the economic worth of the system.

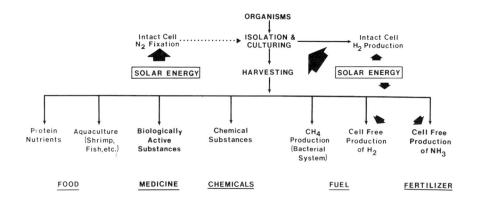

Figure 13   Utilization of Tropical Photosynthetic Marine Organisms

A.)   Hydrogen Producing Algae as a Source of Food and Fertilizer.

Strains may be isolated which are nutritionally rich (high protein, essential amino acids and vitamin content) and easier to process (digestion, extraction) than other already commercially tested strains (e.g. Chlorella). These strains could be used directly for human or animal consumption.

One of the primary considerations in any discussion of food sources is protein content. Previous studies have shown that the concentration of protein varies considerably in living organisms. In our laboratory, a wide variety of marine algae (both macro and microalgae) and photosynthetic bacterial species are being tested for protein concentration. Many blue-green algae and photosynthetic bacteria are characteristically high in protein (i.e., 20-60%) (68). This leads us to the conclusion that further attempts at economic scale mass culturing of these organisms might be profitable.

Of course, protein content is not the only important consideration in the evaluation of "food value". Therefore, a survey of amino acid composition of the protein and the general chemical composition (lipid, carbohydrate, vitamin and mineral content) of numerous algal groups should be made.

From another standpoint many blue-green algae fix nitrogen and therefore could be cultured economically and harvested for use as a combined nitrogen source, i.e., fertilizer.

B.)   Applications to Aquaculture.

One of the factors which determines the success or failure of ventures in aquaculture is the proper choice of primary food sources. Green algae and diatoms have already been successfully employed, as the base of the food chain, to feed zooplankton, which in turn are utilized by the larval and adult stages of the organisms being exploited (i.e., crustaceans, shellfish and fish). Dr. Kobayashi's laboratory in Japan has successfully used freshwater or soil photosynthetic bacteria (cf. Rhodospirillum species) for aquaculture purposes. However, other marine photosynthetic bacteria and blue-green algae have yet to be tested for their applicability as food in aquaculture. For example, preliminary indications support the notion that marine photosynthetic bacteria and blue-green algae could be used in aquaculture. The question might arise as to why marine photosynthetic bacteria and blue-green algae should be used instead of presently developed food sources. There are several advantages which favor the former as a food resource: 1) From the author's laboratory experience in the biochemistry of a wide variety of algae and bacteria, it is clear that the cell walls of photosynthetic bacteria and blue-green algae are much softer and easier to digest than those of green algal species; 2) Some marine photosynthetic bacteria and blue-green algae have a high nutritional value as mentioned above; and 3) The growth rate of some marine photosynthetic bacteria and blue-green algae is considerably faster than many of the presently used food organisms. Mass cultures of H_2-producing algal strains could provide food for the aquaculture of shrimp, crabs, shellfish, and fish (either directly or through the culture of zooplankton).

C.)   Methane Production.

Several years ago, Oswald and his coworkers (85) reported that methane production using sewage and algae might be economically feasible. Some H_2 producing photosynthetic strains may prove to be an economical source of carbohydrate material for bacteria-mediated methane production.

D.)   Use of Marine Algae in Drug and Other Medical Research.

The surveying and research of metabolically active substances produced by marine algae has received some attention in the past few years. However, it remains a potentially profitable area for investigation.

Blue-green algae occupy a unique taxonomic position among living organisms, between procaryotic bacteria and eucaryotic plants. However, being procaryotic they are more closely related to bacteria than any other major classification of organisms. The cellular structure and biochemistry of blue-green algae differs significantly from that found in other living forms. Some species produce substances which are highly toxic to higher animal forms (e.g.   blue-green algae may have been implicated in the recent Biscayne Bay, Miami fish kill). Unfortunately, our knowledge of marine blue-green algal biochemistry is far inferior to our understanding of bacterial systems.

New metabolically active substances may therefore be found in many of these as yet relatively unexplored species. Blue-green algae found in subtropical coastal waters makes it an ideal region to undertake a search for new metabolically active substances. It is possible to hypothesize that some of these uncharacterized substances produced by the H_2-producing algae may have a heretofore unobserved effect on the growth, cell division, metabolism and immunological response of cells.

E.)   The Development and Utilization of Chemical Products of Algae.

There is an enormous range of chemical substances produced by algae which are of interest to man. In a project such as this in which there is a potential for developing systems which will continuously produce large quantities of algal material, it is important to explore ways of exploiting this resource from a chemical point of view.

F.)   Use of Marine Photosynthetic Organisms for the Development of Cell Free Hydrogen Production and Nitrogen Fixation Systems (72)

As discussed above (Section V) the technology of cell free hydrogen production could be developed. A similar approach could be used for the cell free nitrogen fixation process (Figure 14). Although it needs long term and continuous research efforts for the development of these systems, it could be a new method for the production of fuel and fertilizer in the future.

These are definitely areas of research which could enhance the overall commercial and economic feasibility of a hydrogen photoproduction system.

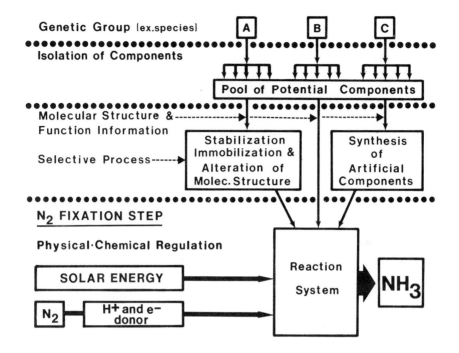

Figure 14   Biochemical Process of N_2 Fixation: Cell Free System

ACKNOWLEDGEMENTS

I would like to thank the following people for their contribution to this   research: Susanna Barciela,   Penny   Dalton,   James Frank, Arnold Goodman, Janet Greenbaum, Laura Haynes, Cheryl Hill, Edward Phlips, Varghese Ponmattam, JoAnn Radway, Diana Rosner, Hein Skjoldal, and Nancy McKeever Targett.

I would also like to thank Edward Phlips and JoAnn Radway for their help on the preparation of this manuscript. This work was supported by the National Science Foundation (AER 75-11171)Gulf Oil Foundation, and Engineering Foundation.

REFERENCES

1. Abeles, F.B. 1964. Cell free hydrogenase from Chlamydomonas. Plant Physiol. 39: 169.
2. Arnon, D.I., A. Mitsui and A. Paneque. 1961. Photoproduction of hydrogen gas coupled with photosynthetic photophosphorylation. Science 134: 1425.
3. Barr, W.J. and F.A. Parker. 1976. The introduction of methanol as a new fuel into United States economy. Foundation for Ocean Research, San Diego.
4. Beck, R.W. 1976. Workshop on solar energy for nitrogen fixation and hydrogen production. University of Tennessee, Knoxville.
5. Ben-Amotz, A. and M. Gibbs. 1974. H_2 photoevolution and photoreduction by algae. Abstract of the Annual Meeting of the American Society of Plant Physiologists. p. 27.
6. Ben-Amotz, A., D.L. Erbes, M.A. Riederer-Henderson, D.G. Peavey and M. Gibbs. 1975. H_2 metabolism in photosynthetic organisms. I. Dark H_2 evolution and uptake by algae and mosses. Plant Physiol. 56: 72.
7. Ben-Amotz, A. and M. Gibbs. 1975. H_2 metabolism in photosynthetic organisms. II. Light-dependent H_2 evolution by preparation from Chlamydomonas, Scenedesmus and Spinach. Biochem. Biophys. Res. Comm. 64: 355.
8. Benemann, J.R., J.A. Berenson, N.O. Kaplan and M.D. Kamen. 1973. Hydrogen evolution by a chloroplast-ferredoxin-hydrogenase system. Proc. Natl. Acad. Sci. U.S.A. 70: 2317.
9. Benemann, J.R. and N.M. Weare. 1974. Hydrogen evolution by nitrogen-fixing Anabaena cylindrica cultures. Science 184: 174.
10. Bishop, N.I. 1966. Partial reactions of photosynthesis and photoreduction. Ann. Rev. Plant Physiol. 17: 185.
11. Bishop, N.I. and H. Gaffron. 1963. On the interrelation of the mechanisms for oxygen and hydrogen evolution in adapted algae. In: Photosynthetic Mechanisms of Green Plants. NAS-NRC Publication 1145. p. 441.
12. Bolton, J.R. (Ed). 1976. Abstracts of International Conference on the Photochemical Conversion and Storage of Solar Energy.
13. Bose, S.K. and H. Gest. 1962. Hydrogenase and light stimulated electron transfer reactions in photosynthetic bacteria. Nature 195: 1168.
14. Bregoff, H.M. and M.D. Kamen. 1952. Studies on the metabolism of photosynthetic bacteria. XIV. Quantitative relations between malate dissimilation, photoproduction of hydrogen and nitrogen metabolism in Rhodospirillum rubrum. Arch. Biochem. Biophys. 36: 202.
15. Calvin, M. 1974. Solar energy by photosynthesis. Science 184: 375.

16.   C.E.Q., E.R.D.A., E.P.A., F.E.A., F.P.C., D.I., N.S.F.   1975.
      Energy alternatives: a comparative analysis. The Science
      and Public Policy Program.   University of Oklahoma,
      Norman, Oklahoma.
17.   Cooper, S.R. and M. Calvin. 1974. Solar energy by photosynthesis:
      manganese complex photolysis. Science 185: 376.
18.   Donovan, P., W. Woodward, F.H. Morse and L.O. Herwig. 1972.
      An assessment of solar energy as a national energy resource.
      NSF/NASA Solar Energy Panel.
19.   Energy Research and Development Administration. 1976. Hydro-
      gen Fuels: A. Bibliography.   Office of Public Affairs,
      Technical Information Center. Washington, D.C.
20.   Feigenblum, E. and Alvin I. Krasna.   1970.   Solubilization and
      properties of the hydrogenase of Chromatium.   Biochem.
      Biophys. Acta 198: 157.
21.   Frenkel, A.W. 1952. Hydrogen evolution by the flagellate green
      alga Chlamydomonas moewusii. Arch. Biochem. Biophys. 38:
      219.
22.   Gaffron, H. and J. Rubin. 1942. Fermentative and photochemical
      production of hydrogen in algae. J. Gen. Physiol. 26: 219.
23.   Gainer, J.L. (Ed). 1976.   Enzyme Technology and Renewable
      Resources.   University of Virginia and National Science
      Foundation (RANN).
24.   Gallon, J.R., T.A. LaRue and W.G. Kurz. 1972. Characteristics of
      nitrogenase activity in broken cell preparations of the blue-
      green alga Gloeocapsa sp. LB795.   Can. J. Microbiol. 18:
      327.
25.   Gest, H.   1951.   Metabolic patterns in photosynthetic bacteria.
      Bacteriol. Rev. 15: 183.
26.   Gest, H. 1963. Metabolic aspects of bacterial photosynthesis. In:
      Bacterial Photosynthesis. (Eds) H. Gest, A. San Pietro and
      L.P. Vernon. The Antioch Press, Yellow Springs. p. 129.
27.   Gest, H.   1971.   Energy conversion and generation of reducing
      power in bacterial photosynthesis.   Advances in Microbial
      Physiology 7:243.
28.   Gest, H., J. Judis and H.D. Peck, Jr.   1956.   Reduction of
      molecular nitrogen and relationships with photosynthesis and
      hydrogen metabolism. In: Inorganic Nitrogen Metabolism.
      (Eds) W.D. McElroy and B.   Glass.   Johns Hopkins Press,
      Baltimore. p. 298.
29.   Gest, H. and M.D. Kamen. 1949.   Studies on the metabolism of
      photosynthetic bacteria. IV.   Photochemical production of
      molecular hydrogen by growing cultures of photosynthetic
      bacteria. J. Bacteriol. 58: 239.
30.   Gest, H. and M.D. Kamen. 1949. Photoproduction of molecular
      hydrogen by Rhodospirillum rubrum. Science 109: 558.
31.   Gest, H. and M.D. Kamen. 1960. The photosynthetic bacteria. In:
      Encyclopedia of Plant Physiology. (Ed) W. Ruhland. Vol. 2.
      Springer Verlag, Berlin. p. 568.

32. Gest, H., M.D. Kamen and H.M. Bregoff. 1950. Studies on the metabolism of photosynthetic bacteria. V. Photoproduction of hydrogen and nitrogen fixation by Rhodospirillum rubrum. J. Biol. Chem. 182: 153.

33. Gest, H., J.G. Ormerod and K.S. Ormerod. 1962. Photometabolism of Rhodospirillum rubrum: Light-dependent dissimilation of organic compounds to carbon dioxide and molecular hydrogen by an anaerobic citric acid cycle. Arch. Biochem. Biophys. 97: 21.

34. Gibbs, M., A. Hollaender, B. Kok, L.O. Krampitz and A. San Pietro (Organizers) 1973. Proceeding of the workshop on bio-solar conversion. NSF-RANN.

35. Gitlitz, P.H. and A. I. Krasna. 1975. Structural and catalytic properties of hydrogenase from Chromatium. Biochemistry 14: 2561.

36. Gray, C.T. and H. Gest. 1965. Biological formation of molecular hydrogen. Science 148: 186.

37. Hall, D.O. 1976. Photobiological energy conversion. FEBS Letters 64: 6.

38. Hartman, H. and A.I. Krasna. 1963. Studies on the "Adaptation" of hydrogenase in Scenedesmus. J. Biol. Chem. 238: 749.

39. Hartman, H., and A.I. Krasna. 1964. Properties of the hydrogenase of Scenedesmus. Biochem. Biophys. Acta. 92: 52.

40. Haystead, A., R. Robinson and W.D.P. Stewart. 1970. Nitrogenase activity in extracts of heterocystous and non-heterocystous blue-green algae. Arch. Microbiol. 74: 235.

41. Healey, F.P. 1970. The mechanism of hydrogen evolution by Chlamydomonas moewusii. Plant Physiol. 45: 153-159.

42. Healey, F.P. 1970. Hydrogen evolution by several algae. Planta 91: 220.

43. Hollaender, A., K.J. Monty, R.M. Pearlstein, F. Shmidt-Bleek, W.T. Snyder, and E. Volkin (Eds). 1972. An inquiry into biological energy conversion. Gatlinburg. NSF-RANN

44. Jones, L.W. and N.I. Bishop. 1976. Simultaneous measurement of oxygen and hydrogen exchange from the blue-green alga Anabaena. Plant Physiol. 57: 659.

45. Kaltwasser, H. and H. Gaffron. 1964. Effects of carbon dioxide and glucose on photohydrogen production in Scenedesmus. Plant Physiol. 39: xiii.

46. Kaltwasser, H., T.S. Stuart and H. Gaffron. 1969. Light-dependent hydrogen evolution by Scenedesmus. Planta 89: 309.

47. Kamen, M.D. and H. Gest. 1949. Evidence for a nitrogenase system in the photosynthetic bacteria. Science 109: 560.

48. Kamen, M.D. and H. Gest. 1952. Serendipic aspects of recent nutritional research in bacterial photosynthesis. In: Phosphorus metabolism. Vol. II. (Eds) W.D. McElory and B. Glass. The Johns Hopkins Press, Baltimore. p. 507.

49. Kelley, J.H. and E.A. Laumann. 1975. Hydrogen Tomorrow. Jet Propulsion Laboratory, California Institute of Technology, Pasadena, California.

50.   Kessler, E. 1962. Hydrogenase und H_2 Stoffwechsel bei Algen. Deut. Bot. Ges. (N.F.) 1: 92.

51.   Kessler, E. and H. Maifarth.   1960.   Vorkommen und Leistungstahigkeit von Hydrogenase bei linigen Grunalgen. Arch. Mikrobiol. 37: 215.

52.   Kessler, E. 1974. Hydrogenase, photoreduction and anaerobic growth. In: Algal Physiology and Biochemistry. (Ed) W.D.P. Stewart. Blackwell Scientific Publication, Oxford. p. 456.

53.   King, D. and M. Gibbs. 1976. Hydrogen metabolism in photosynthetic organisms: the role of carbon metabolism. Abstract of the Annual Meeting of the American Society of Plant Physiologists. p. 60.

54.   Kok, B. 1973. Photosynthesis. In: Proceedings of the Workshop on Bio-Solar Conversion. (Ed) M. Gibbs, A. Hollaender, B. Kok, L.O. Krampitz, and A. San Pietro. NSF-RANN Report. p. 22.

55.   Kok, B., C.F. Fowler, H.H. Hardt, and R.J. Radmer.   1976. Biological solar energy conversion: approaches to overcome yield stability and product limitations.   In: Enzyme Technology and Renewable Resources. (Ed) J.L. Gainer. University of Virginia and NSF-RANN and Oral Presentation by B. Kok.

56.   Krampitz, L. O. 1973. Hydrogen production by photosynthesis and hydrogenase activity.   NSF-RANN Report No. HA1, HA3, HA5. N-73-013. Biophotolysis of water. NSF-RANN Report No. HA2. N-73-014.

57.   Krasna, A. I. and D. Rittenberg.   1956.   A comparison of the hydrogenase activities of different microorganisms. Proc. Natl. Acad. Sci. 42: 180.

58.   Krasna, A. I. 1976. Bioconversion of solar energy. In: Enzyme Technology and Renewable Resources. (Ed.)   J. L. Gainer. University of Virginia NSF-RANN. p. 61.

59.   Kumazawa, S. and A. Mitsui. 1976; Kumazawa, S., S. Barciela and A. Mitsui. 1976. Manuscripts in preparation.

60.   Kumazawa, S., J. Frank, H. R. Skjoldal and A. Mitsui.   1976. Hydrogen production by tropical marine blue-green algae and photosynthetic bacteria.   Abstract of the Annual Meeting of the American Society of Plant Physiologists. p. 61.

61.   Kwei-Hwang Lee, J. and M. Stiller. 1967. Hydrogenase activity in cell-free preparations of Chlorella. Biochem. Biophys. Acta. 132: 503.

62.   Lien, S. and A. San Pietro. 1975. An inquiry into biophotolysis of water to produce hydrogen. Indiana University and NSF.

63.   Livingston, R. S. and B. McNeill (Eds). 1975. Beyond Petroleum. Stanford University, Stanford.

64.   Losada, M., M. Nozaki and D. I. Arnon. 1961. Photoproduction of molecular hydrogen from thiosulfate by Chromatium cells. In: Light and Life. (Eds) W. D. McElroy and B. Glass. Johns Hopkins University Press, Baltimore. p. 570.

65. Mitsui, A. 1967. Physiological role of algal ferredoxin: Relation to photoproduction of hydrogen gas, photoreduction of NADP, photoreduction of nitrite, photofixation of nitrogen and photophosphorylation. In: Studies of mechanism in photosynthesis I. (Ed) A. Takamiya. Tokyo University, Tokyo, Japan. p. 53. (In Japanese).

66. Mitsui, A. 1975. Utilization of solar energy for hydrogen production by cell free system of photosynthetic organisms. In Hydrogen Energy, Part A. (Ed) T. N. Veziroglu, Plenum Publishing Co. , New York. p. 309.

67. Mitsui, A. 1975. Photoproduction of hydrogen via microbial and biochemical processes. In: Proceedings of Symposium-Course "Hydrogen Energy Fundamentals." (Ed) T. N. Veziroglu, University of Miami. S-2, 31.

68. Mitsui, A. 1975. Multiple utilization of tropical and subtropical marine photosynthetic organisms. In: The Proceedings of the Third International Ocean Development Conference. Seino Printing Co. 3: 11.

69. Mitsui, A. 1975. Photoproduction of hydrogen via photosynthetic processes. In: Proceeding of US-Japan Joint Seminar, "Key Technologies for the Hydrogen Energy System." (Ed) T. Ohta. Yokohama National University. p. 75. (Revised form of the symposium proceeding "Hydrogen Energy Fundamentals").

70. Mitsui, A. 1976. Bioconversion of solar energy in salt water photosynthetic hydrogen production system. In: Proceedings of the First World Hydrogen Energy Conference. (Ed) T. N. Veziroglu, University of Miami. 2: 4B-77.

71. Mitsui, A. 1976. Solar energy bioconversion by marine blue-green algae. In: Abstracts of the International Conference on the Photochemical Conversion and Storage of Solar Energy A2-3. Middlesex College, The University of Western Ontario.

72. Mitsui, A. 1976. Long-range concepts: Application of photosynthetic hydrogen production and nitrogen fixation research. In: Proceedings of a Conference on Capturing the Sun Through Bioconversion. The Washington Metropolitan Studies. Washington, D. C. p. 653.

73. Mitsui, A. 1976. A survey of hydrogen producing photosynthetic organism in sub-tropical and tropical marine environments (Abstract). In: Enzyme Technology and Renewable Resources. (Ed) J. L. Gainer. University of Virginia and NSF-RANN. p. 39.

74. Mitsui, A. 1976. A survey of hydrogen producing photosynthetic organisms in tropical and subtropical marine environment. NSF/RANN Annual Report.

75. Mitsui, A. and D. I. Arnon. 1962. Photoproduction of hydrogen gas by isolated chloroplasts in relation to cyclic and non-cyclic electron flow. Plant Physiol. 37S: IV.

76. Mitsui, A. 1974. The association of photosynthetic organisms with debris of macroalgae, seagrasses and mangrove leaves. Abstract of 1974 Annual Meeting of American Society for

Microbiology. p. 24.
77. Mitsui, A. and S. Kumazawa. Manuscript in preparation.
78. Mitsui, A., A. Paneque and D. I. Arnon. 1962. Photoreduction of methylviologen by isolated chloroplasts. (Manuscript) Photoproduction of hydrogen by chloroplast-methylviologen-hydrogenase system. (Manuscript) Photoproduction of hydrogen and cyclic and non-cyclic phosphorylation by chromatophores of blue-green algae. Quoted in following review papers by D. I. Arnon: Photosynthetic Mechanisms of Green Plants. (Eds) B. Kok and A. T. Jagendorf. Pub. No. 1145, Washington, D. C. NAS-NRC. p. 195. (1962) and Science 194: 1460 (1965).
79. NAS. 1973. Water scarcity may limit use of western coal. Science. 181: 525.
80. Neil, G., D. J. D. Nicholas, J. O'M. Bockris and J. F. McCann. 1976. The photosynthetic production of hydrogen. International J. of Hydrogen Energy. 1: 45.
81. Newton, J. W. 1976. Photoproduction of molecular hydrogen by a plant-algal symbiotic system. Science 191: 559.
82. Ohta, T. (Ed). 1975. U.S.-Japan Joint Seminar, Key Technologies for Hydrogen Energy System. U.S. National Science Foundation and Japan Society for Promotion of Science. Yokohama National University, Yokohama.
83. Ormerod, J. G. and H. Gest. 1962. Hydrogen photosynthesis and alternative metabolic pathways in photosynthetic bacteria. Bacteriol. Rev. 26: 51.
84. Ormerod, J. A., K. S. Ormerod and H. Gest. 1961. Light-dependent utilization of organic compounds and photoproduction of molecular hydrogen by photosynthetic bacteria: relationships with nitrogen metabolisms. Arch. Biochem. Biophys. 94: 449.
85. Oswald, W. J. and C.G. Golueke. 1960. Biological transformation of solar energy. Adv. in Applied Microbiol. 2: 223.
86. Peters, G. A., W. R. Evans and R. E. Toia, Jr. 1976. Azolla-Anabaena azollae relationship. IV. Photosynthetically driven, nitrogenase-catalyzed H_2 production. Plant Physiol 58: 118.
87. Radway, J. and A. Mitsui. Manuscript in preparation.
88. Radway, J., D. Rosner, J. Greenbaum and A. Mitsui. 1976. Association of blue-green algae and photosynthetic bacteria with macrophytes in the subtropical marine environment. Abstracts of the Annual Meeting of the American Society for Microbiology. p. 128.
89. Rao, K. K., L. Rosa and D. O. Hall. 1976. Prolonged production of hydrogen gas by a chloroplast biocatalytic system. Biochem. Biophys. Res. Comm. 68: 21.
90. Rosner, D., J. Radway and A. Mitsui. 1976. Isolation and growth physiology of blue-green algae from the tropical marine environment of the Atlantic ocean. Abstracts of the Annual Meeting of the American Society of Plant Physiologists. p. 106.

91. Schick, J. J. 1971. Interrelationship of nitrogen fixation, hydrogen evolution and photoreduction in Rhodospirillum rubrum. Arch. Mikrobiol. 75: 102.

92. San Pietro, A. and R. K. Togasaki. 1976. Bio-solar conversion: Search for algal hydrogenase with greater oxygen resistance. In: Enzyme Technology and Renewable Resources. (Ed) J. L. Gainer. University of Virginia and NSF-RANN, p. 45.

93. Spruit, C. J. P. 1954. Photoproduction of hydrogen and oxygen in Chlorella. In: Proceedings of First International Photobiology Congress. Amsterdam. p. 323.

94. Spruit, C. J. P. 1958. Simultaneous photoproduction of hydrogen and oxygen by Chlorella. Meded. Landbouwhogesch. Wageningen 58: 1.

95. Spruit, C. J. P. 1962. Photoreduction and anaerobiosis. In: Physiology and Biochemistry of Algae. (Ed.) R. A. Lewin. Academic Press, New York and London. p. 47.

96. Stuart, T. S. 1971. Hydrogen production by photosystem I of Scenedesmus: Effect of heat and salicylaldoxime on electron transport and photophosphorylation. Planta 96: 81.

97. Stuart, T. S. and H. Gaffron. 1971. The kinetics of hydrogen photoproduction by adapted algae. Planta 100: 228.

98. Stuart, T. S. and H. Gaffron. 1972. The mechanism of hydrogen photoproduction by several algae. I. The effect of inhibitors of photophosphorylation. Planta 106: 91.

99. Stuart, T. S. and H. Gaffron. 1972. The mechanism of hydrogen photoproduction by several algae. II. The contribution of photosystem II. Planta 106: 101.

100. Stuart, T. S. and H. Gaffron. 1972. The gas exchange of hydrogen-adapted algae as followed by mass spectrometry. Plant Physiol. 50: 130.

101. Stuart, T. S., E. W. Herold, Jr. and H. Gaffron. 1972. A simple combination mass spectrometer inlet and oxygen electrode chamber for sampling gases dissolved in liquids. Anal. Biochem. 46: 91.

102. Stuart, T. S. and H. Kaltwasser. 1970. Photoproduction of hydrogen by photosystem I of Scenedesmus. Planta 91: 302.

103. Tamiya, H. 1955. Growing Chlorella for food and feed. Proceedings of world symposium on applied solar energy. Phoenix, Arizona. p. 231.

104. Targett, N. and A. Mitsui. Manuscript in preparation.

105. Veziroglu, T.N. (Ed.). 1975. Symposium Proceedings: Hydrogen Energy Fundamentals. University of Miami, Miami.

106. Veziroglu, T.N. (Ed.). 1975. Hydrogen Energy, Parts A and B. Plenum Press, New York.

107. Veziroglu, T.N. (Ed.). 1976. Conference Proceedings of the 1st World Hydrogen Energy Conference. University of Miami, Miami.

108. Wang, R., R.P. Healy and J. Myers. 1971. Amperometric measurement of hydrogen evolution in Chlamydomonas. Plant Physiol. 48:108.

109.   Ward, M.A.   1970a.   Whole cell and cell-free hydrogenases of algae. Phytochemistry 9:259.
110.   Ward, M.A.   1970b.   Adaptation of hydrogenase in cell-free preparations from Chlamydomonas. Phytochemistry 9:267.
111.   Watanabe, A.   1970.   Studies on the application of Cyanophyta in Japan. Schweizerische Zeitschrift fur Hydrologie. 32:566.
112.   Whatley, F.R. and B.R. Grant.   1963.   Photoreduction of methyl viologen by spinach chloroplasts. Federation Proceedings. p. 227.
113.   Yagi, T.   1976.   Separation of hydrogenase-catalyzed hydrogen-evolution system from electron-donating system by means of enzymic electric cell techniques. Proc. Natl. Acad. Sci. U.S.A. 73:2947.

CATALYTIC AND STRUCTURAL PROPERTIES OF
THE ENZYME HYDROGENASE AND ITS ROLE IN
BIOPHOTOLYSIS OF WATER

ALVIN I. KRASNA

Department of Biochemistry
Columbia University
College of Physicians and Surgeons
New York, New York 10032

During the past few years there has been considerable research interest in the biophotolysis of
water by solar energy to produce hydrogen. Since this
reaction derives its required energy from the sun, it
has important practical implications since it provides
a means from the bioconversion of solar energy. The
hydrogen produced is an ideal energy source since its
only combustion product is water, free of the common
environmental pollutants. The water is returned to
the biosphere completing the cycle.
   The biological system used for hydrogen production consists of chloroplasts and the enzyme hydrogenase. The photosynthetic component catalyzes the
photolysis of water by sunlight via photosystems II
and I to produce oxygen and reduced ferredoxin, and
hydrogenase catalyzes the evolution of hydrogen gas
from reduced ferredoxin. Such a system is found naturally in intact hydrogenase-containing algae or can
be constructed by coupling isolated chloroplasts to
hydrogenase via a suitable electron carrier. The research being carried out in many laboratories focuses
on the development of a stable continuous system
capable of sustained production of large quantities of
hydrogen. Since hydrogenase is an integral component
of such systems, a knowledge of its structure and
catalytic properties is essential for its efficient
coupling to the photosynthetic component.
   Hydrogenase is an enzyme, found in many microorganisms, which reversibly activates molecular hydrogen. It is presumably present whenever hydrogen is
taken up or produced by organisms, with the exception

of the evolution of hydrogen by nitrogenase under certain conditions. The substrate, hydrogen, is the
simplest stable molecule and its existence in different isotopic and spin forms affords a convenient means
of studying its interaction with hydrogenase. The
catalytic activity of the enzyme can be assayed by any
method which measures its interaction with substrate.
There are four general methods most commonly employed
which are suitable for studying the enzyme in whole
cells, crude preparations, or in pure form.

The most commonly used method involves reduction of an acceptor by hydrogen

$$H_2 + \text{oxidized acceptor} \rightarrow \text{reduced acceptor}. \quad (1)$$

A large variety of acceptors have been used including
dyes, inorganic and organic compounds, as well as biological electron carriers. With the exception of dyes,
it appears that the reduction of an acceptor is the
resultant of two reactions; activation of hydrogen by
hydrogenase and transfer of the activated hydrogen to
the acceptor. The latter reaction often requires
other enzymes and electron carriers in addition to
hydrogenase.

Evolution of hydrogen from a reduced acceptor
of low potential according to the equation

$$\text{reduced acceptor} \rightarrow \text{oxidized acceptor} + H_2 \quad (2)$$

is also a convenient assay for hydrogenase. Viologen
dyes, ferredoxin, cytochrome c_3, NADH, and organic
compounds have been used in this reaction. With the
exception of viologen dyes, it appears that other
enzymes and carriers are required in addition to hydrogenase.

Direct activation of hydrogen by hydrogenase,
independent of added electron carriers, can be assayed
by measuring the enzymatic catalysis of the exchange
reaction between hydrogen and deuterated or tritiated
water,

$$H_2 + HDO \rightleftharpoons HD + H_2O \quad (3)$$

$$H_2 + HTO \rightleftharpoons HT + H_2O . \quad (4)$$

The direct activation of hydrogen by the enzyme can
also be studied by measuring the enzymatic catalysis
of the conversion of para hydrogen to ortho hydrogen,

$$pH_2 \ (\downarrow\uparrow) + H_2O \rightleftharpoons oH_2 \ (\uparrow\uparrow) + H_2O . \quad (5)$$

Detailed studies on the exchange and conversion reactions have established the mechanism of

hydrogen cleavage by hydrogenase. Conversion is observed in H_2O, but not in D_2O. In D_2O, the exchange
reaction produces HD and DD, with HD being produced
five times faster than DD. Since homolytic cleavage
of H_2 would not produce HD, its appearance suggests
that hydrogen is cleaved heterolytically to form an
enzyme hydride and a proton,

$$E + H_2 \rightleftharpoons EH^- + H^+ . \tag{6}$$

Conversion in H_2O can be represented as

$$E + pH_2 \ (\uparrow\downarrow) \rightleftharpoons EH^- + H^+ \tag{7}$$

$$EH^- + H^+ \rightleftharpoons E + oH_2 \ (\uparrow\uparrow) . \tag{8}$$

Clearly if the conversion is run in D_2O, reaction 8 would produce HD and no pH_2 would be converted to oH_2. The production of HD and DD in the
exchange reaction according to the reactions below

$$E + H_2 \rightarrow EH^- + H^+ \tag{6}$$

$$EH^- + D^+ \rightarrow E + HD \tag{9}$$

$$E + HD \rightarrow ED^- + H^+ \tag{10}$$

$$ED^- + D^+ \rightarrow E + DD \tag{11}$$

requires that HD be formed before DD.

The degree of hydride formation, as measured
by the ratio of HD formation to DD formation, differs
among different hydrogenases. Table 1 gives this
ratio for the hydrogenases from a number of microorganisms. Even for the hydrogenase from P. vulgaris,

Table 1

Ratio of HD to DD Formation for Different Hydrogenases

Organism	HD/DD
Proteus vulgaris	5.0
Chromatium	2.5
Clostridium pasteurianum	2.2
Scenedesmus	2.0
Desulfovibrio desulfuricans	0.9

where HD is formed five times faster than DD, the rate
of DD formation is greater than would be expected if
HD is an obligatory intermediate for DD formation,
and if the hydride on the enzyme is not exchangeable
with water. The other hydrogenases listed give even

more DD than expected. This could be due to exchange
of the hydride on the enzyme (which would be different
for different hydrogenases) or to a molecular cage
effect by which HD formed by the enzyme immediately
reacts again to form DD before the HD diffuses away
from the enzyme. Though a molecular cage effect has
not been proven, a diffusion cage effect has been
demonstrated at high enzyme concentration where diffusion of hydrogen gas becomes rate limiting.

In contrast to hydrogenase which cleaves hydrogen heterolytically, platinum cleaves hydrogen
homolytically. Platinum catalyzes the exchange reaction with HDO or HTO and the conversion reaction in
H_2O, but not in D_2O. The exchange reaction in D_2O
yields DD predominantly with only small amounts of HD.
This establishes that hydrogen is cleaved homolytically and the reaction can be represented as

$$H_2 \rightleftarrows 2H \cdot \tag{12}$$

$$2H \cdot \rightleftarrows 2H^+ + 2e \tag{13}$$

or as

$$H_2 + Pt \xrightarrow{k_1} HPtH \xrightarrow[k_3]{D^+} HPtD \xrightarrow[k_3']{D^+} DPtD \tag{14}$$

with the downward reactions:
$HPtH \xrightarrow{k_2} H_2 + Pt$, $HPtD \xrightarrow{k_2'} HD + Pt$, $DPtD \xrightarrow{k_2''} D_2 + Pt$

with k_3 and k_3' being much greater than k_2, k_2' and
k_2''.

Hydrogenase is an iron enzyme and is inhibited reversibly by CO. The inhibition is not reversed
by light. Of particular interest with respect to biophotolysis is the inhibition of hydrogenase by oxygen.
With many hydrogenases the inhibition is reversible and
due to the formation of an inactive oxygen complex.
With other hydrogenases the inhibition leads to irreversible loss of catalytic activity and cannot be
reversed by removal of oxygen. In either case, it
is clear that oxygen produced during photolysis of
water by chloroplasts would have an adverse effect on
hydrogenase.

Depending on the microorganism used as a
source of enzyme, hydrogenase can be obtained in
soluble or particulate form which can be solubilized
by various means. The enzymes from Desulfovibrio

vulgaris, Clostridium pasteurianum, and Chromatium
have been purified to homogeneity. The molecular
weight of the enzyme from the first two organisms is
60,000 and the molecular weight of Chromatium hydrogenase is 100,000. The respective isoelectric points
are 6.25, 5.0, and 4.3. The Desulfovibrio and
Chromatium enzymes are reported to be composed of two
subunits. All three hydrogenases contain non-heme
iron and acid-labile sulfide, generally in equimolar
quantities, and visible and EPR spectra establish that
hydrogenase is an Fe-S protein in which the structure
of the chromophore is similar to that found in the
ferredoxins.
        Certain species of photosynthetic algae con-
tain hydrogenase activity which appears only after a
dark anaerobic "adaptation" period. This "adaptation",
not requiring protein synthesis, involves conversion
of an inactive form of hydrogenase to an active one.
Complete removal of oxygen is insufficient to activate
the hydrogenase and the process requires the intact
cell, since freezing and thawing cells inhibits
adaptation. Adaptation is inhibited by iron chelating
or sulfhydryl and disulfide reagents, which have no
effect on hydrogenase after adaptation. Adapted
algae carry out hydrogen photosynthesis at low light
intensities. In this reaction carbon dioxide is reduced by hydrogen in the light. In the absence of
carbon dioxide, adapted algae photoproduce hydrogen.
At high light intensities, hydrogenase is inactivated
by oxygen produced during normal photosynthesis. Inactivation of algal hydrogenase by oxygen is not reversed by removal of oxygen. A period of dark anaerobic adaptation is required to reactivate the enzyme.
        From the properties of hydrogenase described
it is apparent that a major difficulty in the biophotolysis scheme is the evolution of oxygen by the
photosynthetic component during photolysis of water.
Oxygen inhibits hydrogenase and oxidizes low potential
electron carriers which couple chloroplasts to hydrogenase. A second major problem is the instability of
biological systems for long term use.
        In this laboratory the requirements for photo-
production of hydrogen by adapted algae was studied in
Scenedesmus 393 and Chlorella 11C. Both organisms
showed good photosynthetic activity, as measured by
oxygen evolution with quinone as acceptor, and hydrogenase activity, as measured by reduction of methylene
blue with hydrogen. In the absence of oxygen removing

systems, freshly harvested cells of both algae produced hydrogen in the light for six hours, with Chlorella being five times more active than Scenedesmus in rate and quantity of hydrogen evolved. The reaction was light dependent and inhibited by DCMU establishing that water was the source of electrons for hydrogen formation.

As the cells aged for one week or more at 4°, the hydrogen photoproduction activity of Scenedesmus cells remained unchanged. However, the high activity of Chlorella decreased from a rate of 2 μmole/min to 1 μmole/min and the total gas evolved in six hours decreased from 200 μmoles to 30 μmoles. (Data are per ml packed cells.) This may be due to the presence of an endogenous oxygen acceptor in freshly harvested Chlorella cells which is depleted on storage, while Scenedesmus cells initially have very little endogenous acceptor.

These cells were used to test the effectiveness of different agents for the removal of oxygen produced during the photoproduction of hydrogen. Sodium dithionite in a center well, which is an effective oxygen removing system, did not stimulate hydrogen production. Glucose and glucose oxidase had only a slight stimulatory effect. These two systems are effective in deoxygenating hydrogenase, and their failure to stimulate photoproduction of hydrogen would suggest that oxygen produced during photosynthesis is not inhibiting hydrogenase but is oxidizing low potential carriers which couple chloroplasts to hydrogenase. Addition of sodium dithionite to the reaction produced a considerable stimulation of hydrogen photoproduction. With Scenedesmus the initial rate increased from 0.3 to 1.5 μmoles/min and the total hydrogen produced increased from 40 μmoles to 285 μmoles, a level not attainable previously with these cells. With Chlorella, the rate increased from 1 to 3 μmoles/min and the total hydrogen production increased from 30 μmoles to 300 μmoles. The stimulation was light dependent and inhibited by DCMU suggesting that dithionite was not acting as an electron donor to the photosystems but as an effective oxygen removing agent. The action of these different oxygen removing systems in stimulating hydrogen photoproduction would suggest that in algae the oxygen produced intracellularly during photosynthesis inhibits hydrogen production before oxygen diffuses out of the cell.

The biophotolysis of water to produce hydrogen was also studied in a coupled system of isolated chloroplasts and bacterial hydrogenase. To determine the rate limiting step in the coupled system, each component was studied separately. Chloroplasts which actively evolved oxygen in the presence of Hill acceptors or consumed oxygen in a Mehler reaction with methyl viologen, reduced methyl viologen anaerobically at a slow rate in the presence of glucose, glucose oxidase, catalase, and ethanol. The maximum concentration of reduced methyl viologen obtainable was 100 µM and could not be increased by raising the total concentration of viologen. Hydrogen production from this steady state concentration of reduced methyl viologen in the dark in the absence of chloroplasts with dithionite as reductant was very rapid with different hydrogenases or with platinum or palladium as catalyst. Increasing the viologen concentration increased the rate of hydrogen production. It is clear that the rate limiting step in a coupled system of chloroplasts and hydrogenase is the photoreduction of the coupling electron carrier by the chloroplast component. It is this reaction which must be increased in order to achieve reasonable levels of continuous hydrogen production in a coupled system.

In developing partial or completely synthetic systems for the photoproduction of hydrogen, the replacement of hydrogenase by heavy metal catalysts was studied. Platinum, palladium, and rhodium catalysts were active in the reduction of methylene blue or benzyl viologen with hydrogen and the evolution of hydrogen from reduced methyl viologen. Often synthetic catalysts showed greater activity than hydrogenase and should be able to replace the enzyme in hydrogen photoproducing systems.

Though there are no efficient synthetic systems for the direct photolysis of water, synthetic photoreducing systems have been described and serve as useful models. The reported photoreduction of methyl viologen by EDTA in the presence of proflavin was studied as a possible synthetic system independent of chloroplasts. This system is most efficient at 450 nm, which is the wavelength of maximal spectral intensity of sunlight. The optimum conditions for photoreduction of methyl viologen were: proflavin, 0.04 µM; methyl viologen, 0.4 or 4 µM; EDTA, 4 or 40 µM; phosphate buffer, 4 mM, pH 6.0. The higher concentrations of methyl viologen and EDTA

gave slightly faster rates of reduction but lower percentage yields. Addition of hydrogenase or synthetic
catalysts to the reduced methyl viologen in the dark
led to its rapid reoxidation. In a continuous system
in the light containing the photoreducing components
and hydrogenase, substantial quantities of hydrogen
were evolved continuously for over six hours. The
system was made completely synthetic by replacement
of hydrogenase with platinum or palladium. These
catalysts evolved hydrogen at a more rapid rate and to
a greater extent than the enzyme. It is clear that
once the carrier is reduced, either chemically in the
dark or by a photoreducing system, synthetic catalysts
can readily replace hydrogenase for hydrogen production.
The use of EDTA as a source of electrons for hydrogen
production is not practical and more suitable electron
donors, preferably available as waste products, must
be sought.

USE OF AN ENZYMIC ELECTRIC CELL AND IMMOBILIZED HYDROGENASE IN THE STUDY OF THE BIOPHOTOLYSIS OF WATER TO PRODUCE HYDROGEN *

TATSUHIKO YAGI

Department of Chemistry, Shizuoka University, Shizuoka 422, Japan

Summary

Recently much attention has been focused on the "biophotolysis of water" in which solar energy is converted to chemical energy stored in the molecules of hydrogen, a non-polluting fuel, to meet future needs of energy supply.

Artificial water-biophotolytic systems so far reported consisted of chloroplasts which photoexcite electrons to reduce a low-potential carrier and hydrogenase which releases hydrogen in the presence of the reduced low-potential carrier. Difficulties encountered in this system arose from the facts that hydrogenase-catalyzed hydrogen-producing system is very much susceptible to oxygen produced by the chloroplast photosynthetic system and that each component of this system is unstable when isolated from the cells.

In this paper, I will report on the use of an enzymic electric cell to separate hydrogenase-catalyzed hydrogen-producing system (cathode reaction) from the electron-donating system (anode reaction) in order to protect hydrogenase from the by-products of electron-donating reactions. Another problem to be overcome is the instability of the isolated components such as chloroplasts and hydrogenase. An attempt to stabilize hydrogenase by immobilization technique will be reported.

Introduction

Recently attention has been focused on the "biophotolysis of water to produce hydrogen" in which solar energy is transformed into chemical

* Supported in part by a grant (No. 111911) from the Ministry of Education, Science and Culture of Japan.

energy stored in the molecules of hydrogen. The principle of this process is based on the earlier observations of Gaffron & Rubin (7) and Gest & Kamen (6) that some green algae and photosynthetic bacteria produced hydrogen when illuminated under certain conditions.

Feasibility of this process is now being investigated by several workers in order for us to prepare to meet future needs of energy supply (9, 12). This process consists of photoreduction of a low-potential electron carrier with concomitant production of oxygen, and the reduction of protons to produce hydrogen using the reduced low-potential carrier by means of hydrogenase. An electron flow in this process is schematically illustrated in Fig. 1. This scheme is so familiar to those who are interested in "biophotolysis of water" that no superfluous explanations will be made here.

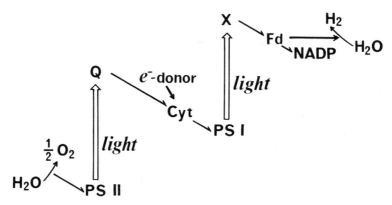

Fig. 1.   Flow of electrons in the "biophotolysis of water."

This process seems feasible at glance, but includes a problem that hydrogenase and its electron carrier are susceptible to oxygen, which is formed by the chloroplast photoelectron-donating reaction.

In order for us to produce hydrogen by biophotolysis of water, protection of hydrogenase and its electron carrier from oxygen seem essential. Immobilization of hydrogenase and chloroplasts confers them stability and resistance to oxygen to some extent (10, 11, 14, 15). I have suggested a use of an enzymic electric cell to separate a hydrogenase-catalyzed hydrogen-evolution system (cathode) from an electron-donating system (anode) in order that hydrogenase and its electron carrier might not be affected by by-products of electron-donating reaction (16).

In this paper, I shall report on the preparations of immobilized hydrogenase which retain activity and resist against inactivation during repeated use, and an application of immobilized hydrogenase in the above mentioned hydrogen-producing system.

Materials and Methods

Apparatus

The enzymic electric cell used in this study was the same as that reported in the previous paper (16). The cell container was connected with a Warburg manometer, and the change in the gas volume was measured by conventional manometric technique.

Hydrogenase

Cytochrome c_3 hydrogenase (hydrogen:ferricytochrome c_3 oxidoreductase, EC 1.12.2.1) was purified from particulate fraction of cells of *Desulfovibrio vulgaris*, Miyazaki, as described before (18). The activity of hydrogenase, either in aqueous solution or in the immobilized state was assayed by the enzymic electric cell method (17).

Results

Immobilization of hydrogenase

Two kinds of immobilized hydrogenase gel-pieces were prepared. (1) Hydrogenase entrapped in normal polyacrylamide gel: (i) a solution containing 750 mg of acrylamide and 40 mg of N, N'-methylenebisacrylamide in 3.7 ml of 0.02 M phosphate buffer, pH 7.0, (ii) an 0.6 ml portion of purified hydrogenase (13.6 units/ml), (iii) 0.5 ml of 5 % N, N, N', N'-tetramethylethylenediamine, and (iv) 0.5 ml of freshly prepared 1.0 % ammonium peroxodisulfate solution, were mixed in this order, and left to stand for gelation. (2) Hydrogenase entrapped in polyacrylamide gel with cross-linker, glutaraldehyde: (i) a solution containing 750 mg of acrylamide and 40 mg of N,N'-methylenebisacrylamide in 3.1 ml of 0.02 M phosphate buffer, pH 7.0, (ii) an 0.2 ml portion of purified hydrogenase (130 units/ml), (iii) 1.0 ml of aqueous 25 % glutaraldehyde solution, (iv) 0.5 ml of 5 % N, N, N', N'-tetramethylethylenediamine solution, and (v) 0.5 ml of freshly prepared 1.0 % ammonium peroxodisulfate solution, were mixed in this order, and left standing for gelation. Both of the hydrogenase-containing gel thus prepared were cut into pieces of 70 mg weight, and the activity was measured. In typical experiments, the activity of two pieces of gel was about 25 milliunits. When these pieces were removed from the reaction mixture, the residual activity was less than 1 milliunits.

Stability of the immobilized hydrogenase during storage

Stability of hydrogenase preparations, either in an aqueous solution
or in the immobilized state were compared, and the results are shown
in Fig. 2.

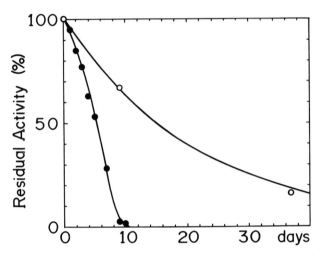

*Fig. 2. Stability of hydrogenase during storage.   Hydrogenase
preparations were left standing at room temperature (26°)   for
several days, and their activities were   assayed   at   the   time
indicated. ○: the immobilized hydrogenase,   and ●: an aqueous
hydrogenase solution.*

Stability of the immobilized hydrogenase during repeated use

The activity of gel-pieces of the immobilized hydrogenase was assayed by the enzymic electric cell method, the gel-pieces were removed
from the cell, stored for one or several days at 4° in 0.1 M phosphate
buffer or in water, and the activity was assayed again. This was repeated several times, and the results are shown in Table I.

Table I. *Relative activity of the reused hydrogenase*

	1st	2nd	3rd	4th	5th
(a)	100 %	96 %	100 %	85 %	94 %
(b)	100 %	61 %	33 %	18 %	

(a) *hydrogenase entrapped in acrylamide gel with glutaraldehyde,*
(b) *hydrogenase entrapped in normal acrylamide gel.*

Cathodic hydrogen evolution catalyzed by the immobilized hydrogenase

Twelve pieces of the glutaraldehyde-treated hydrogenase-gel were placed in the cathode container of the enzymic electric cell described in the previous paper, in which hydrogenase-catalyzed hydrogen-evolution reaction was separated from the electron-donating reaction (16), and the rate of hydrogen evolution and the short-circuit current were measured. The results are illustrated in Fig. 3.

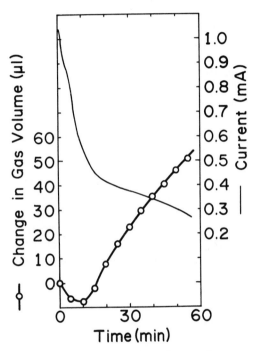

Fig. 3.   Time-course curves for the change in gas volume and in the short-circuit current of the enzymic electric cell. The reaction mixture in the cathode contained 12 pieces of the glutaraldehyde-treated hydrogenase gel and 0.5 ml of the methylviologen-phosphate stock solution* in 3.0 ml.   The anode was a zinc electrode inserted in ammonium chloride solution.   Gas phase was nitrogen.   The reaction was started by closing the circuit as described previously (16), and the change in gas volume and current were recorded. No gas-evolution was observed in the absence of hydrogenase.   (* a mixture of 3 volumes of 0.2 M phosphate buffer, pH 7.0, and 1 volume of 0.04 M methylviologen)

Discussion

Since the discovery of Gaffron & Rubin (7) and Gest & Kamen (6) on the photoevolution of hydrogen from photosynthetic microorganisms, several papers had dealt with photoproduction of hydrogen by photosynthetic algae, bacteria, or cell-free systems thereof. Arnon et al. (1) observed photoproduction of hydrogen from a mixture containing spinach chloroplasts and clostridial hydrogenase (EC 1.12.7.1) in the presence of cysteine which acted as an electron donor for photosystem I. Mitsui (13), Ben-Amotz & Gibbs (2), and Kitajima & Butler (10) also succeeded in photoproduction of hydrogen from mixtures consisting of chloroplasts and hydrogenase or immobilized forms thereof in the presence of an electron donor for photosystem I such as ascorbate or dichlorophenolindophenol. No hydrogen production was observed in the absence of the electron donor. The addition of an electron donor for photosystem I, however, almost nullified the effectiveness of this process as a means for obtaining industrial energy from solar energy, since much industrial energy must have been supplied to produce these electron donors.

Benemann et al. (3) observed production of hydrogen in the absence of externally added electron donors by illuminating a chloroplast-ferredoxin-hydrogenase system, in which water was acting as an electron donor. The rate of hydrogen production, however, was extremely low, and the system deteriorated rapidly. Rao et al. (15) made similar observations with chloroplasts fixed by glutaraldehyde-treatment. The rate of hydrogen production in these cases were much improved when these reaction systems were supplemented with glucose and glucose oxidase (EC 1.1.3.4) which acted as a scavenger of oxygen. In fact, the addition of a scavenger of oxygen was essential to maintain the activities of these systems. However, this will merely cause competition between industrial energy and food source for agricultural products. We must, therefore, try to produce hydrogen without the aid of any electron donors or scavengers of oxygen.

Benemann & Weare (4) succeeded in simultaneous photoproduction of hydrogen and oxygen from water under the atmosphere of argon using actively growing cultures of *Anabaena cylindrica*. In this case, photoexcited electrons were transferred to protons to produce hydrogen via nitrogenase system instead of hydrogenase, and therefore, the reaction was not sensitive to oxygen. This advantage was, however, counterbalanced by the fact that hydrogen production was strongly inhibited by nitrogen. Their observations that hydrogen evolution by nitrogen-starved cultures in the dark was severalfold higher than acetylene reduction in the dark suggested the contribution of endogenous electron donors in the hydrogen-evolution reaction. The extent of the contribution by purely biophotolytic process in the production of hydrogen was not estimated, nor any cell-free systems have been obtained. The activities of the growing cultures might be controlled by unexpected regulating mechanisms which might inhibit continuous production of hydrogen. Thus, protection of hydrogenase and its electron donors from oxygen, and preparation of the cell-free system seem essential to produce hydrogen by biophotolysis of water. Lappi et al. (11) had a limited success in preparing stable immobilized hydrogenase particles.

The immobilized hydrogenase preparations reported in the present paper are resistant to atmospheric oxygen, and maintain their activities during repeated use. In other words, the stability of hydrogenase was strengthened to a level of that of inorganic heavy-metal catalysts. Preference of the stable hydrogenase preparations to inorganic catalysts is obvious. Any heavy-metal catalysts, when deteriorated, might cause pollution, whereas the immobilized hydrogenase which consisted of mainly organic matters with a trace of iron, when deteriorated, could be incinerated without fear of pollution.

Before we can succeed, however, a stable photoelectron-donating system must be prepared. Recent observations by Fong & Winograd (5) on photogalvanic effects of chlorophyll-quinhydrone half-cell reactions, or the use of semiconductors such as TiO_2 as suggested by Fujishima & Honda (6), could be helpful in producing hydrogen by biophotolysis of water in the enzymic electric cell.

I am deeply indepted to Miss Yoko Mizukami for her excellent technical assistance.

References

1. Arnon, D. I., A. Mitsui, & A. Paneque. 1961 Photoproduction of hydrogen gas coupled with photosynthetic phosphorylation. Science 134: 1425.
2. Ben-Amotz, A. & M. Gibbs. 1975. H_2 metabolism in photosynthetic organisms. II. Light-dependent H_2 evolution by preparations from Chlamidomonas, Scenedesmus and spinach. Biochem. Biophys. Res. Commun. 64: 355-359.
3. Benemann, J. R., J. A. Berenson, N. O. Kaplan, & M. D. Kamen. 1973. Hydrogen evolution by a chloroplast-ferredoxin-hydrogenase system. Proc. Nat. Acad. Sci. USA 70: 2317-2320.
4. Benemann, J. R. & N. M. Weare. 1974. Hydrogen evolution by nitrogen-fixing Anabaena cylindrica cultures. Science 184:174-175.
5. Fong, F. K. & N. Winograd. 1976. In vitro solar conversion after the primary light reaction in photosynthesis. Reversible photogalvanic effects of chlorophyll-quinhydrone half-cell reactions. J. Amer. Chem. Soc. 98: 2287-2289.

6. Fujishima, A. & K. Honda. 1972. Electrochemical photolysis of water at a semiconductor electrode. Nature 238: 37-38.
7. Gaffron, H. & J. Rubin. 1942. Fermentative and photochemical production of hydrogen in algae. J. Gen. Physiol. 26: 219-240.
8. Gest, H. & M. D. Kamen. 1949. Studies on the metabolism of photosynthetic bacteria. IV. Photochemical production of molecular hydrogen by growing cultures of photosynthetic bacteria. J. Bacteriol. 58: 239-245.
9. Gibbs, M., A. Hollaender, B. Kok., L. O. Krampitz, & A. San Pietro. organizers. 1973. Proc. Workshop Bio-Solar Conversion, Dept. Plant Sci., Indiana Univ.
10. Kitajima, M. & W. L. Butler. 1976. Microencapsulation of chloroplast particles. Plant Physiol. 57: 746-750.
11. Lappi, D. A., F. E. Stolzenbach, N. O. Kaplan, & M. D. Kamen. 1976. Immobilization of hydrogenase on glass beads. Biochem. Biophys. Res. Commun. 69: 878-884.
12. Lien, S. & A. San Pietro. 1975. An Inquiry into Biophotolysis of Water to Produce Hydrogen. Dept. Plant Sci., Indiana Univ.
13. Mitsui, A. 1975. The utilization of solar energy for hydrogen production by cell free system of photosynthetic organisms. *In* T. N. Veziroglu, ed., Hydrogen Energy. Part A. Plenum Publ. Corp. New York. pp. 309-316.
14. Packer, L. 1976. Problems in the stabilization of the in vitro photochemical activity of chloroplasts used for hydrogen production. FEBS Lett. 64: 17-19.
15. Rao, K. K., L. Rosa, & D. O. Hall. 1976. Prolonged production of hydrogen gas by a chloroplast biocatalytic system. Biochem. Biophys. Res. Commun. 68: 21-28.
16. Yagi, T. 1976. Separation of hydrogenase-catalyzed hydrogen-evolution system from electron-donating system by means of enzymic electric cell technique. Proc. Nat. Acad. Sci. USA 73: 2947-2949.
17. Yagi, T., M. Goto, K. Nakano, K. Kimura, & H. Inokuchi. 1975. A new assay method for hydrogenase based on an enzymic electrode reaction. The enzymic electric cell method. J. Biochem. 78: 443-454.
18. Yagi, T., K. Kimura, H. Daidoji, F. Sakai, S. Tamura, & H. Inokuchi. 1976. Properties of purified hydrogenase from the particulate fraction of Desulfovibrio vulgaris, Miyazaki. J. Biochem. 79: 661-671.

THE MECHANISM OF HYDROGEN PHOTOEVOLUTION IN PHOTOSYNTHETIC ORGANISMS

DAN KING, DAVID L. ERBES, AMI BEN-AMOTZ[1], and MARTIN GIBBS

Institute for Photobiology of Cells and Organelles
Brandeis University, Waltham, Mass. 02154

Most studies of H_2 photoevolution in photosynthetic organisms have been carried out with in vivo systems. Recently, several reports have been presented utilizing cell-free systems. Hydrogen formation in a preparation of Chlamydomonas eugametos has been reported by Abeles (1). Benneman et al. (4) has reported a reconstituted spinach chloroplast-clostridial hydrogenase system capable of light-dependent H_2 evolution from water. Roa et al.(12) reported similar results using spinach or lettuce chloroplasts and E.coli or clostridial hydrogenase. We have also presented some results utilizing a cell-free, hydrogenase-containing system from Chlamydomonas reinhardi (3). The cells were broken by extended sonication in a medium of high osmolarity. The resulting preparations, while capable of photoevolution of H_2, were not completely satisfactory with respect to reproducibility. In this report, we have adapted a method of brief sonication of algal cells in a medium of low osmolarity (6) to obtain stable, sustained rates of H_2 photoevolution. With these preparations, we have undertaken to evaluate and characterize the participation of the photosystems in H_2 evolution. The effects of ferredoxin, electron donors, and inhibitors are also presented.

MATERIALS AND METHODS

Cultures of Chlamydomonas reinhardi (Wt and F60[2]) were grown on the TAP medium of Gorman and Levine (7). Cells were harvested by centrifugation and resuspended in 10mM KH_2PO_4, 20mM KCl, 2.5mM $MgCl_2$, pH 7.4, at a final chlorophyll concentration of 0.5-1.0 mg/ml. Cells were adapted under N_2 and sonicated anaerobically with a Branson Sonifer 200, 75W, for 15-20 sec at 0 C in the dark.

[1]Present address: Israel Oceanographic & Limnological Research, Ltd., Haifa Laboratories, Haifa, 21 Hativat Golani Road.
[2]F60 is a mutant strain of C. reinhardi deficient in phosphoribulokinase (11).
[3]Abbreviations: Asc, Ascorbic acid; DBMIB, Dibromothymoquinone; DCMU, Dichlorophenylmethylurea; DPIP, Dichlorophenolindolphenol; DSPD, Disalicylidenepropanediamine; DTT, Dithiothreitol; FCCP, Carboxycyanide p-trifluoromethoxyphenylhydrazone; PMS, Phenazine methosulfate; PSI, PSII, Photosystem I, Photosystem II.

The resulting suspensions were centrifuged at 500g for 10 min
under N_2. Supernatants were transferred anaerobically into serum
bottles. Three milliliters were used for manometric assays.
Problems incurred with oxygen evolution during particle preparation were eliminated by the use of the mutant F60.

C. reinhardii hydrogenase was prepared from mass cultures.
About 200g of wet cell paste were suspended in 50mM Tris-Cl, pH
8.0, containing 10µM DCMU[3], to give a final volume of 1.2 liters.
Five minutes of evacuation followed by H_2 flushing was repeated
three times and the suspension was frozen in liquid N_2. The
thawed suspension was subjected to successive ammonium sulfate
precipitations anaerobically and centrifugation yielded a precipitate containing both ferredoxin and hydrogenase. The pellet was
dissolved in appropriate buffers for use or further purification.

Hydrogenase activity was monitored during purification and
handling by methyl viologen reduction. The appearance of reduced
methyl viologen was followed at 605nm and an extinction coefficient of $9.3mM^{-1}cm^{-1}$ was used to calculate the rate of reduction.

Spinach chloroplasts were prepared according to Avron and
Gibbs(2). C. reinhardi ferredoxin was prepared as a byproduct
of hydrogenase preparations. Chlorella pyrenoidosa ferredoxin was
a gift of A. Schmidt and J. Schiff. Clostridium pasteurianum
ferredoxin was kindly supplied by R.H. Burris.

NADP photoreduction was performed with an Eppendorf fluor-
imeter attached to a Varian recorder and modified for illumination of the cuvette. Actinic light was provided by a 500 W pro-
jection lamp filtered through a Baird-Atomic, Inc., No. 6400 in-
terference filter.

All operations were carried out under conditions of strict
anaerobicity. Nitrogen was used only after being assayed at less
than 5µl/1 O_2. Hydrogen was purified by passage through Deoxo
cartridges and/or through a 120cm column of BASF R3-11 catlayst
heated to 155 C.

RESULTS

Cell-free preparations of anaerobically adapted and sonicated algae demonstrated light-dependent H_2 evolution upon the
addition of hydrogenase and suitable electron donors (Table I).
Longer sonication periods resulted in a loss of photoevolutionary
capacity. Dithiothreitol and NADH served as the most effective
electron donors. A low concentration of PMS (7µM) eliminated H_2
photoevolution as did the presence of DSPD, an inhibitor of ferredoxin (14). The plastoquinone inhibitor, DBMIB (5), eliminated
evolution from NADH, but only partially inhibited the flow of
electrons from DTT.   The presence of NADP had no effect on evolution from NADH while NADPH was inhibitory. NADPH would not serve
as a source of electrons for H_2 evolution.

Figure 1 records the effects of the presence of 10µM DCMU
on H_2 evolution utilizing NADH or DTT as electron donors. Gas
evolution was reduced by 50%. Light intensities were held at low
levels (50-100 footcandles) since in the absence of DCMU and

TABLE I
Stimulation of H2 Photoevolution by Reduced Compounds Using
Sonicated Chlamydomonas reinhardi

The complete reaction mixture (3.2ml) contained an aliquot
of hydrogenase, the indicated electron donor, 10^{-5}M DCMU, and
supernatant fraction of broken particles at 250-300µg Chl/ml.

Donor	H2 Evolution µMol/mg Chl/hr	Donor	H2 Evolution Mol/mg Chl/hr
NADH, 2mM	6.1	DTT, 2mM	7.4
", + 7µM PMS	0.2	", + 7µM PMS	0.0
", + 1mM DSPD	0.0	", + 1mM DSPD	0.0
", + 10µM DBMIB	0.1	", + 10µM DBMIB	4.0
", + 10mM NADP	5.8	Asc, 10mM, + 30µM DPIP	0.1
", + 2mM NADPH	2.3	Cysteine, 10mM	1.6
NADPH, 2mM	0.2	Glutathione, 10mM	0.5
Minus donor	0.0-1.0	Dark	0.0-0.5

despite the presence of alkaline pyrogallol, higher intensities
inhibited gas evolution in these algal systems. Fig. 1 also demonstrates the apparent uncoupling of photophosphorylation from
H2 evolution in the sonicated algae. Addition of the uncoupler,
FCCP, not only stimulated the rate of H2 evolution selectively
with NADH relative to DTT, but also resulted in a linear time
course. Hydrogen formation in the presence of DTT was already
enhanced, presumably due to the uncoupling action of DTT (10).
The lag in H2 evolution from NADH may have been due to a period
required for uncoupling phosphorylation from electron flow within the particles. Sensitivity to uncouplers was limited to particles prepared with shorter sonication times. The presence or
absence of a lag period may also represent the degree of uncoupling of particles due to variations in preparation.
The question of the role of ferredoxin in the mechanism of
H2 photoevolution was studied in two respects. Particles were
prepared and washed anaerobically. When these particles were reconstituted with a ferredoxin-free hydrogenase and the suitable
donors were added, no H2 photoevolution was observed. The addition of ferredoxin to this system restored activity. Secondly,
we compared the apparent Km values of various ferredoxins in dark
H2 evolution from dithionite using a hydrogenase prepared from C.
reinhardi. These apparent Km values were compared with the Km
values for NADP photoreduction catalyzed by spinach chloroplast
membranes (Table II). The data indicate a basic compatibility
of ferredoxins and a much more efficient transfer of electrons
to NADP reduction than to protons for hydrogen production.

CONCLUSIONS
The preparation of anaerobic particles by brief sonication
resulted in cell-free preparations capable of reasonable, repro-
ducible, and stable rates of H2 photoevolution. The particles
demonstrated a requirement for exogenous electron sources and a

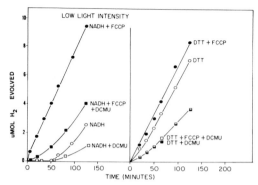

Figure 1. Effects of FCCP and DCMU on Hydrogen Photoevolution by Sonicated Chlamydomonas.

dependency upon ferredoxin.   The best donors are those compounds which are poor acceptors of electrons.

It would appear that both photosystems are involved in the flow of electrons for hydrogen photoevolution.   Inhibition by DCMU indicated that the in vitro system mimics the in vivo algal cell with respect to photosystem involvement.   Whether water is split under our conditions to provide electrons for PS II remains an open question.   It is clear that electrons are donated to the oxidizing side of PS II.   At very low light intensities, the rate of O_2 evolution may be below the rate of respiratory uptake.

The stimulation of H_2 evolution by FCCP indicates that the oxidation of NADH was coupled to an energy-conserving site.   In

TABLE II

Apparent Km Values of Algal and Bacterial Ferredoxins for Hydrogen Evolution and NADP Photoreduction

The reaction mixture(3ml) for dark H_2 evolution contained: 25mM Na, K phosphate, pH 7.3; 2mM $MgCl_2$, 20mM dithionite; required ferredoxin; and 9mg/ml (2.85 ml) protein of crude Chlamydomonas hydrogenase prepared as in Methods and Materials. The reaction mixture (1ml) for NADP photoreduction contained: 25mM Na, K phosphate, pH 7.3; 2mM KCl, 0.1mM NADP, ferredoxin as required; and spinach chloroplasts containing 30µg chlorophyll.

Reaction	Ferredoxin	Specific Activity µmol/mg Prot./hr	Apparent Km µM
Dark H_2 evolution	Chlamydomonas	0.4	20
with Chlamydomonas	Clostridium	0.2	50
hydrogenase	Clorella	0.18	62
NADP Photoreduction	Chlorella	36	0.7
with spinach chloro-	Clostridium	33	0.8
plasts	Chlorella	40	0.8

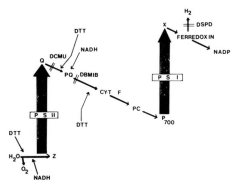

Figure 2. Suggested Mechanism for Hydrogen Photoevolution In
Chlamydomonas reinhardi

some preliminary experiments, a molar ratio of ATP to H_2 of 0.8
± 0.15 (five determinations) was observed, but until highly purified hydrogenase of Chlamydomonas is available, stoichiometric
measurements of light-dependent H_2 evolution, ATP formation, and
NADH oxidation must be considered tentative. Another complication
is the possibility of a cyclic pathway of photophosphorylation
involving the formation of reduced pyridine nucleotide by PS I and
its subsequent reoxidation by plastoquinone.

The requirement for ferredoxin appears to be established and
the basic compatibility of ferredoxins from various sources with
Chlamydomonas hydrogenase denies specificity with regard to transfer of electrons from the photosystems to hydrogenase. The much
lower apparent Km for NADP photoreduction relative to the ferredoxin Km for dark hydrogen evolution apparently reflects the
efficiency with which the photosystems shuttle electrons to various acceptors.

The mechanism shown in Fig. 2 is an interpretation of the
available data. Photosystems are viewed as photoevolving H_2 from
reduced pyridine nucleotide by the introduction of electrons
either into plastoquinone or into the oxidizing site of photosystem II. Non-specific donors could donote electrons anywhere a
proper redox potential existed. Water may be photooxidized and
compete with other donors for the site of donation in PS II. DCMU
would block any PS II-dependent activity and DBMIB would interfere
with any plastoquinone-dependent activity. The transfer of electrons from ferredoxin through hydrogenase to protons is thought
to be the mechanism of hydrogen generation.

Finally, the source of reduced pyridine nucleotide must be
considered. Studies with higher plant chloroplasts (8,13) indicate
that the pyridine nucleotides, oxidized or reduced, cannot cross
the plastid envelope, resulting in the concept of a glyceralde-
hyde-3-P/glycerate-3-P shuttle to transfer reducing equivalents
from plastid to cytoplasm. Since, in intact cells, H_2 evolution
is accompanied by the formation of CO_2 (9), a reduced carbon

substance may well serve as the source of both gases. Thus we envisage a shuttle with cytoplasmic glyceraldehyde-3-P formed during the fermentation of cellular substances as the substrate for oxidation by the chloroplast-containing glyceraldehyde-3-P dehydrogenase.

SUMMARY

NADH and DTT supported light-dependent H_2 evolution using cell-free extracts of Chlamydomonas reinhardi. Light-dependent H_2 evolution was eliminated by 7μM PMS or 1mM DSPD while 10μM DCMU provided only 50% inhibition. DBMIB (10μM) partially inhibited H_2 evolution from DTT and completely blocked evolution from NADH. These data indicate that DTT and NADH can donate electrons to both the oxidizing and reducing sides of PS II. Beyond the site of DCMU action electrons from NADH are funnelled into or near PQ while DTT contributes electrons to a number of sites.

Reduced ferredoxins from Chlamydomonas, Chlorella, and Clostridium were able to act as electron donors to C. reinhardi hydrogenase. The apparent Km's for ferredoxin were much higher for H_2 evolution than for NADP photoreduction.

Acknowledgements; This research was generously supported by National Science Foundation Grant BMS 71-00978 and Energy Research and Development Administration Grant ET(11-1)3231.

LITERATURE CITED

1. Abeles, F.B. 1964. Cell-free hydrogenase from Chlamydomonas Plant Physiol. 39:169-176.
2. Avron, M. and M. Gibbs.   1974. Properties of phosphoribulokinase of whole chloroplasts. Plant Physiol. 53:136-139.
3. Ben-Amotz, Ami and M. Gibbs.   1975. H_2 metabolism in photosynthetic organisms II. Light-dependent H_2 evolution by preparations from Chlamydomonas, Scenedesmus, and spinach. Biochim. Biophys. Res. Commun. 64:355-359.
4. Bennemen, J.R., J.A. Berenson, N.O. Kaplan and M.D. Kamen. 1973. H_2 evolution by a chloroplast-ferredoxin-hydrogenase system. Proc. Nat.Acad.Sci. 70:2317-2320.
5. Bohme, H.,S. Reimer and A. Trebst. 1971. The effect of dibromothymoquinone, an antagonist of plastoquinone, on noncyclic and cyclic electron flow systems in isolated chloroplasts. Z. Naturforsch.26b:341-352.
6. Brand, J.J., V.A. Curtis, R.K. Togasaki and A. San Pietro. 1975. Partial reactions of photosynthesis in briefly sonicated Chlamydomonas. Plant Physiol. 55: 187-191.
7. Gorman, D.S. and R.P. Levine. 1965. Cytochrome f and plastocyanin: Their sequence in the photosynthetic electron transport chain of Chlamydomonas reinhardi. Proc. Nat. Acad. Sci. 54:1665-1669.

8. Heber, U.W. and K.A. Santarius. 1965. Compartmentation and reduction of pyridine nucleotides in relation to photosynthesis. Biochim. Biophys. Acta 109: 390–408.

9. Kessler, E. Hydrogenase, photoreduction and anaerobic growth. In.W.D.P.Stewart, ed. Algal Physiology and Biochemistry. Blackwell(Oxford)pp. 456–473.

10. Marchant, R.H. 1974. An analysis of the effect of added thiols on the rate of O2 uptake during Mehler reactions which involves superoxides production. Proc. Third Inter. Cong. Photosynthesis. pp. 637–643.

11. Moll, B. and R.P. Levine. 1965. Characterization of a Photosynthetic mutant strain of Chlamydomonas reinhardi deficient in phosphoribulokinase activity. Plant Physiol. 46: 576–580.

12. Roa, K.K., L.Rosa and D.O. Hall. 1976. Prolonged production of hydrogen gas by a chloroplast biocatalytic system. Biochem. Biophys. Res. Commun.68 (1):21–28.

13. Stocking, C.R. and S. Larson. 1969. A chloroplast cytoplasmic shuttle and the reduction of extraplastid NAD. Biochem. Biophys. Res. Commun. 37: 278–282.

14. Trebst, A. and M. Burba. 1967. Über die Hemmung photosynthetisch Reaktionen in isolierten Chloroplasten und Chlorella durch Disalicylidenepropanediamine Z. Pflanzenphysiol. 57: 419–433.

MUTATIONAL ANALYSIS OF *CHLAMYDOMONAS REINHARDI*: APPLICATION
TO BIOLOGICAL SOLAR ENERGY CONVERSION.*

A. CHARLES MCBRIDE, STEPHEN LIEN, ROBERT K. TOGASAKI AND ANTHONY
SAN PIETRO

*Department of Plant Sciences, Indiana University, Bloomington,
Indiana 47401*

Abstract

        Three possible applications of mutant strains of
Chlamydomonas reinhardi to Biological Solar Energy Conversion
are described. First, hydrogen photoevolution was studied
using anaerobically adapted wild type cells and two photosynthetically incompetent mutant strains, one with a lesion
on the oxidizing side of PSII and the other on the reducing
side. Based on analysis of the initial rate of hydrogen photoevolution and the effect of the herbicide, DCMU, it appears
that water is the primary hydrogen source for the rapid initial
photoevolution. Secondly, a selection program was developed
for isolation of mutant strains with a more oxygen resistant
hydrogenase than wild type. The mutant strains exhibited the
expected *in vivo* properties, evidenced by kinetic analysis of
hydrogen photoevolution and the oxyhydrogen reaction. Thirdly,
a selection program was developed for isolation of mutant
strains with greater resistance towards DCMU inhibition of
photosynthesis. Some isolates exhibited DCMU resistance only at
the level of intact cell photosynthesis while others showed
resistant oxygen photoevolution under both *in vivo* and *in
vitro* conditions. One isolate exhibited a differential
resistance toward DCMU and simazine.

        Because of its aptness for genetic and biochemical analy-
sis and manipulations (10, 11, 12), the photosynthetic green alga
Chlamydomonas reinhardi is a very useful organism with which to
study many aspects of Biological Solar Energy Conversion.

---

*Supported by grants (BMS 75-03415 to A.S.P. and BMS 75-19643 to
R.K.T. and AER 75-16962 to A.S.P. and R.K.T.) from the National
Science Foundation, U.S.A.

Much of the information obtained with this organism is
potentially applicable to higher plant systems. A classic
example is the elucidation of the sequential arrangement of
cytochrome f and plastocyanin in the photosynthetic electron
transport pathway. The sequence was deduced initially from
Chlamydomonas mutational data and later confirmed for higher
plant systems using inhibitors such as mercury and cyanide
(6, 7, 8, 9). In this paper we report on some projects currently
under investigation in our laboratories that have relevance to
Biological Solar Energy Conversion. The first two projects
focus on phenomena unique to algal systems; the third has
potential application for higher plant systems.

Hydrogen Photoevolution

During the past year we have been investigating the photoevolution of hydrogen by anaerobically adapted wild type cells
and a number of photosynthetically incompetent mutant strains
(13). Cells were cultured in the tris-acetate-phosphate (TAP)
medium described by Gorman and Levine (b), harvested, washed
twice with 1 mM potassium phosphate buffer, pH 6.8, and
resuspended to a final chlorophyll concentration of 10 µg/ml
in the same buffer. Twelve ml of cell suspension was adapted
for at least one hour under anaerobic conditions and then
transferred to a reaction vessel with three openings. Electrodes
for the measurement of O_2 (Clark type electrode) and H_2 (electrode
modified according to Wang *et al* (22)) concentration were
inserted through the side openings. The tips of the electrodes
were positioned well below the surface of the cell suspension.
Both openings were sealed gas tight with teflon tape and
parafilm. The top opening was plugged with a rubber stopper
through which three hypodermic needles were inserted. One long
needle, reaching nearly to the bottom of the vessel, was used
to flush the cell suspension with gas of a desired composition.
Two short needles, neither of which reached the surface of the
cell suspension, were used to maintain the vessel anaerobic;
one needle served as entry and the other as exit for a stream
of nitrogen. The cell suspension was continuously stirred
with a magnetic stirrer and kept dark at 25°C. The electrodes
were connected to a dual pen recorder (Fisher Recordall Series
5000). Each electrode was calibrated with a standard gas
mixture (1% H_2/99% N_2 and 1% O_2/99% N_2) obtained from Matheson
Co.

When anaerobically adapted wild type cells are illuminated,
the initial rapid rise in H_2 concentration ceased within a few
minutes as the O_2 concentration rose. The initial rate of H_2
photoevolution varied greatly, from 15 to 110 µmoles H_2/mg
CHL·hr. In all cases, the increase in H_2 concentration ceased

as the O_2 concentration approached about 1%. DCMU (10 μM)
inhibited this initial rate by 70% to 90%. Typical data for
H_2 photoevolution by wild type cells and two mutant strains,
pet 20-8 and pet 10-2, are given in Table I. With wild type
cells the initial rate of 70 μmoles H_2/mg Chl·hr is decreased
80% in the presence of 10 μM DCMU. Mutant strain pet 20-8,
whose photosynthetic electron transport pathway is interrupted
at a site(s) on the reducing side of PSII, photoevolved hydrogen
at a rate equivalent to 10% that of wild type. Further, the
rate with this mutant was not affected by 10 μM DCMU.

Table I. *Initial rates of photohydrogen evolution by wild type
and mutant strains of Chlamydomonas reinhardi.*[a]

Strain	DCMU (10^{-5}M)	μmole H_2/mg Chl·hr	% of wild type
wild type	–	70	100
wild type	+	14	20
pet 20-8	–	7.7	11
pet 20-8	+	7.0	10
pet 10-2	–	32	47
pet 10-2	+	9.6	14

[a]Reaction mixture (12 ml) contained whole cell equivalent to
12 μg chl/ml in 1 mM PO_4 buffer at pH 7. Incident white light
intensity from a tungsten lamp was 800 foot candles, and the
reaction mixture was kept at 25°C.

Mutant strain pet 10-2, whose photosynthetic electron
transport pathway is interrupted at some site(s) on the
oxidizing side of PSII, photoevolved hydrogen at an initial
rate of 32 μmoles/mg Chl·hr. Interestingly, this rate declined
rapidly despite the near absence of oxygen inside the reaction
vessel. The rapid initial rate could be restored by an
additional 30 minute further dark anaerobic incubation of the
cell suspension. Further, the rapid initial H_2 photoevolution
by adapted pet 10-2 is quite susceptible to inhibition by 10 μM
DCMU as seen in Table I. Clearly, the major electron source for
this phenomenon is not water but may be a pool of some internal
electron donor to the photosystem II reaction center which is
rapidly depleted in the light and slowly restored in the dark.

If this hypothesis is correct, interruption of the photosynthetic transport pathway by mutation on either the oxidizing or reducing side of PSII should result in a drastic decrease in H_2 photo-evolution. The data presented in Table I are consistent with the view that water serves as the major hydrogen donor during the rapid initial photohydrogen evolution. To establish this point more firmly, we must achieve in future a sustained H_2 photoevolution in parallel to and stoichiometric with oxygen photoevolution. Obviously, we need a mutant organism whose hydrogenase is more resistant to inhibition by oxygen than is the protein in wild type cells.

Hydrogenase

Chlamydomonas wild type cells catalyze photoreduction; i.e., light-dependent CO_2 fixation under a hydrogen atmosphere in the presence of DCMU (2). It is clearly established that CO_2 assimilation under photoreductive conditions is mediated by the photosynthetic carbon cycle. The radioactive products from such fixation are very similar to those for photosynthetic CO_2 fixation (5, 18). Under photoreductive conditions, ATP is provided by cyclic photophosphorylation while the reducing power derives from H_2 via hydrogenase. Clearly, photoreduction is hydrogenase dependent. Thus, if culture conditions were established such that the survival of cells depended solely on photoreduction, the survival of cells having an O_2 sensitive hydrogenase would be endangered simply by introduction of oxygen to the culture (21). In actual practice, sufficient oxygen is provided so that wild type cells will not survive. Mutagenized cells which survive this treatment should contain an O_2 resistant hydrogenase. In fact, when wild type cells of Chlamydomonas reinhardi 137c (+) were incubated in the light in the DCMU supplemented (100 μM) liquid minimal medium (20), in the light under an atmosphere of 5% O_2, 25% N_2 and 70% H_2, the cells died within four days. Experimentally, wild type cells were mutagenized by nitrosoguanidine treatment (4), washed three times with minimal medium, resuspened in the same medium containing 100 μM DCMU and subjected to the selective condition described above. After eight days incubation under the selective condition, aliquots of the cell suspension were plated out on TAP plates, and incubated under normal photosynthetic conditions for an additional 10 days. Cells from the several colonies which formed were cultured in liquid TAP medium and then subjected to another round of mutagenesis and selection. The oxygen concentration during the second round of selection was increased to 8% and surviving cells were isolated on TAP plates. Survivors of the two consecutive rounds of selection are currently being analyzed for their ability to photoevolve hydrogen. Typical data for one isolate 1B3 are shown in Figure 1 together with comparable data for wild type cells.

Figure 1. H_2 and O_2 metabolism in Chlamydomonas reinhardi[a]

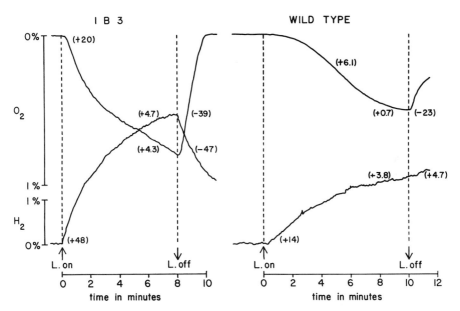

[a]For experimental conditions, see legends for Table 1.

     Upon illumination of wild type cells, H_2 evolution commenced immediately; the initial rate of 14 μmoles H_2/mg Chl·hr decreased substantially after ten minutes. Oxygen evolution, on the other hand, showed a distinct lag period followed by a period of evolution which terminated completely after ten minutes. Upon cessation of illumination, there was no decrease in the H_2 concentration whereas there was a rapid decrease in O_2 concentration. It appears that after ten minutes of exposure to the indicated amount of O_2 in the light, wild type cells are incapable of carrying out the oxyhydrogen reaction, which is dependent on an active hydrogenase, even though they are still capable of respiration. We believe that this phenomenon reflects the inactivation of hydrogenase by molecular oxygen. In contrast, mutant 1B3 catalyzed a very active oxyhydrogen reaction, even after eight minutes of exposure with light to a higher level of O_2 than for the wild type cells. The initial rate of H_2 photoevolution by mutant cells was 48 μmoles H_2/mg Chl·hr; the rate after eight minutes was 4.7 μmoles H_2/mg Chl·hr. The latter value was corrected for the concurrent

H_2 consumption via the oxyhydrogen reaction using the rate of postillumination H_2 consumption of 47 μmole H_2/mg Chl·hr as the measure of the oxyhydrogen reaction rate. Using this correction, a H_2 photoevolution rate of 51.7 μmoles H_2/mg Chl·hr, at the end of eight minutes, was calculated. Thus, within the time span of the experiment, there was no indication of O_2 inactivation of hydrogenase activity in this mutant strain. This datum alone is insufficient to prove that mutant 1B3 has an altered hydrogenase, i.e., more resistant to O_2 inhibition than wild type. However, we can say that a hydrogenase dependent reaction catalyzed by this organism (1B3) is more O_2 resistant than when catalyzed by wild type cells. Clearly, if the mutation produced an organism with an oxygen resistant hydrogenase, this is indeed the phenotype one would expect.

Thus, our selection procedure yielded some putative mutants with the desired *in vivo* characteristics. We plan to continue this selection program to obtain mutant strain with an even greater tolerance towards oxygen. Further, individual mutant strains will be analyzed to establish both the genetic and biochemical basis of the oxygen resistance.

Herbicide Sensitivity

We are trying to alter the sensitivity of *Chlamydomonas* towards several herbicides which are potent inhibitors of Hill activity. There are a number of herbicides with different chemical structures which affect photosynthetic reactions rather similarly (1, 15). However, the mode of their inhibition is not presently known. More knowledge about the mechanisms of herbicides, such as DCMU, may allow for practical application. Thus, we selected initially for mutant strains with greater resistance toward DCMU inhibition of photosynthesis, and have successfully isolated several such mutant strains (16).

Wild type cells were mutagenized by either ultraviolet light or nitrosguanidine treatment (16). Mutagenized cells were then subjected to selection in the light on gradient plates (17, 19) of solid minimal medium containing 10 μM DCMU in the bottom layer and 1 μM DCMU in the top layer. Colonies which formed on these initial plates were transferred to fresh plates of minimal media containing 10 μM DCMU. They were finally transferred to and maintained on minimal media plates (20). Isolates demonstrating good growth on DCMU (10 μM) supplemented minimal media plates were cultured in liquid TAP medium and tested for the effect of DCMU on their photosynthetic activity.

The effect of DCMU (6.6×10^{-7}M) on oxygen evolution by whole cells of wild type and two isolates, DR-18 and DR-65a, is given in Table II (16). Both isolates exhibited substantially greater resistance towards DCMU inhibition than did wild type.

Table II. *Effect of DCMU on oxygen evolution by whole cells*[a]

Strain	μmole O_2 evolved/mg chl·hr.		
	No DCMU	6.6×10^{-7}M DCMU	% control
Wild type	86	25	29
DR-65a	91	46	51
DR-18	87	88	101

[a]Reaction mixture (1.5 ml) contained whole cell equivalent to 23-25 μg Chl/ml, 6.6 mM $NaHCO_3$, 27 mM PO_4 buffer at pH 7. For further detail, see reference 16.

The effect of DCMU (3.3×10^{-7}M) on oxygen evolution by sonicated cell fragments (3) of wild type cells and 5 isolates, including DR-18 and DR-65a, is shown in Table III (16).

Table III. *Effect of DCMU on oxygen evolution by cell fragments*[a]

Strain	μmole O_2 evolved/mg chl·hr.		
	No DCMU	3.3×10^{-7}M DCMU	% control
Wild type	71	1	1
Dr-65a	104	2	2
DR-18	105	87	83
DR-25	98	67	68
DR-43	76	61	80
DR-67	67	40	60

[a]Reaction mixture (1.5 ml) contained sonicated cell equivalent to 20-25 μg chl/ml, 0.66 mM p-benzoquinone, 25.2 mM PO_4 buffer at pH 7. For further detail, see reference 16.

Whereas wild type and DR-65a were nearly completely inhibited, all other isolates demonstrated a high degree of DCMU resistance. DR-18, exhibited the highest level of DCMU resistance and was further analyzed. The effects of three herbicides, DCMU, 3-(4-chlorophyenyl)-1,1-dimethylurea (CMU), and 2-chloro-4,6-bis (ethylamino)-s-triazine (simazine), upon whole cell oxygen evolution by wild type and DR-18 are presented in Figure 2 (14).

Figure 2.   Inhibition of the steady rate of photosynthesis with whole cell suspensions of the wild type and mutant strain DR-18 by the inhibitors.[a]

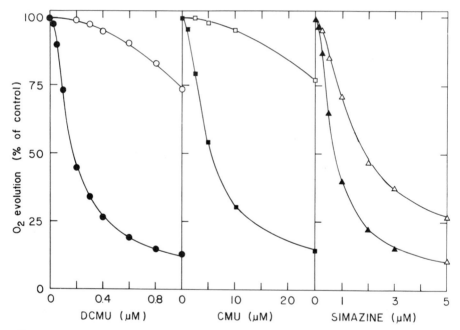

[a]The solid and blank symbols denote experiments using the wild type strain and mutant strain DR-18, respectively. Reaction mixture (2 ml) contained whole cell equivalent to 10 μg Chl/ml, in TAP media. For further detail, see reference 13.

It is clear that the resistance of DR-18 towards CMU, a close chemical analogue of DCMU, is very similar to that observed with DCMU; in contrast, its resistance towards simazine is much less. In other words DR-18 cells can distinguish between simazine on the one hand and DCMU and CMU on the other. This data opens exciting new possibilities for introduction of differential resistance towards herbicides of different chemical classes by means of mutation. Aside from providing basic information with respect to interaction of the herbicide with

the photosynthetic apparatus of green plants, this finding may
lead to very practical application. Rather than utilizing individual chemical specificity toward plants, we may be able to
engineer our crop plant(s) such that they will have specific
resistance toward wide spectrum herbicides. When we consider
the enormous complexity of herbicidal chemicals and green plants,
it may be much more reasonable to use the latter as the vehicle
to provide herbicidal specificity. In the past, naturally
occurring specific herbicidal resistance in a few plants has
been effectively exploited (1, 15); for example, the resistance
of maize toward simazine and the resistance of rice toward
dimethyl tetrachloroterephthalate (DCPA). The approach
described in this paper, in combination with plant tissue
culture and plantlet regeneration, might well provide a more
systematic and rapid means to understand herbicidal action and
contribute to plant productivity, a vital consequence of
biological solar of energy conversion.

Literature Cited

1. Ashton, F. M., and A. S. Crafts. 1973. Mode of action of
    herbicides. A Wiley-Interscience Publication, New York.
2. Bishop, N. I. 1966. Partial Reactions of Photosynthesis
    and Photoreduction. Annu. Rev. Plant Physiol. 17:
    185-208.
3. Curtis, V. A., Brand, J. J., and R. K. Togasaki. 1975.
    Partial reactions of photosynthesis in briefly sonicated
    Chlamydomonas. Plant Physiol. 55: 183-186.
4. Gillham, N. W. 1965. Induction of chromosomal mutation
    in *Chlamydomonas reinhardi* N-methyl-N'-nitro-N'nitro-
    soguanidine. Genetics. 52: 429-537.
5. Gingras, G., Goldsby, R. A., and M. Calvin. 1963.
    Carbon dioxide metabolism in hydrogen-adapted *Scenedesmus*.
    Arch. Biochem. Biophys. 100: 178-184.
6. Gorman, D. S., and R. P. Levine. 1966. Cytochrome f and
    plastocyanind: their sequence in the photosynthetic
    electron transport chain of *Chlamydomonas reinhardi*.
    Proc. Natl. Acad. Sci. U.S. 54: 1665-1669.
7. Gorman, D. S. and R. P. Levine. 1966. Photosynthetic
    electron transport chain of *Chlamydomonas reinhardi*.
    VI. Electron transport in mutant strains lacking either
    cytochrome 553 or plastocyanin. Plant Physiol. 41:
    1648-1656.
8. Izawa, S., Kraayenhof, R., Ruuge, E. K., and D. Devault.
    1973. The site of KCN inhibition in the photosynthetic
    electron transport pathway. Biochim. Biophys. Acta 314:
    328-339.

9. Kimimura, M. and S. Katoh. 1972. Studies on electron transport associated with photosystem I. I. Functional site of plastocyanin: Inhibitory effects of $HgCl_2$ on electron transport and plastocyanin in chloroplasts. Biochim. Biophys. Acta. 283: 279-292.

10. Levine, R. P., and W. T. Ebersold. 1960. The genetics and cytolohy of *Chlaymdomonas*. Annu. Rev. Microbiol. 14: 197-216.

11. Levine, R. P. 1969. Analysis of photosynthesis using mutant strains of algae and higher plants. Annu. Rev. Plant Physiol. 20: 523-540.

12. Levine, R. P., and U. W. Goodenough. 1970. The genetics of photosynthesis and of the chloroplast in *Chlamydomonas reinhardi*. Annu. Rev. Genetics 4: 397-408.

13. Lien, S., Togasaki, R. K., and A. San Pietro. 1974. Electron transport activity and CHLA fluorescence induction of photosynthetic mutants of *Chlamydomonas reinhardi*. Plant Physiol. 53: S-121.

14. Lien, S., McBride, J. C., McBride, A. C., Togasaki, R. K., and A. San Pietro. A comparative study of photosystem II specific inhibitors: The differential action on a DCMU resistant mutant strain of *C. reinhardi*. Plant Cell Physiol. In press.

15. Matsunaka, S. 1976. Introduction to plant toxicology (in Japanese). U. Tokyo Press, Tokyo.

16. McBride, J. C., McBride, A. C., and R. K. Togasaki. Isolation of *Chlamydomonas reinhardi* mutants resistant to the herbicide, DCMU. Plant Cell Physiol. In press.

17. Nakamura, K., and C. S. Gowans. 1964. Nicotinic acid-excreting mutants in *Chlamydomonas*. Nature 202 (4934): 826-827.

18. Russell, G. K. and M. Gibbs. 1968. Evidence for the participation of the reductive pentose phosphate cycle in photoreduction and the oxyhydrogen reaction. Plant Physiol. 43: 649-652.

19. Scherr, G. H., and M. E. Rafelson (Jr.). 1962. The directed isolation of mutants producing increased amounts of metabolites. J. Appl. Bact. 25: 187-194.

20. Sueoka, N. 1960. Mitotic replication of deoxyribonucleic acid in *Chlamydomonas reinhardi*. Proc. Natl. Acad. Sci. U.S. 46: 83-91.

21. Togasaki, R. K. 1973. A proposal to search for mutant strains carrying oxygen resistant hydrogenase. Proceedings of the workshop on bio-solar conversion held 5-6 September 1973 at Bethesda, Maryland, supported by the NSF-RANN. pp. 60-61.

22. Wang, R., Healey, F. P., and J. Myers. 1971. Amperometric measurement of hydrogen evolution in *Chlamydomonas*. Plant Physiol. 48: 108-110.

HYDROGEN PHOTOPRODUCTION FROM WATER

Satoru Kurita, Kenji Toyoda, Toshikatsu Endo, Naoki Mochizuki,
Masura Honya and Takashi Onami

Research Institute, Ishikawajima-Harima Heavy Industries Co.,
Ltd. Toyosu 3-1-15, Koto-ku, Tokyo 135-91, Japan

Summary

An attempt has been made to photoproduce hydrogen from
water using a hydroquinone-quinone redox system. The rate of
hydrogen photoproduction from hydroquinone increased upon
addition of either potassium ferricyanide, potassium ferrocyanide
or methylviologen. The generation of hydroquinone was verified
by absorption spectrometry.

In further experiments with the same redox system using a
photoelectrochemical cell, the rate of hydrogen evolution was
found to increase with the intensity of incident sunlight.

Introduction

Studies were conducted as early as 1937 on the photoproduction of hydrogen from water by biological solar energy conversion
(1-17). In that year, R. Hill first reported on biophotolysis,
which later came to be known as the "Hill reaction." Hill's
name is also associated with the requirement for oxidizing agents
for water photolysis, such as potassium ferric oxalate, which
are today universally recognized to be effective with plant
chloroplasts. Warburg has further shown that p-benzoquinone
also can function as a Hill reagent.

The present authors undertook a study on hydrogen photoproduction from water with the hydroquinone-benzoquinone redox
system, which represents a simple practical model based on the
analogy of hydrogen production by natural photosynthesis.

The system operates in several successive reaction stages to split water:

$$2Q + 2H_2O \xrightarrow[\text{(PSII)}]{h\nu} 2QH_2 + O_2 \uparrow \tag{1}$$

$$2QH_2 \xrightarrow[\text{(PSI)}]{h\nu} 2Q + 2H_2 \uparrow \tag{2}$$

$$2H_2O \xrightarrow{h\nu} 2H_2 \uparrow + O_2 \uparrow \tag{3}$$

or

$$2H_2O + 2Q \xrightarrow[\text{(PSII)}]{h\nu} 2QH_2 + O_2 \uparrow \tag{4}$$

$$2H_2Q + X \longrightarrow 2Q + H_2X \tag{5}$$

$$2H_2X \xrightarrow[\text{(PSI)}]{h\nu} X + 2H_2 \uparrow \tag{6}$$

$$2H_2O \xrightarrow{h\nu} 2H_2 \uparrow + O_2 \uparrow \tag{7}$$

where Q   : Quinone
       QH_2 : Hydroquinone
       X   : Third compound

The light-induced reactions (1) and (4) produce hydroquinone from quinone and correspond to photosystem II (PSII); reactions (2) and (6) produce protons from hydroquinone and correspond to photosystem I (PSI). These reactions demand a total quantum energy of about 2.0 eV, equivalent to that of light of 620 nm wavelength.

The direct photodecomposition of water to OH and H requires at least 5.08 eV, equivalent to light of 244 nm or shorter, which would mean practically below 190 nm considering absorption (see also Appendix 1).

Use of a suitable catalyst to induce water decomposition in the form

$$2H_2O \xrightarrow{h\nu} 2H_2 + O_2$$

should reduce the quantity of energy required by half and permit
hydrogen to be produced by photolysis of water by illumination
with green light (below 507 nm) contained in sunlight.

This possibility holds promise for the hydroquinone-quinone
system as an energy-sparing alternative to direct photodecomposition for producing hydrogen.

Experiments were conducted first on hydrogen photoproduction
in a Warburg vessel. This was followed by trial continuous photoproduction in a photoelectrochemical cell.

Results and Discussion

Hydrogen Evolution From Hydroquinone

Two identical reaction mixtures were prepared and placed
in Warburg vessels. One of them was wrapped in aluminum foil
to seal off all light, while the other was exposed to light
from a fish lamp, emitting cold light (Toshiba "Fishlux" model
FL40S BRF/NL). In each vessel, the reaction mixture composed
of 4 µmoles of methylviologen and 4 µmoles of either potassium
ferricyanide or ferrocyanide, dissolved in 2 ml of 0.1 N
sulfuric or hydrochloric acid solution, was placed in the main
compartment; 0.012 mg of pyrogallol dissolved in 20% sodium
hydroxide was in the center well.

The side arm held 5 µmoles of hydroquinone solution and the
gas phase was nitrogen. The reaction in the illuminated vessel
was initiated by tilting the vessel to mix the hydroquinone
solution with the reaction mixture in the main compartment.
The subsequent change of gas pressure was measured manometrically.
The gaseous product formed in the vessel was analyzed by gas
chromatography and verified to be hydrogen.

The process of hydrogen evolution is depicted in Table I.
It is seen that the addition of either ferricyanide or ferrocyanide ions, or else of methylviologen, contributed to enhance
hydrogen evolution, but that by far the most striking results
were obtained when hydroquinone was added. This effect might
be attributed to the contribution of the ferricyanide or ferrocyanide ions to hydrogen photoproduction in a hydroquinone-p-benzoquinone redox system according to the scheme (also see Appendix
2):

	ΔG^1 (Kcal/mole)	
	pH 1.0	pH 7.0
$QH_2 + 2Fe(CN)_6^{-3} \rightarrow Q + 2H^+ + 2Fe(CN)_6^{-4}:$	12.86	-3.73
$2Fe(CN)_6^{-4} + 2H^+ \rightarrow 2Fe(CN)_6^{-3} + H_2 \uparrow$	19.36	35.95
$QH_2 \longrightarrow Q + H_2 \uparrow$	32.22	32.22

In the absence of hydroquinone, ferricyanide or ferrocyanide ions may enhance water decomposition to produce hydrogen through the scheme

$$H_2O + 2Fe(CN)_6^{-3} \longrightarrow 2Fe(CN)_6^{-4} + 2H^+ + \frac{1}{2}O_2$$

$$2H^+ + 2Fe(CN)_6^{-4} \longrightarrow 2Fe(CN)_6^{-3} + H_2 \uparrow$$

$$H_2O \longrightarrow H_2 + \frac{1}{2}O_2$$

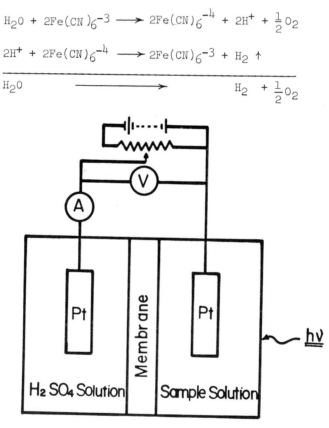

Fig. 1   Photoelectrochemical cell used for measuring reaction mixture photocurrent.

Table 1. Hydrogen evolution from reaction mixtures in Warburg test under illumination

Reaction mixtures		1	2	3	4	5	6	7	8	9
Methylviologen	(0.02 M x 0.2 ml)	+	+	+	+	−	+	+	+	−
$K_3Fe(CN)_6$	(0.02 M x 0.2 ml)	+	−	+	+	−	−	+	−	−
$K_4Fe(CN)_6$	(0.02 M x 0.2 ml)	+	+	−	+	+	+	−	−	−
Hydroquinone	(0.01 M x 0.5 ml)	+	+	+	−	+	−	−	+	−
In H_2SO_4 solution	(1.9 ml)	+	+	+	+	+	+	+	+	+
Hydrogen evolved (µl)		28.0	19.6	14.0	7.0	12.6	5.9	5.0	6.0	0.3

TABLE II. Photocurrent observed in reaction mixtures

Reaction mixture*	1 (1)	2 (4)	3	4	5	(9)
Methylviologen	+	+	−	+	−	−
$K_3Fe(CN)_6$	+	+	−	−	+	−
$K_4Fe(CN)_6$	+	+	−	−	−	−
Hydroquinone	+	−	+	+	−	−
0.5 M H_2SO_4	+	+	+	+	+	+
Photocurrent						
−Without external potential (μA)	25	16	4	6	16	6
−With external potential						
−Applied potential (V)	1.00	1.35	0.87	0.90	1.13	1.25
−Photocurrent (mA)	3.3	−	4.0	6.0	0.8	1.4

*Reaction mixture numbers given in brackets refer to reaction mixtures given in Table I.

Using the photoelectrochemical cell shown in Fig. 1, the photocurrents generated by the reaction mixtures described earlier were measured and the results are shown in Table 2.

In the absence of an externally applied potential, the photocurrent showed a tendency reminiscent of the hydrogen evolution in the preceding Warburg test. But the behavior was markedly different upon application of an external potential. To elucidate the characteristics of hydrogen evolution by photoelectrochemical methods, this procedure was used in subsequent experiments.

Hydroquinone Photoformation From p-Benzoquinone

Direct Photoformation: Tests on hydroquinone photoformation were undertaken with a 0.01 M aqueous solution of p-benzoquinone exposed to sunlight at 20°C. Almost all the p-benzoquinone was converted into hydroquinone after 6 hours. The product was verified by absorption spectrometry (Fig. 2).

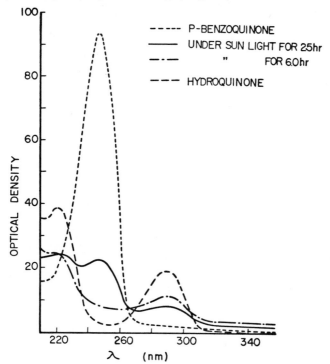

Fig. 2   *Absorption spectra for hydroquinone photoformation by reaction between p-benzoquinone and water.*

Photoelectrochemical Formation: The photochemical cell
shown in Fig. 3 was used as reaction vessel.   It is comprised of
two chambers, separated from each other by a cation exchange
membrane.   One of the chambers carries an n-type titanium dioxide
semi-conductor electrode immersed in 500 ml of 0.5 M sodium
hydroxide solution.   The electrode is 2 mm thick, with diameters
of about 16 mm for xenon lamp illumination and 10 mm for sunlight.   It was prepared from single crystal wafer (001) crystal
surface and treated at 600°C under hydrogen gas.   The other
chamber is equipped with a platinum electrode and contains 500
ml of 0.5 M sulfuric acid solution, either (a) alone or
(b) containing p-benzoquinone.   The two electrodes are connected
to each other.

The reaction was initiated by illuminating the system and
the results are shown in Fig. 4.   In the absence of p-benzoquinone
(Fig. 4, Case A), direct photoevolution of hydrogen and oxygen
occurred.   Illumination of the titanium dioxide electrode
generated an electron flow toward the platinum electrode,
leaving a positive hole on the titanium oxide surface.   Oxygen
was evolved from this electrode.   In the neighboring solution,
protons were generated which migrated through the membrane
toward the platinum electrode where reduction occurred and
hydrogen gas was evolved.   This phenomenon was first discovered
by K. Honda (17).

When p-benzoquinone was present (Fig. 4, Case B), hydroquinone was formed at the platinum electrode; hydrogen evolution
did not occur.

The photocurrent intensity was twice that noted for
direct photoevolution without p-benzoquinone (Fig. 4).   These
data suggest the possibility of enhancing the rate of hydrogen
photoproduction by inclusion of quinone in the system.

*Photoelectrochemical generation of H_2 from H_2O with
quinone system:*   A photoelectrochemical cell was used as
reaction vessel.   As shown in Fig. 5, the vessel consists of
two sections I and II; each is separated into two 50 ml
compartments separated by a cation exchange membrane permeable
to protons.

Reactions involved in oxygen evolution and in the production of hydroquinone from p-benzoquinone occur in section II;
whereas, in section I hydrogen evolution from hydroquinone
occurs.   Circulation of the hydroquinone solution by means of
pumps C_1 and C_2 allowed this reaction to proceed continuously.
Illumination for the n-type titanium dioxide electrode was
provided through a quartz window in compartment F_1.

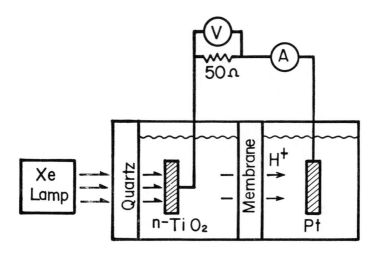

Fig. 3 Photoelectrochemical cell with n-TiO₂ electrode.

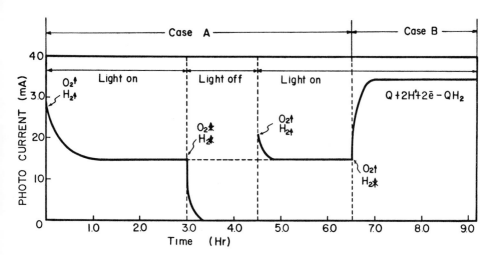

Fig. 4 Photocurrent observed upon hydrogen evolution (Case A) and hydroquinone formation by photolysis of water (Case B).

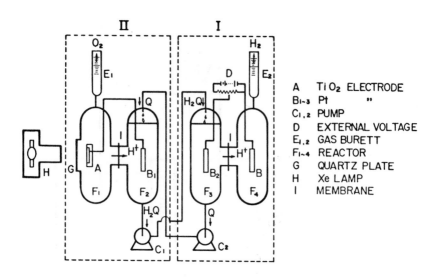

A   Ti O$_2$ ELECTRODE
B$_{1-3}$ Pt   "
C$_{1,2}$ PUMP
D   EXTERNAL VOLTAGE
E$_{1,2}$ GAS BURETT
F$_{1-4}$ REACTOR
G   QUARTZ PLATE
H   Xe LAMP
I   MEMBRANE

Fig. 5 Photoelectrochemical cell apparatus for photoproduction of hydrogen with quinone-hydroquinone system.

The reaction was started by switching on a 500 W xenon lamp or by admitting sulight. The gaseous products from sections I and II were confirmed by gas chromatography to be hydrogen and oxygen, respectively.

An external potential of 0.8 -1.0 V was applied to maintain a suitable rate of continuous reaction according to the scheme:

$$QH_2 \xrightarrow{h\nu} Q + 2H^+ + 2e^- \quad \text{(in photoelectrochemical cell)}$$

$$2e^- + 2H^+ \longrightarrow H_2 \quad \text{(with externally applied potential)}$$

$$\overline{QH_2 \longrightarrow Q + H_2}$$

Hydrogen was evolved at the rate of 2.55 ml/3 hr and oxygen at 1.1 ml/3 hr by photoreaction with xenon lamp illumination, giving a hydrogen/oxygen ratio of 2.3. The corresponding values obtained with sunlight were 1.54 ml/hr for hydrogen and 0.72 ml/ hr for oxygen.

The photocurrent was 3.8 mA under daylight in fine weather and 0.5 mA under similar conditions in cloudy weather (Figs. 6A and 6B). The rate of hydrogen and oxygen evolution is directly proportional to photocurrent intensity.

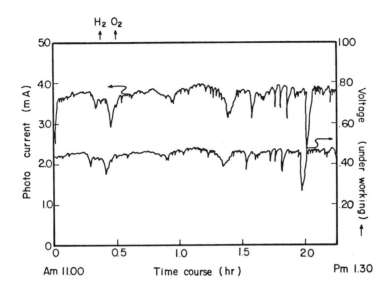

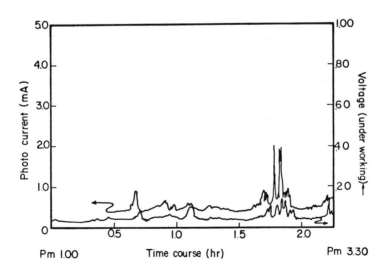

Fig. 6 Photocurrent observed upon hydrogen evolution from water with quinone-hydroquinone system under daylight in fine weather (a) and in cloudy weather (b).

The most serious difficulty foreseen for practical application of the present method is to realize a durable membrane permeable only to protons.

Other possibilities for developing the method may lie in replacement of the externally applied potential by a catalyst such as hydrogenase for reactions (2) and (6).

Conclusion

Using either ferricyanide, ferrocyanide or methylviologen as catalyst, it was established that hydrogen formation occurs by photoproduction from hydroquinone, which in turn is obtained from p-benzoquinone and water by photoformation.

As a consequence, it should be possible to obtain hydrogen from water by photoproduction via p-benzoquinone and hydroquinone, through the scheme

$$Q + H_2O \xrightarrow{\text{h}\nu} QH_2 + \frac{1}{2}O_2$$

$$QH_2 \longrightarrow O_2 + H_2$$

$$\overline{H_2O \longrightarrow H_2 + \frac{1}{2}O_2}$$

With a view to enhancing the reaction rates, tests were conducted on a photoelectrochemical procedure whereby hydrogen was obtained by photoproduction with a hydroquinone-quinone redox system in a photoelectrochemical cell. The method proved to be twice as effective as the direct photoproduction of hydrogen from water. The rate of hydrogen evolution was found to increase with the intensity of incident sunlight.

The feasibility of a simple hydroquinone-p-benzoquinone redox system, analogous to natural photosynthesis, was thus established. A more practical procedure embodying this principle is under development.

Acknowledgment

The present study was supported by a research grant from the Agency of Industrial Science and Technology, Ministry of International Trade and Industry, Government of Japan. The authors are indebted for kind permission to publish these results.

Several helpful discussions with Dr. A. San Pietro are gratefully acknowledged.

References

(1) R. Hill: Nature 139, 881 (1937).
(2) O. Warburg: Naturwissenschaften 32, 301 (1944).
(3) H. Gaffron and J. Rubin: Gen. Physio. 26, 219 (1942).
(4) H. Gest and M. D. Kamen: Science 109, 558 (1949).
(5) A. Mitsui: Proc. Sympos.-Course "Hydrogen Energy Fundamentals," Univ. Miami S-2, 31 (1975).
(6) A. Hollaender, K. J. Monty, R. M. Pearstein, F. Schmidt-Bleek, W. T. Snyder and E. Volkin: "An Inquiry into Biological Energy Conversion," Gatlinburg (1972).
(7) P. Donovan, W. Woodward, F. H. Morse and L. O. Herwig: NSF/NASA Solar Energy Panel, "An assessment of solar energy as a national energy resource" (1972).
(8) M. Gibbs, A. Hollaender, B. Kok, L. O. Krampitz and A. San Pietro: "Proc. Workshop on Bio-solar Conversion" (1973).
(9) T. N. Veziroglu: Proc. "Hydrogen Economy Miami Energy Conf." (1974).
(10) T. Ota: Proc. NSF-JSPS Sympos. "Key Technologies for Hydrogen Energy System" (1975).
(11) T. N. Veziroglu: Proc. Sympos. "Hydrogen Energy Fundamentals" Miami Beach (1975).
(12) R. Beck: Rep. on NSF-RANN Workshop "Solar Energy for Nitrogen Fixation and Hydrogen Production" Gatlinburg (1975).
(13) A. Mitsui: Proc. "Hydrogen Economy Miami Energy Conf." Univ. Miami A 309 (1974).
(14) A. Mitsui: Proc. U.S.-Japan Joint Seminar "Key Technologies for Hydrogen Energy System" 75 (1975).
(15) A. Mitsui: Proc. Third Int. Ocean Development Conf., 3, 11 (1975).
(16) S. Lien and A. San Pietro: NSF-RANN Rep. on "An Inquiry into Biophotolysis of Water to Produce Hydrogen" (1975).
(17) K. Honda and A. Fujishima: Nature 37 238 (1972).

Appendix 1

RELATION OF REDOX POTENTIAL Vs P^H

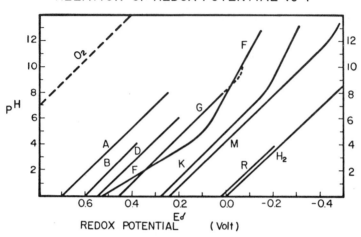

REDOX POTENTIAL E^d (Volt)

A : Quinone. B : p-Xyloauinone. C:m-Bromoindophenol. D : l, 2 Napthoquinone.
E : l-Napthol-2 sulphonic acid indophenol. F : Methylene blue. G: l-l Napthoquinone.
H : Indigo tetrasulphonate, K : Indio disniphonate. M : l-5 Anthraquinoe disulphonate.
N : Anthraquinone diulphonate. R : Ti^{+++} - Ti^{++++}.

Appendix 2

	PH				
	1	3	7	10	12
$Q + 2Fe(CN)_6^{4-} + 2H_2O$ $\rightarrow QH_2 + 2Fe(CN)_6^{3-} + 2OH^-$	-13.42 Kcal/mol	7.88	3.23	1148	15.08
$2Fe(CN)_6^{3-} + 2OH^-$ $\rightarrow 2Fe(CN)_6^{4-} + H_2O + \frac{1}{2}O_2\uparrow$	37.85	32.32	21.21	12.96	7.43
$QH_2 + 2Fe(CN)_6^{3-}$ $\rightarrow Q + 2H^+ + 2Fe(CN)_6^{4-}$	12.86	7.33	-3.73	-12.03	-15.63
$2Fe(CN)_6^{4-} + 2H^+$ $\rightarrow 2Fe(CN)_6^{3-} + H_2\uparrow$	19.36	24.90	35.96	44.26	49.79

THE MOLECULAR MECHANISMS OF PHOTOSYNTHETIC
HYDROGEN AND OXYGEN PRODUCTION

Elias Greenbaum

The Rockefeller University
New York, New York   10021

The key experimental approach in attempting to elucidate the
molecular mechanism of photosynthetic oxygen evolution is the
technique of single turnover flashes. By combining this technique with the oxygen polarograph, Joliot and his co-workers
were able to demonstrate that the oxygen evolved from dark
adapted *Chlorella* or isolated chloroplasts undergoes a damped
oscillation with periodicity four (1). Further work along these
lines led to the linear, four step model of Photosystem II
proposed by Kok and his co-workers (2) which is the widely
accepted model of photosynthetic oxygen evolution. We have developed, in our laboratory, a new technique using a flow system
and a zirconium oxide high temperature electrode as the sensing
element (3). Unlike the oxygen polarograph, used in all previous flash work, the flow apparatus is capable of absolute calibration and in addition has sufficient sensitivity to detect the
oxygen and/or hydrogen evolved from photosynthetic organisms
illuminated with single-turnover, saturating flashes. The
recent results on oxygen evolution show disagreement with the
accepted model (2) and a new model has been proposed (4) which is
summarized below. The newly developed flow apparatus has also
been used to perform the first experiments in which the flash
pattern of *hydrogen* evolution from photosynthetic organisms
illuminated with single turnover saturating flashes of light is
determined (5). The results on both hydrogen and oxygen will be
summarized here.

Fig. 1. summarizes a set of experimental results. The tri-
angular points refer to hydrogen evolution. The left ordinate
indicates the individually resolved, absolute yield of hydrogen
per mole of chlorophyll per flash. The abscissa indicates the
flash number. It is clearly seen in Fig. 1 that when dark
anaerobically adapted *Chlorella* are illuminated with single-
turnover flashes of light in the absence of added redox reagents,
the yield of hydrogen on the first flash is non-zero and is
essentially equal to the hydrogen yield per flash for the subsequent flashes. The fact that algae can evolve molecular
hydrogen under anaerobic conditions was first reported by
Gaffron and Rubin (6).

For the anaerobic conditions of Fig. 1 (hydrogen evolution)
the components of the electron transport chain linking
Photosystem I and Photosystem II are fully reduced (7). Consequently, oxygen evolution is not observed for anaerobically

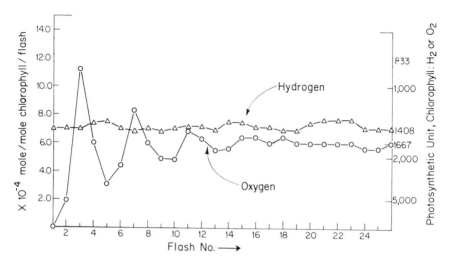

Fig. 1. *The yield of oxygen or hydrogen per mole of chlorophyll per flash. All experiments done in a helium atmosphere containing 10 ppm O_2. The time between flashes is 10 sec. The concentration of p-benzoquinone for the oxygen curve is 1.0 mM. Chlorella vulgaris were grown on a mineral medium at a light intensity of 3×10^3 erg/cm^2/sec. The gas detection system consisted of a flow apparatus designed and built in our laboratory. The sensing element is a zirconium oxide high temperature Nernst electrode. Absolute calibration of the apparatus is achieved by placing an electrolysis cell in tandem with the reaction cuvette containing the algae. The flash lamps were two General Radio Stroboslaves type 1539A. These delivered saturating flashes with $t_{1/2}$=4 μsec. The timing for the flashlamps was provided by Tektronix 160 series pulse generators. The chlorophyll content of the algae was determined spectrophotometrically by extraction into methanol.*

adapted algae in the absence of chemical oxidants. If p-benzoquinone is added to the anaerobic *Chlorella* suspension, the same experimental procedure measures the oxygen yield per flash. These results are indicated by the circular data points of the lower oxygen curve. Under these oxidizing-conditions (although still strictly anaerobic at 10 ppm O_2) there is no evidence for evolution of hydrogen.

Returning to the oxygen curve of Fig. 1 we see that the yield of oxygen on the first flash, is zero, unlike the corresponding observation for hydrogen. There is a relatively small yield of O_2 on the second flash which can be increased by adding ionic oxidants or increasing the flash frequency (3). The third flash has maximal O_2 yield and the subsequent yields can be described as a damped oscillation with periodicity four. Oxygen oscillations have formed the experimental underpinning of the theories

concerning the molecular mechanism of oxygen evolution (1-4).
These oscillations indicate a role for photoproduced, sequential,
metastable intermediates which are serially involved in the formation of oxygen. Apparently the four oxidizing equivalents
necessary to produce a molecule of oxygen from water cannot be
produced in a single flash. It requires at least two flashes to
produce some oxygen. Such is not the case for hydrogen. One
flash is sufficient. Thus, the molecular mechanism of hydrogen
evolution must be such that no photoproduced, metastable intermediates are serially involved as they are in the case of oxygen
evolution. Moreover, since two reducing equivalents are required to make one molecule of hydrogen it follows that in the
photoproduction of molecular hydrogen the reducing equivalents
from at least two photosystems are fed into a common pool (probably ferredoxin) following the common assumption that the photosystems form one equivalent per single turnover flash,   as
indicated in Fig. 2.

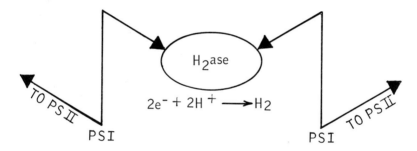

*Fig. 2. Reducing equivalents from at least two Photosystems I
are fed into a common pool that produce hydrogen via hydrogenase.*

     The size of the photosynthetic unit for hydrogen evolution in
Chlorella is chlorophyll:$H_2 \approx 1400:1$. The photosynthetic unit size
for oxygen evolution (the "Emerson and Arnold Unit") for
Chlorella coupled to benzoquinone is chlorophyll:$O_2 \approx 1700:1$. These
data allow a determination of the ratio of hydrogen to oxygen,
both normalized to the chlorophyll content of the algae. The
ratio of H_2 to O_2 is 1.2. Were the movement of electrons by
chlorophyll equally efficient for hydrogen as for oxygen then
the stoichiometric ratio would be two. These data indicate that
with regard to the photophysical apparatus of photosynthesis,
the ability to utilize absorbed visible quanta for the light
driven reaction of photosynthetic hydrogen evolution is at least
60% as efficient as the ability to utilize absorbed visible
quanta for the light driven reaction of photosynthetic oxygen
evolution. Conceivably, for other algae, the figure could be
higher.
     According to Kok's model the oxidizing side of photosystem II

consists of four relatively stable redox states, S_n states
$(n = 0,1,2,3)$. Each successive S state is more oxidized than its
predecessor by one electron equivalent. The state S_4 is a cum-
mulation of 4(+) oxidizing equivalents. S_4 spontaneously
evolves O_2 with the system cycling back to S_0. The experimental
underpinning of Kok's model is the observation that when dark
adapted *Chlorella* or isolated chloroplasts are illuminated with
brief saturating flashes, the oxygen evolved follows a pattern of
damped oscillations with periodicity four. According to the S
state model the significance of periodicity four is the fact that
four oxidizing equivalents in four steps are required to make a
molecule of oxygen from water:

$$2H_2O \longrightarrow O_2 + 4H^+ + 4e^- \tag{1}$$

In order to account for the observed damping, parameters α
(misses) and β (double hits) were introduced. Numerical values
for these parameters are obtained by fitting the theory to the
experimental results. These parameters vary from 5% to 20% of
the centers per flash. Moreover, in order to explain the fact
that the maximal yield of oxygen occurred in the third flash it
was necessary to hypothesize the existence of two stable redox
ground states S_0 and S_1 and that S_1 is favored in the dark. It
was later postulated (8) that molecular oxygen is an exclusive
oxidant of the S states and that the steady state ratio $S_1:S_0$ is
determined by a kinetic competition between O_2 and endogenous
reductants. However, experiments (3) on the yield of oxygen per
flash of *Chlorella* coupled to chemical oxidants (such as p-benzoquinone) under anaerobic conditions show that the hypothesis of
molecular oxygen as an exclusive oxidant of the S states is not
tenable.

In Figure 3 the oxygen yield per flash of *Chlorella*
under a wide variety of redox conditions is presented. It is
clearly evident that under highly oxidizing conditions the
oscillations are clear and well defined. However, as reductant
is progressively added to the cells, the oscillations fade out
and are replaced by a monotonic climb to the steady state. It
as been discovered (4) that this systematic behavior (including
the oscillations) can be accounted for with one simple assumption:

$$Y_{k+4} \propto Y_k \tag{2}$$

i.e. the yield of oxygen on the $(k+4)^{th}$ flash is proportional to
the yield on the kth. The simplest equation that can be written
with this assumption is:

$$Y_{k+4} = C_1 Y_k + C_2 \tag{3}$$

with solution (see reference 4)

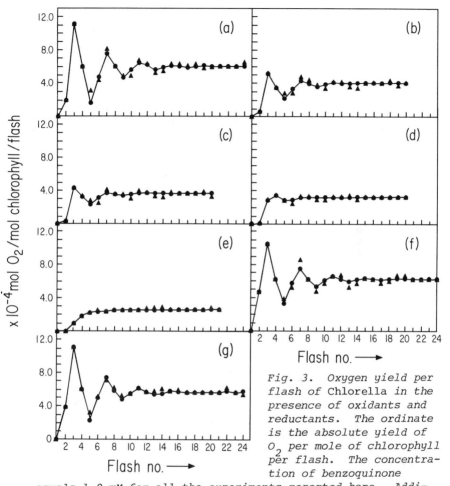

Fig. 3. *Oxygen yield per flash of* Chlorella *in the presence of oxidants and reductants. The ordinate is the absolute yield of O_2 per mole of chlorophyll per flash. The concentration of benzoquinone equals 1.0 mM for all the experiments reported here. Additional concentrations are as follows: a) BQ the only added redox component; b) HQ = 0.5 mM; c) HQ = 1.0 mM; d) HQ = 1.5 mM; e) HQ = 2.0 mM; f) $Fe(CN)_6^{3-}$ = 3.0 mM; g) $IrCl_6^{2-}$ = 3.0 mM.* ▲ *Experimental data.* ● *Theoretical points.* Chlorella *were suspended in 20 mM Pi buffer at pH = 7.*

$$Y_k = (\sqrt[4]{C_1})^k \{K_1 + A\cos(\tfrac{1}{2}k\pi + B)\} + 1 \qquad (4)$$

All of the data in Fig. 3 have been fit with equation 4 in the following uniform way: A trial value of C_1 was chosen. Then the constants K_1, A and B were determined from the values Y_2, Y_3

and Y_4 by solving three simultaneous equations in the three unknowns. Final values were determined by the C_1 that gave the best value for damping. It is indeed remarkable as well as intriguing that such a simple assumption as (2) can describe the oxygen yields under such a wide variety of conditions.

One further point that is worth mention is the yield, Y_2, of oxygen on the second flash. According to Kok's model $Y_2 \neq 0$ is due to "double hitting", i.e., two photochemical events that occur within the envelope of the flash profile that has $t_{1/2}=4$ μsec and a 10 μsec tail. It can be seen in Fig. 3 that the value of Y_2 depends on the redox environment of the algal cells. In order to test the hypothesis of double hitting, the experiment of Fig. 4 was performed. In this experiment Y_2 is measured as a function

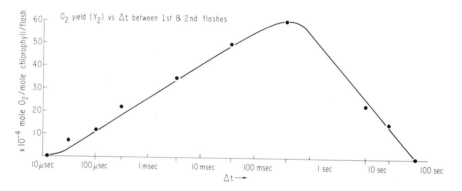

Fig. 4. *The yield of oxygen on the second flash, Y_2 versus spacing between flashes.*

of, Δt, the spacing between the flashes. It is seen that the yield, Y_2 of oxygen in the 10-30 μsec time spacing is <5% of the maximun yield at 0.3 sec. Therefore, the non-zero values of Y_2 cannot be due to double hitting when this term is used in the conventional sense.

In conclusion, it has been shown that the flash pattern yield of oxygen or hydrogen from *Chlorella* are markedly different from each other and that each pattern is rich in information regarding molecular mechanism. From the point of view of biological solar energy conversion, however, it is most encouraging that the photosynthetic unit based on hydrogen evolution is comparable in size to that based on oxygen evolution.

Acknowlegements The author would like to thank Prof. D. Mauzerall for valuable discussions and for critically reading the manuscript and Ms. Mary Ellen Genther for typing the manuscript. This research was supported by National Science Foundation grant PCM74-11747

References

1. P. Joliot, G. Barbieri and R. Chabaud, Photochem. Photobiol., *10*, 309-329 (1969); P. Joliot and B. Kok in Bioenergetics of Photosynthesis (Edited by Govindjee) pp. 387-412, Academic Press, New York (1975).
2. B. Kok, B. Forbush and M.P. McGloin, Photochem. Photobiol. *11*, 457-475 (1970); *ibid, 14*, 307-321 (1971).
3. E. Greenbaum and D. C. Mauzerall, Photochem. Photobiol. *23*, 369-372 (1976).
4. E. Greenbaum, Photochem. Photobiol. (1976) in press.
5. E. Greenbaum, Science (1976), submitted for publication.
6. H. Gaffron and J. Rubin, J. Gen. Physiol. *26*, 219-240 (1942).
7. B. Diner and D. C. Mauzerall, Biochim. Biophys. Acta *305*, 329-352 (1973).
8. B. Kok, R. Radmer and C. F. Fowler, Proceedings of the Third International Congress on Photosynthesis (Edited by M. Avson) pp 485-496, Elsevier, Amsterdam (1974)

DEVELOPMENT OF OXYGEN-EVOLVING SYSTEM
AS DESCRIBED IN TERMS OF THERMOLUMINESCENCE

YORINAO INOUE AND KAZUO SHIBATA

Laboratory of Plant Physiology,
The Institute of Physical and Chemical Research,
Wako-shi, Saitama, 351, japan

Summary

The present paper reviews the characteristics of thermoluminescence bands of angiosperm(wheat) and gymnosperm(spruce) leaves and
green algal(*Scenedesmus*) cells measured in previous studies together with some new data. Attention was concentrated to the development of various luminescence bands during development of photosynthetic apparatus. The five luminescence bands generally
observed for mature leaves and cells were characterized and some
of them were correlated to the oxygen-evolving system. The multi-
quantum activation of the latent oxygen-evolving system in dark-
grown or flashed leaves or dark-grown or Mn-deficient *Scenedesmus*
cells was demonstrated in terms of thermoluminescence as well as
by the measurements of delayed emission and photosynthetic
activity.

Introduction

The development of photosynthetic apparatus proceeds differently in different organisms. In angiosperms, the development is
strictly dependent on light. Photoconversion of protochlorophyl-
lide to chlorophyllide triggers the rapid biosynthesis of pigments
and proteins necessary for formation of photosynthetic membranes
(43). On illumination of the etiolated wheat(angiosperm) leaves,
various photosynthetic activities appear at early stages of
greening and the two photosystems including the oxygen-evolving
system are completely formed in 10 to 15 hrs of illumination (7,8).
Under intermittent illumination with flashes at intervals of
more than several minutes, however, the development of angiosperm
chloroplasts proceeds differently. More chlorophyll *a* is formed
without lag to yield leaves with a higher *a/b* ratio (1). The
chloroplasts in such intermittently illuminated leaves are devoid
grana, and only primary thylakoids are differentiated from pro-
lamellar bodies (39). Such chloroplasts are capable of bringing

about PS-I photoreactions and the Hill reaction in the presence
of diphenylcarbazide as electron donor, but are incapable of
bringing about certain PS-II photoreactions such as oxygen evolution, delayed light emission and fluorescence variation (17,29,
35, 40). It has been revealed recently that the oxygen-evolving
system in such etioplasts remains latent unless activated through
sequential photoreactions by illumination with continuous light
or short interval flashes (19,33).

As opposed to angiosperms, the gymnosperm leaves are capable
of biosynthesizing chlorophylls independent of light, so that
the spruce seedlings grown in complete darkness are deeply green
and contain a considerable amount of chlorophylls (30). The
chloroplasts isolated from such dark-grown spruce leaves have a
well developed lamellar system and are capable of reducing Hill's
reagents by light in the presence of diphenylcarbazide as electron donor. Similarly to the intermittently illuminated wheat
leaves, the chloroplasts in these leaves are specifically devoid
of the water-splitting activity, which remains latent unless
activated by continuous light or short interval flashes (22,30).

Chlorophyll formation and development of photosynthetic apparatus in green algae depend on the species and culture conditions.
In the case of *Scenedesmus obliquus* used in this study, the cells
grown heterotrophically in darkness were deeply green and contained a sufficient amout of chlorophylls. In this organism, the
chlorophyll formation is independent of light in heterotrophic
culture as in the case of gymnosperms. It was reported that all
the photosynthetic apparatus are developed completely in the dark
grown cells (26), since the cells show a high activity of oxygen
evolution. It is, however, likely that the light used for the
measurement of oxygen evolution may have photoactivated the
latent oxygen-evolving system in this experiment. According to
the study by Cheniae and Martin (15), the oxygen-evolving system
in dark-grown *Chlorella* cells remains latent unless activated by
light.

The pigment content and composition in *Scenedesmus* cells are
not affected by Mn-deficiency, but the oxygen evolution is depressed heavily (10). The deficient cells absorb Mn ions rapidly
into their photosystems. The depressed oxygen-evolving activity
is restored when incubated in the light in the presence of Mn.

This article reviews the glow curves observed for three photosynthetic organisms of wheat leaves, spruce leaves, and *Scenedesmus* cells. Thermoluminescence from plant materials was first observed by Arnold (6) who observed four luminescence bands emitted
at different temperatures which were denoted as Z, A, B and C
bands (4). Similar measurements were carried out on intact algal
cells, isolated chloroplasts and photosynthetic particles by
various investigators (3,6,16,28,36,38), and it is generally accepted that these emission peaks result from recombination of
electrons and positive holes generated by chlorophyll photoreactions and stabilized in frozen states (2). Of these bands,
Arnold's Z band was identified to be an emission from the triplet

state of chlorophyll *a* (37), and is not involved directly in photosynthetic reactions. Since the other A, B and C bands are emitted from PS-II particles (18,28), they are considered to originate from the back reaction on the electron transport chain between PS I and PS II to donate electrons to PS II. Arnold and Azzi (5) stated that two quanta are required for charging the A or B band and suggested that the emission originates from the holes in the oxygen-evolving center in which the charge transfer proceeds stepwise (24,25). In spite of these observations on the site involved in the thermoluminescence emission, the chemical nature of the site responsible for the emission is unknown. We have measured the glow curves of the above various organisms during development of their photosynthetic apparatus. It was indicated that glow curves provide useful informations about the state of the oxygen-evolving system, in particular, of the Mn catalyst which is inferred to be involved in the water-splitting reactions.

Materials and Methods

Thermoluminescence measurement

Thermoluminescence from leaves and algal cells was measured by the method described previously (18), which is briefly described below. Leaves, chloroplasts or algal cells placed on the holder was cooled in a Dewar bottle to a desired temperature below 0°C, illuminated for 1 min with red light($\geqq$630 nm, 600 μW/cm^2), cooled down to liquid nitrogen temperature and, then, heated at a rate of 0.5°C/sec. The photons emitted from the sample during heating was measured with a Jasco photon counter model KC-200 equipped with an EMI photomultiplier model 9659QB. The photo count integrated in every 1 sec was recorded on an X-Y recorder against the sample temperature monitored with a thermocouple.

Preparation of sample leaves

Seven-day old etiolated leaves of wheat(*Triticum aestivum* L.) were cut into segments and spread on moist filter paper to be illuminated for greening with coninuous light(750 μW/cm^2) or with intermittent flashes(1 msec flash/5 min dark, $1.5 \cdot 10^4$ergs/cm^2/ flash). The intermittently flashed leaves obtained after exposure to 300 flashes were futher illuminated with continuous light (680 $\pm$ 5 nm, 14 μW/cm^2) for photoactivation of their latent water-splitting system.

Spruce seeds(*Picea abies* L.) were germinated and grown on moist vermiculite in darkness, and green needle-like leaves were harvested from 20-day old seedlings. The dark-grown leaves were subjected to activation of the latent water-splitting system by illumination with continuous red light($\geqq$630 nm, 28 μW/cm^2) or with 2-μsec Xe flashes($2 \cdot 10^2$ ergs/cm^2/flash) repeated at varied intervals.

Preparation of sample algal cells

Scenedesmus obliquus were cultured heterotrophically in darkness for 3 days in an organic medium containing yeast extract and
polypepton or autotrophically in the light for 14 days in an inorganic MDM medium for blue green algae. The Mn-deficient cells
were obtained by cultivating the cells in a similar MDM medium.
All the ingredients of the medium except Ca, Mg and heavy metals
were filtered through a 30-cm long Chelex column. Ultra-pure Ca
and Mg salts were added separately and the heavy metals other
than iron and Mn were added as a Mn-free A5 solution. An iron
solution was prepared by dissolving pure Fe metal in ultra-pure
HCl and then neutralized with ultra-pure NaOH. The Ultra-pure
chemicals were purchased from Ventron(U.S.A.). The dark-grown or
Mn-deficient cells in suspension were illuminated with the same
light source as used for illumination of higher plant leaves.

Results

Development of thermoluminescence bands in angiosperms

The changes of the glow curve during development of photosyn-
theticapparatus in angiosperm leaves are summarized in Fig. 1.

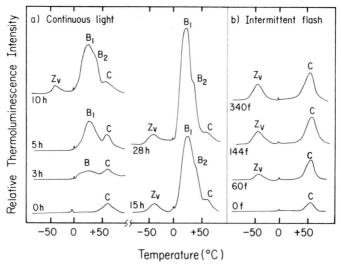

Temperature (°C)

*Fig. 1. Development of thermoluminescence bands during greening
of wheat leaves under continuous(a) and intermittent(b) illumination. Excitation was made by 1-min illumination with red light
at -60°C. Figures on each glow curve stand for the time in hr of
continuous illumination and the number of flashes given in every
5 min for (a) and (b), respectively.*

The glow curves of fully greened mature leaves measured after excitation at -60°C showed three distinct luminescence peaks. The
peaks at -45 and +55°C were denoted as Z_v and C bands, respectively, and the main peak around +30°C with a shoulder around +40°C
is a composite of B_1 and B_2 bands(upper curves on the middle column). The designation of these bands was made according to the
nomenclature by Arnold (4) with partial modification (18). The
emission temperature of the Z_v band was variable depending on the
illumination temperature. Mature leaves emitted the A band in
addition to these bands, when the excitation was made at a higher
temperature of -20°C. The A band and the B bands(B_1 and B_2)
appeared alternatively: under the excitation condition to yield
a strong A band, the B bands were weak, whereas the A band was
emitted weakly under the excitation condition to yield strong B
bands.

On illumination of the etiolated wheat leaves with continuous
light, the B_1 and B_2 bands appeared first at 3 hrs, and the Z_v
band at 10 hrs(Fig. 1a). The B_1 and B_2 bands were intensified
during prolonged illumination. When the glow curve was measured
by excitation at a higher temperature of -20°C, on the other hand,
the A band was developed instead of the B bands.

When the etiolated leaves were greened under intermittent flash
illumination at 5-min interval, the Z_v band fist appeared at 5 hrs,
and was gradually intensified during further illumination with
340 flashes. However, neither the B bands nor the A band appeared
even after 28 hr of illumination with 340 flashes(Fig. 1b).

On exposure of such intermittently illuminated leaves to continuous light, a rapid chnage of the glow curve was observed as
shown by Fig. 2. The B bands were markedly enhanced after a brief
exposure of the leaves to weak red light, and the enhancement
continued during the prolonged illumination to be as high as those
observed for mature leaves. This development of either the B
bands or the A band was accompanied by generation of the Hill
activity. The band development proceeded parallel with the generation of the Hill activity (21).

Development of thermoluminescence bands in gymnosperms

Fig. 3 shows the glow curves of dark-grown and light-grown
spruce leaves measured by excitation at three different temperatures. As mentinoned above, the glow curve was greatly dependent
on the excitation temperature. The glow curves obtained for
light-grown leaves(solid curves) by excitation at -75°C shows a
strong band at +20°C with a shoulder at +30-40°C and two weak
bands at -65 and +70°C, respectively. The broad band is a compo-
sit of the B_1 and B_2 bands. The small band at -65°C is the Z_v
band. The glow curves obtained at a higher excitation temperature
of -45°C shows the variable Z_v band at a higher emission temperature of -25°C, a strong B_2 band with a B_1 shoulder and a flat C
band. At an even higher excitation temperature of -20°C, the A
band newly appeared around -5°C and the B_1 and B_2 bands remarkably

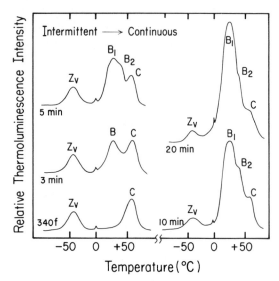

Fig. 2. *Development of thermoluminescence bands during photoacti-*
vation of the latent water-splitting system. Wheat leaves
greened by 340 flashes in 28 hrs were exposed to continuous red
light for the period in min indicated on each curve. Excitation
conditions were the same as in the experiment of Fig. 1.

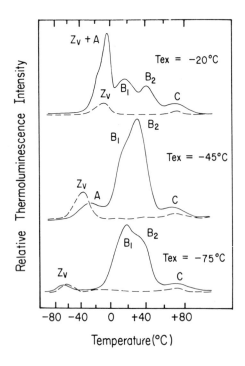

Fig. 3. *Glow curves of*
light-grown (solid curves)
and dark-grown (broken
curves) spruce leaves
measured at three dif-
ferent excitation temp-
eratures of -75, -45 and
-20°C, indicated as T_{ex}
on each group of curves.

decreased in height.

The glow curves for dark-grown spruce leaves(broken curves) were greatly different from those of mature light-grown leaves. They are composed of a distinct Z_v band and a very weak C band, but are completely devoid of the A, B_1 and B_2 bands. At -45°C excitation, the intensity of the Z_v band for the dark-grown leaves was considerably stronger than that for the light-grown leaves.

When such dark-grown spruce leaves were illuminated with continuous red light, the glow curve changed remarkably. Fig. 4 shows the effect of this illumination. At the excitation temperature of -45°C(curves on the left side), the broad B_2 band appeared after a brief exposure for 2 min, and this band was intensified progressively during prolonged illumination until 180 min. During this development of the B_2 band, the Z_v band was not intensified but rather decreased in height. When similar leaves under development were examined by excitation at -20°C(curves on the right side), the A, B_1 abd B_2 bands comparable in relative height appeared. The glow curve obtained after 180 min of illumination was very similar to that of mature leaves excited at the same temperature shown by the top curve in Fig. 3.

This change of thermoluminescence was accompanied by generation of the Hill activity with water as electron donor, enhancement of delayed emission and appearance of fluorescence variation (22,

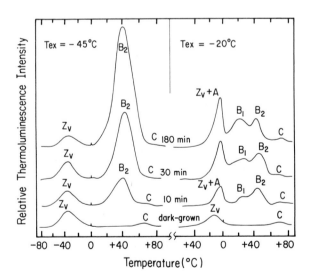

Fig. 4. Development of glow peaks during continuous illumination of dark-grown spruce leaves. The dark-grown spruce leaves were illuminated with red light for the period in min indicated commonly for each glow curve in the middle. The glow curves on the left and right sides are those measured by excitation at -45 °C and -20°C, respectively.

30). These observations are in good agreement with those obtained for the angiosperm leaves greened under intermittent illumination with long interval flashes (19, 41), and suggested strongly that the A, B_1 and B_2 bands originates from energy storage in
the water-splitting system in PS II, whereas the other two bands
are emitted from some other systems.

The activation of the latent water-splitting system is a multi-
quantum process. The process previously analyzed for intermittently illuminated wheat leaves by measuring delayed emission enhancement (19) showed that the process is made up of three consecutive photoreactions. The parallelism between the development
of the characteristic thermoluminescence bands and the generation
of the Hill activity in both angiosperm and gymnosperm leaves
suggested that the band development also requires sequential
photoreactions. This inference was confirmed by the following
experiment. The dark-grown leaves were exposed to 600 flashes
repeated at various dark interval(t_d) before subjected to thermoluminescence measurement. As seen from the glow curves in Fig. 5,
the flashes at an interval of 1 sec enhanced the B_2 band maximally, while the flashes at longer or shorter intervals were less
effective for the enhancement. This is more clearly demonstrated
by the experiment of Fig. 6, in which the height of the B_2 band
thus induced was plotted against the dark interval between
flashes(solid circles). A similar experiment was made by measuring the Hill activity as a function of t_d, and the result showed
the same dependency of the activity yield on t_d(open circles).
This dependency also agrees with the dependency found on photoactivation of the Hill activity in flashed wheat leaves (19).
It was thus found that the development of the B_2 band is a multi-
quantum process involving at least two photoreadtions. In addition to the B_2 band, the A and B_1 bands appeared when the glow
curves were measured by excitation at higher temperatures.

Development of thermoluminescence bands in green algae

Heterotrophic culture of *Scenedesmus obliquus* cells in darkness yields deeply green cells containing normal amout of chlorophylls. The study by Cheniae and Martin (15) revealed that the
water-splitting system in such dark-grown algal cells remains
latent unless exposed to light. We have measurd the glow curves
of such dark-grown cells during this photoactivation process.
The glow curve obtained for light-grown cells was composed of two
bands(Fig. 7), a weak but distinct Z_v band and a strong B band,
but the A and C bands were not observed. As opposed to the data
obtained for higher plant leaves, the glow curves of algal cells
were practically not dependent on the excitation temperature.
The A and C bands could not be observed at any excitation temperatures between -80 and -10°C, and the B band did not split into
components, the B_1 and B_2 bands. The glow curve for dark-grown
cells was greatly different from this profile of light-grown
ells. The B band was absent but the Z_v band was more distinct.

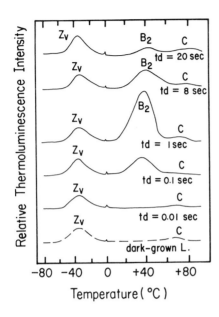

Fig. 5. Induction of glow peaks in spruce leaves by short interval flashes. Dark-grown spruce leaves were exposed at room temperature to 600 flashes at various intervals indicated as t_d on each glow curve. Excitation was made at $-45°C$.

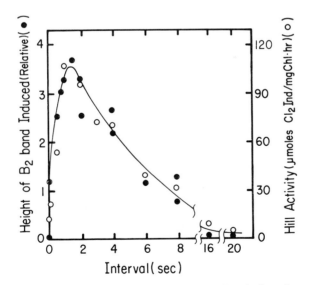

Fig. 6. Effects of the flash interval on the induction of glow peaks and the generation of the Hill activity in dark-grown spruce leaves during photoactivation by flashes. Solid and open circles along the bell-shaped curve are the relative height of B_2 band induced and the Hill activity generated by 600 flashes repeated at various intervals.

The B band was rapidly developed when the dark-grown cells were exposed to continuous light or to short interval repetitive flashes. Fig. 8 shows the glow curve obtained for dark-grown cells measured after exposure to 300 flashes repeated at varied intervals. The optimal interval for generation of the B band was about 2 sec which roughly equal to the optimal interval determined for higher plant leaves. It is evident from this result that the induction of the B band in algal cells is a multi-quantum process involving more than two consecutive photoreactions.

A similar experiment was carried out for Mn-deficient *Scenedesmus* cells obtained by cultivating the cells autotrophical-ly in a Mn-depleted medium. The growth in the deficient medium was suppressed at a level which is about 15% of the level as measured as chlorophyll content. The Mn content in such deficient cells was less than 5% of that in normal cells, and the molar ratio of chlrophylls to Mn was 890 and 40 for the deficient and normal cells, respectively. The photosynthetic activity of the Mn-deficient cells measured in terms of oxygen evolution was about 20% of the normal cells.

The bottom curve on the left side of Fig. 9 shows the glow curve obtained for such Mn-deficient cells, in which the B band at +35°C is extremely low as compared with that of normal cells shown by the upper curve in the same figure. Four glow curves

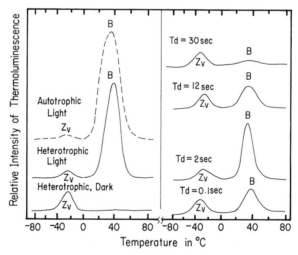

Fig. 7(left curves). Glow curves of Scenedesmus obliquus cells cultured under different conditions indicated on each profile. Heterotrophic light cells were obtained by overnight illumination of the dark-grown cells with white light. The concentration of the cell suspension was 10 μg Chl/cm².

Fig. 8(right curves). Development of glow peaks in dark-grown Scenedesmus cells induced by exposure to 300 flashes repeated at various intervals indicated as t_d on each curve.

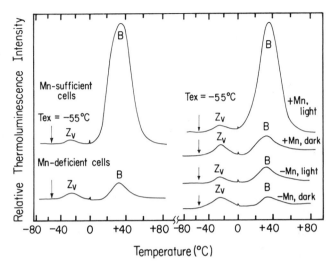

Temperature (°C)

Fig. 9. Glow curves of Mn-deficient(lower curve) and Mn-suffici-
ent(upper curve) Scenedesmus cells on the left side, and the
enhancement of the B band by the addition of Mn^{2+} on the right.
Four glow curves on the right side were measured for Mn-deficient
cells after 12-hr incubation in the presence or absence of 50 μM
of $MnCl_2$ in the light or dark, as specified on each glow curve.

on the right side are the data obtained after 12 hrs of incubation
of such Mn-deficient cells in the presence or absence of 50 μM
of $MnCl_2$ in the light or in the dark. As seen from the solid
curves on the top, the B band was greatly enhanced by the illumination in the presence of Mn^{2+}, while the incubation under other
conditions did not alter the profile at all. These data indicate
that photoreactions are involved in the recovery of the missing
luminescence band.

The light conditions required for the enhancement of the B
band was further examined by flash illumination. The Mn-deficient cells were pre-incubated in darkness for 2 hrs in a medium
containing 50 μM of Mn^{2+} and, then, exposed to Xe flashes repeated
300 times at uniform intervals(t_d). The glow curves measured
after exposure to flashes showed a marked enhancement of the B
band. The optimal interval for the enhancement was 2-4 sec, as
seen from the result shown by curve A with solid circles in Fig.
10. This dependency of the band enhancement observed for algal
cells is similar to that of the activity dependency observed for
angiosperm leaves (19) which is shown by curve B with open triangles in the same figure for comparison. The optimal interval
for angiosperm leaves was about 1 sec which is slightly shorter
than that for algal cells. This dependency also agrees with that
found for the development of the B_2 band in dark-grown spruce
leaves (23), and also with that for the photoactivation of the

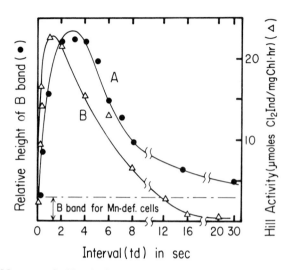

Fig. 10. Effects of flash interval on the B-band enhancement in
Mn-deficient cells induced by repetitive flashes(curve A with
solid circles) and those on the Hill-activity generation in
intermittently flashed wheat leaves(curve B with open triangles).
Mn-deficient cells were exposed to 300 flashes after 2-hr incubation in a medium containing 50 μM MnCl$_2$. The intermittently
flashes leaves were exposed to short intervals flashes for 100
times before measurement of thermoluminescence. The interval was
varied between 10 msec and 30 sec. The broken line indicates the
original height of the B band observed for Mn-deficient cells.

oxygen-evolving activity in Mn-extracted algal cells observed
previously by Radmer and Cheniae (33). It is indicated that the
Mn catalyst plays an important role in the energy storage for
thermoluminescence.

Discussion

The glow curves obtained for mature leaves are composed of 4 to 6
luminescence peaks. They are the Z, Z_v, A, B_1, B_2 and C bands.
The five bands except the Z_v band were emitted at constant temperatures, but the emission temperature of the Z_v band depended
on the excitation temperature, being usually higher by 10-30°C
than the excitation temperature between -140 and -10°C.
   Table 1 compares the thermoluminescence bands designated by
various investigators for different plant materials. The number
of observed bands and their emission temperatures observed by
different workers do not completely agree with each other. This
is mainly due to the difference in excitation condition. The
glow curves obtained after continuous irradiation during cooling

Table 1. *Correspondence of thermoluminescence bands observed by various investigators*

Author[sample]*	Luminescence bands (emission temperature)				
Arnold & Azzi[1] (1968)	Z (-155°C)	A (-15°C)	B (+30°C)		C (+55°C)
Rubin & Venediktov[2] (1969)		peak (-15°C)	peak (+10°C)	peak (+25°C)	peak (+55°C)
Shuvalov & Litvin[3] (1969)	C_{II} (-160°C)	C_{III} (-15°C)	C_{IV} (+20°C)		
Lurie & Bertsch[4] (1974)		$peak_1$ (-10°C)	$peak_2$ (+30°C)	$peak_3$ (+40°C)	
Sane et al[5] (1974)	peak (118°K)	peak (254°K)	peak (257°K)	peak (290°K)	peak (320°K)
Ichikawa *et al* (1975) Inoue *et al* (1976) [6]	Z (-160°C) + Z_v (variable)	A (-10°C)	B_1 (+25°C) *(alternative)*	B_2 (+40°C)	C (+55°C)

*[1] spinach leaves, [2] *Chlorella* cells, [3] higher plant leaves and algal cells, [4] spinach chloroplasts, [5] spinach leaves [6] spinach and wheat leaves.

from room temperature to liquid nitrogen temperature was considerably different from those obtained by excitation at a fixed temperature, and the glow curve obtained by a fixed excitation temperature varied depending on the excitation temperature. In general, the excitation at a higher temperature around -20°C yielded glow curves with a strong A band around -10°C, while the excitation at a low temperature around -50°C yielded glow curves with strong B bands around +35-40°C. The alternative nature found between these A and B bands suggests that these two types of bands are emitted from similar energy storage sites under different conditions.

The glow curves of intermittently flashed wheat leaves and dark-grown spruce leaves demonstrated clearly that the A, B_1 and B_2 bands originate from the water-splitting system. These bands were completely absent in the profiles of the flashed wheat leaves and the dark-grown spruce leaves, but they were rapidly developed on exposure of these leaves to continuous light (Figs. 1-4)(ref. 21-23). The chloroplasts isolated from flashed or dark-grown leaves are incapable of bringing about the Hill reaction of oxygen evolution but are capable of photoreducing Hill's reagents in the presence of diphenylcarbazide. It is, therefore, evident that these chloroplasts are specifically devoid of the water-

Table 2. *Thermoluminescence bands of various samples of leaves,
chloroplasts, subchloroplast particles and algal cells*

Sample	Luminescence bands (emission temperature, °C)				
	Z (variable)	A (-10)	B_1 (+25)	B_2 (+40)	C (+55)
Mature spinach leaves	+	+	+	+	+
Heat-treated spinach leaves	+	-	-	-	-
Flashed wheat leaves	+	-	-	-	+
Etiolated wheat leaves	-	-	-	-	+
Light-grown spruce leaves	+	+	+	+	+
Dark-grown spruce leaves	+	-	-	-	+
Whole chloroplasts	+	+	+	±	±
Tris-washed chloroplasts	+ T*	-	-	-	+
Triton PS-II particles	+ T*	-	-	-	+
Triton PS-I particles	-	-	-	-	±
Light-grown *Scenedesmus*	+	-	+**		-
Dark-grown *Scenedesmus*	+	-	-		-
Mn-deficient *Scenedesmus*	+	-	+**		-
Chlamydomonas PET 101***	+	-	-		-
Chlamydomonas PET 201***	+	-	+**		-
Chlamydomonas PET 301***	+	-	+**		-

* Observed around -20°C for Tris-treated chloroplasts or PS-II
   particles derived by treatment with Triton X-100.
** The B band of algal cells do not split into the component
   bands of B_1 and B_2.
*** Mutants isolated by Dr. Togasaki of Indiana University.

splitting system, while the PS-II reaction centers have been completely developed (19,20,31). The chloroplasts isolated from the leaves briefly exposed to continuous light are active to split water to evolve oxygen. The time course of generation of this Hill activity agreed well with the course of development of the three thermoluminescence bands described in the present paper. This parallelism strongly suggested a close correlation between these two phenomena.

This inferrence was further confirmed by the experiment with short interval flashes(Fig. 4). The enhancement of the thermo-luminescence band plotted against flash interval showed a bell-shaped curve with a maximum around 1-2 sec, which indicated that the process is made up of more than two light reactions with a rate-limiting dark reaction (23). The slope at shorter intervals indicates that 1-2 sec of dark period is necessary for the inter-mediate formed on the first photoreaction to be converted in darkness to another intermediate which is to be photoactivated to the final active state, and the slope at longer intervals indi-cates that the photosensitive intermediate decays in darkness. The dependency on the flash interval found for the band induction agrees not only with that for generation of the Hill activity in the same leaves, but also with those previously observed by us (19) for the Hill activity generation in intermittently flashed wheat leaves, and with the results obtained by Cheniae and Martin (15) and by Radmer and Cheniae (33) for Mn-extracted algal cells. A similar dependency on the flash interval was observed for the band enhancement in dark-grown or Mn-deficient algal cells. These similar bell-shaped dependencies observed for dif-ferent organisms indicate that the requirement cf sequential light reactions is a universal characteristic of this activation, and confirms the above conclusion that the three bands originate from the water-splitting system .

The Z_v band is emitted from intermittently flashed wheat leaves (18,21), dark-grown spruce leaves(22) and algal cells in which the water-splitting systme remains latent. The hypothesis that this band is an emission from free or light harvesting chlorophylls may be excluded, since the band is observed for PS-II particles but not for PS-I particles (18). The C band is emitted also independent of the water-splitting system. This band seems not related to photosynthetic activities, since it is observable in the glow curve of etiolated leaves(Fig. 1)(ref. 21). This band is very low or not observed for mature leaves but becomes distinct when the leaves were poisoned with DCMU.

We have recently measured the glow curves of isolated chloro-plasts during inactivation of the oxygen-evolving system by Tris treatment and during reactivation of the Tris treated chloro-plasts by DCIP-ascorbate treatment followed by illumination in the presence of Mn^{2+} and Ca^{2+} (ref. 44). On treating the chloroplasts with Tris, the A and B bands disappeared, whereas the Z_v band was intensified markedly. On reducing the Tris-treated chloroplasts with DCIP-ascorbate, the intensified Z_v band was lowered to the normal height, but the A and B bands did not appeare at this stage of reactivation. These two bands appeared and were intensified on further treatment of the chloroplasts with Mn^{2+} and Ca^{2+} under illuminateion (unpublished data by Inoue & Yamashita). These data reconfirm the close correlation of the three bands to the oxygen-evolving system.

There is another type of thermoluminescence band, which is Arnold's Z band emitted around -160°C. This babd is excited more effectively by blue light and is observed even for boiled algal cells (4). This indicates that this band is not concerned with photosynthesis. Sane et al (37) from its emission spectrum concluded that the band is an emission from the triplet state of chlorophylls.

These thermoluminescence bands in various leaves and algal cells are summarized in Table 2. These data indicate that the above band characterization applies to glow curves of algal cells, although they lack the A and C bands as opposed to those of higher plants. The Z_V band is observed in all algal cells examined but the B band was absent in the glow curves of dark-grown Scenedesmus or Chlamydomonas mutant PET 101, isolated by Dr. Togasaki, both of which are incapable of evolving oxygen because of some defects in the water-splitting system.

The glow curves of Mn-deficient cells seem to provide a new insight into the mechanism of energy storage. As shown by the experiments of Figs. 9 and 10, the B band depressed by shortage of Mn in the medium was recovered by illumination after addition of Mn ions, and the band enhancement is driven by sequential photoreactions. This suggests that a Mn complex formed from Mn^{2+} by photoreactions is involved directly in the emission of the B band or indirectrly in activating a neighbouring site needed for the emission. It seems very probable that the positive holes generated by PS-II photoreaction at low temperature are stabilized in the oxygen-evolving system as cations of activated Mn catalyst.

It has been reported that light is required for incorporation of Mn into the photosynthetic apparatus of Mn-depleted algal cells or chloroplasts (12-14,34,42), but we don't know why two or more photoreactions are necessary. Radmer and Cheniae (33) proposed valence changes of Mn during photoactivation, stepwise oxidation of Mn^{2+} to Mn^{4+} via Mn^{3+}. Attemps to detect such valence changes by means of ESR measurement are made by several investigators (9,32). It has also been proposed that one oxygen-evolving center contains three Mn atoms, two of which work in the water splitting reaction and the remaining one works as electron carrier. According to Loach and Calvin, the redox potential of Mn^{4+} is high enough to oxidize water. Taking these into considerations, one might assme a binuclear Mn complex, $(Mn^{2+}-Mn^{2+})$, as the latent Mn catalyst formed in the Mn-deficient cells with added Mn in the dark. Two sequential photoreactions will convert this complex to $(Mn^{3+}-Mn^{3+})$ via an intermediate $(Mn^{3+}-Mn^{2+})$, and this $(Mn^{3+}-Mn^{3+})$ complex will be excited to $(Mn^{4+}-Mn^{4+})$

through stepwise oxidation. The energy collected by this excitation will be released at room temperature to oxidize water to
evolve oxygen or may be stored at low temperature for thermoluminescence.

Acknowledgements

This work was supported by a research grant for the study of
"*Life Sciences*" at the Institute of Physical and Chemical Research
(Rikagaku Kenkyusho). We wish to thank Miss A. Suzuki for her
excellent secretarial help in the preparation of this manuscript.

References

1. Akoyunoglou, G., Argyroudki-Akoyunoglou, J.H., Michel-Wolwertz
   M.R. and Sironval, C. 1966 Effect of intermittent and continuous light on chlorophyll formation in etiolated plants.
   Physiol. Plant. 19: 1101-1104
2. Arnold, W. 1965 An electron-hole picture of photosynthetis.
   J. Phys. Chem. 69: 788-791
3. Arnold, W. 1966 Light reaction in green plant photosynthesis:
   A method of study. Science 154: 1046-1049
4. Arnold, W. and Azzi, J. 1968 Chlorophyll energy levels and
   electron flow in photosynthesis. Proc. Natl. Acad. Sci. U.S.
   61: 29-35
5. Arnold, W. and Azzi, J. 1971 The mechanism of delayed light
   production by photosynthetic organisms and a new effect of
   electric field on chloroplasts. Proc. Natle. Sci. U.S.
   14: 233-240
6. Arnold, W. and Sherwood, H. Energy storage in chloroplasts
   J. Phy. Chem. 63: 2-4
7. Boardman, N.K., Anderson, J.M., Hiller,R.G., Kahn, A., Roughan
   P.G. Treffry T.E. 1971 Biodynthesis of photosynthetic apparatus during chloroplast development in higher plants. In:
   G. Forti, M. Avron and A, Melandri eds. Proc. 2nd Congr.
   Dr. Jung publishers, The Hague pp. 2265-2286
8. Boardman, N.K., Anderson, J.M., Kahn, A., Thorne, S.W. and
   Treffry, T.E. 1970 Formation of photosynthetic membranes
   during chloroplast development. In : N.K. Boardman, A.W.
   Linnane and R.M. Smillie eds. Autonomy and biogenesis of
   mitochondria and chloroplasts. North-Holland, Amsterdam,
   pp. 70-84
9. Blankenship. R.E., Babcock, G.T. and Sauer, K. 1975 Kinetic
   study of oxygen-evolution parameters in Tris-washed, reactivated chloroplasts. Biochim. Biophys. Acta, 387: 165-175
10. Cheniae, G.M. and Martin, I.F. 1968 Site of mananese function
    in photosynthesis. Biochim. Biophys. Acta, 153: 819-837
11. Cheniae, G.M. and Martin, I.F. 1969 Photoreactivation of
    manganese catalyst in photosynthetic oxygen evolution. Plant
    Physiol. 44: 351-360

12. Cheniae, G.M. and Martin, I.F. 1970 Studies of function of manganese within photosystem II. Roles in O_2 evolution and system II. Biochim. Biophys. Acta, 197: 219-239

13. Cheniae, G.M. and Martin, I.F. 1971 Photoactivation of the manganese catalyst of O_2 evolution I. Biochemical and kinetic aspect. Biochim. Biophys. Acta, 253: 163-181

14. Cheniae, G.M. and Martin, I.F. 1972 Effect of hydroxylamin on photosystem II. II. Photoreversal of the NH_2OH destruction of O_2 evolution. Plant Physiol. 50: 87-94

15. Cheniae, G.M. and Martin, I.F. 1973 Absence of oxygen-evolution capacity in dark-grown Chlorella. Photochem. Photobiol. 11: 441-457

16. Desai, T.S., Sane. P.V. and Tatake, V.G. 1975 Thermoluminescence studies on spinach leaves and Euglena. Photochem. Photobiol. 21: 345-350

17. Dujardin, E., de Kouchkovsky, Y. and Sironval, C. 1970 Properties in leaves grown under a flash regime. Photosynthetica 4: 223-227

18. Ichikawa, T., Inoue, Y. and Shibata, K. 1975 Characteristics of thermoluminescence bands of intact leaves and isolated chloroplasts in relation to the water-splitting activity in photosynthesis. Biochim. Biophys. Acta, 408, 228-239

19. Inoue, Y. 1975 Multiple-flash activation of the water-photolysis system in wheat leaves as observed by delayed emission. Biochim. Biophys. Acta, 396: 402-413

20. Inoue, Y., Kobayashi, Y., Sakamoto, E. and Shibata, K. 1974 Action spectrum for photoactivation of the water-splitting system in plastids of intermittently illuminated leaves Physiol. Plant. , 32: 228-232

21. Inoue, Y., Ichikawa, T. and Shibata, K. 1976 Development of thermoluminescence bands during greenin of wheat leaves under continuous and intermittent illumination. Photochem. Photobiol. 23: 125-130

22. Inoue, Y., Furuta, S., Oku, T. and Shibata, K. 1976 Light-dependent development of thermoluminescence, delayed emission and fluorescence variation in dark-grown spruce leaves. Biochim. Biophys. Acta, in press

23. Inoue, Y., Oku, T., Furuta, S. and Shibata, K. 1976 Multiple-flash development of thermoluminescence bands in dark-grown spruce leaves. Biochim. Biophys. Acta 440: 772-776

24. Joliot, P., Barbieri, G. and Chabaud, R. 1969 Un nouveau modele des centres photochimiqus du system II. Photochem. Photobiol., 10: 309-329

25. Kok, B., Forbush, B and McGloin, M. 1970 Cooperation of charges in photosynthetic O_2 evolution I. A linear four step mechanism. Photochem. Photobiol. 11: 457-475

26. Kulandaiveln, G. and Senger, H. 1976 Changes in the reactivity of the photosynthetic apparatus in heterotrophic aging culture of Scenedesmus obliquus I. Changes in the photochemical activities. Physiol. Plant. 36: 157-164

27. Loach, P.A. and Calvin. M. 1963 Oxidation state of maganese

hematoporphyrin IX in aqueous solution. Biochemistry 2:361-371
28. Lurie, S. and Berstch, W. 1974 Thermoluminescence studies
    on photosynthetic energy conversion I. Evidence of three types
    of energy storage by photosystem II. Biochm. Biophys. Acta,
    357: 429-438
29. Michel, J.M. and Sironval, C. 1972 Evidence for induction of
    photosystem II in primary thylakoids when illuminated with
    continuous light for a short period. FEBS Lett. 27; 231-234
30. Oku, T., Sugawara, K. and Tomita, G. 1974 Functional develop-
    ment of photosystems I and II in dark-grown pine seedlings.
    Plant & Cell Phsiol. 15: 175-178
31. Oku, T. and Tomita, G. 1976 Photoactivation of oxygen-evo-
    lution system in dark-grown spruce seedlings. Physiol. Plant.
    in press.
32. Phung Nhu Hung, S., Houlier, S. and Moyse, A. 1976 Mn content
    changes in wheat etioplast membranes during greening under
    intermittent or continuous light. Plant Sci. Lett. 6:243-251
33. Radmer, R. and Cheniae, G.M. 1971 Photoactivation of the
    manganese catalyst of O_2 evolution II. A two-quantum mechanism
    Biochim. Biophys. Acta 253: 182-186
34. Radmer, R. and Kok, B. 1975 Energy capture in photosynthesis:
    Photosystem II. In: E. Snell, P.D. Boyer, A. Meister and
    C.C. Richardson eds. Ann. Rev. Biochem. vol. 44 Anual Reviwes
    INC. Palo Alto, pp. 409-433
35. Remy, R. 1973 Appearance and development of photosynthetic
    activities in wheat etioplasts greene under continuos or
    intermittent light: Evidence for water side photosystem II
    deficiency after greening under intermittent light.
    Photochem. Photobiol. 18: 409-416
36. Rubin, A.B. and Venediktov, P.S. 1969 Storage of the light
    energy by photosynthetic organisms at low temperature.
    Biofiz. 14: 105-109
37. Sane, P.V., Tatake, V.G. and Desai, T.S. 1974 Detection of the
    triplet state of chlorophylls in vivo. FEBS Lett. 45:290-294
38. Shuvalov, V.A. and Litvin. F.F. 1969 Mechanism of prolonged
    after luminescence of plant leaves and energy storage in the
    photosynthetic reaction center. Molek. Biol. USSR 3:59-73
39. Sironval, C., Michel, J.M., Bronchart, R. and Dujardin, E.
    1969 On the "primary" thylakoids of chloroplasts grown under
    a flash regime. In: H. Metzner ed. Progress in Phtosyn. Res.
    vol. 1 H. Laupp Jr., Tübingen, pp. 47-54
40. Strasser, R.J. and Sironval, C. 1972 Induction of photosystem
    II activity in flashed leaves. FEBS Lett. 28: 56-59
41. Strasser, R.J. and Sironval, C. 1973 Correlation between the
    induction of PS II activity and the induction of variable
    fluorescence in flashed bean leaves by weak green light.
    FEBS Lett. 29: 286-288
42. Takahashi, M. and Asada, K. 1976 Removal of Mn from spinach
    chloroplasts by sodium cyanide and binding of Mn^{2+} to Mn-dep-
    leted chloroplasts. Eur. J. Biochem. 64: 44-52
43. Virgin. H.I. 1972 Chlorophyll biosynthesis and phytochrome

activity. In: K. Mitrakos and W.Shlopshier Jr. eds.
Phytochrome, Academic Press. London & New York. pp.437-479
44. Yamashita, T. and Tomita, G. 1976 Light reactivation of
(Tris-washed)-DPIP-treated chloroplasts: Manganese incorporation chlorophyll fluorescence, action spectrum and oxygen
evolution. Plant & Cell Physiol. 17: 571-582

INHIBITION OF ELECTRON TRANSPORT ON THE OXYGEN-EVOLVING SIDE OF PHOTOSYSTEM II BY AN ANTISERUM TO A POLYPEPTIDE ISOLATED FROM THE THYLAKOID MEMBRANE

Dedicated to Professor Dr. Hans Gaffron at the occasion of his 75th anniversary

GEORG H. SCHMID, WILHELM MENKE, FRIEDERIKE KOENIG AND ALFONS RADUNZ
Max-Planck-Institut für Züchtungsforschung (Erwin-Baur-Institut) 5 Köln-Vogelsang, West-Germany

In 1962 Kreutz and Menke have concluded from results obtained by small angle X-ray scattering that the structure of the thylakoid membrane was essentially asymmetric with proteins located preponderantly towards the outer surface of the thylakoid membrane and lipids mainly towards the inner surface (1). This concept was in contradiction to the general view of that time which was mainly based on evidence obtained from electron microscopy and which proposed that the structure of the thylakoid membrane should be symmetric (2). Later, through investigations which were incited by Mitchell's chemiosmotic theory of photophosphorylation the idea of an asymmetric structure of the thylakoid membrane regained some interest.

Since many years we are working with serological methods on the elucidation of the molecular structure of the thylakoid membrane (3,4,5,6). The preparation of the protein antigens for these investigations is difficult because the major part of the proteins of the thylakoid membrane is insoluble in aqueous solvents. However, the proteins can be solubilized by means of various detergents. Among these sodium dodecylsulfate is the detergent which solubilizes the proteins of the thylakoid membrane in the completest way. As shown by our laboratory (7) sodium dodecylsulfate leads to conformational changes of the polypeptides which are further increased upon removal of the detergent by anion exchange chromatography (Fig. 1). Despite the fact that the secondary and tertiary structure of these polypeptides is greatly altered we were able to show that immunization of rabbits with these polypeptides yields antisera which agglutinate suspensions of stroma-freed chloroplasts and affect photosynthetic electron transport. Consequently, the antisera contain antibodies which are directed towards native antigenic determinants of the thylakoid membrane. For the eluci-

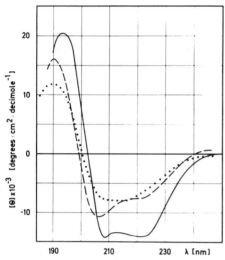

Fig. 1. Far ultraviolet circular dichroism spectra of fragments of the thylakoid membrane (solid line), of the polypeptide mixture obtained from the thylakoids dissolved in 0.22 % dodecyl sulfate-containing phosphate buffer pH 7.2 (dashed line) and the polypeptide mixture after removal of dodecyl sulfate (dotted line). θ is the mean residue ellipticity.

dation of the correlation between the molecular structure and function antisera have in comparison to other reagents the advantage of their high specificity. This, however, is only the case if the antisera are monospecific. Results obtained with antisera which are not monospecific lead to interpretations of restricted value.

In the following we report on an antiserum to a polypeptide fraction which exhibits an apparent molecular weight 11000. However, the molecular weight determined in the ultracentrifuge is considerably lower (Craubner et al. to be published). The fractionation of the polypeptide mixture was achieved by gel permeation chromatography in the presence of dodecylsulfate. The success of the fractionation is verified by means of dodecylsulfate polyacrylamide gel electrophoresis which shows that the polypeptide used for immunization migrates as a fast running single band (Fig. 2).

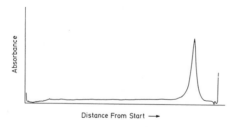

Fig. 2. Optical scan of the dodecylsulfate polyacrylamide gel electropherogram of the polypeptide 11000. Buffer pH 8.8.

Despite this apparent uniformity, the preparation is probably a mixture of several polypeptides with the

same or only slightly differing molecular weights. Four
amino terminal amino acids have been found as dansyl
derivatives. However, the question concerning the final
number of components is not solved yet.
    An antiserum to this polypeptide fraction aggluti-
nates stroma-freed chloroplasts from Antirrhinum and
tobacco. Consequently, antigenic determinants towards
which the antiserum is directed are located in the outer surface of the thylakoid membrane.
    The antiserum to polypeptide 11000 inhibits photo-
synthetic electron transport in stroma-free swellable
chloroplasts from tobacco. The degree of inhibition
depends on the pH and has its optimum at pH 7.4 (Fig.3).

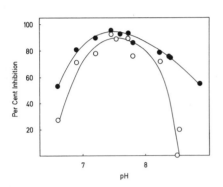

Fig. 3. pH-Dependence of
the degree of inhibition
of photosynthetic elec-
tron transport in tobacco
chloroplasts by the anti-
serum to polypeptide
11000. ●, Reaction in the
system tetramethyl benzi-
dine/ascorbate→anthra-
quinone-2-sulfonate.
O, Reaction with water as
the electron donor and
anthraquinone-2-sulfonate
as the electron acceptor.

At the optimal pH 7.4 the degree of inhibition depends
on the amount of the added antiserum (Fig. 4).

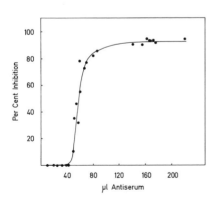

Fig. 4. Dependence of
the degree of inhibition
on the amount of added
antiserum to polypeptide
11000 at pH 7.4. Reac-
tion in the system te-
tramethyl benzidine/as-
corbate→anthraquinone-
2-sulfonate.

The curve shape is sigmoidal which hints at a cooperative effect. The calculation of the Hill interaction
coefficient yields 10. It should be noted that sigmoidal curve shapes are also obtained by the antisera to

polypeptide 33000 (5), to plastocyanin (6) and to cytochrome f. At the optimal pH and in the presence of saturating amounts of antiserum the inhibition site by
the antiserum is localized in the electron transport
scheme. The anthraquinone-2-sulfonate Hill reaction is
inhibited by the antiserum (Fig. 5).

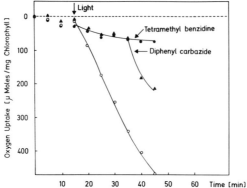

Fig. 5. Electron transport reaction measured as
oxygen uptake in an anthraquinone-2-sulfonate mediated Mehler reaction in wild type tobacco chloroplasts. O, Hill reaction in the presence of
control serum; ●, Hill reaction in the presence
of antiserum, addition of tetramethyl benzidine
does not relieve inhibition. ▲, Hill reaction in
the presence of antiserum, diphenyl carbazide
relieves inhibition.

Addition of tetramethyl benzidine which is an artificial electron donor to photosystem II (8) does not relieve the inhibition. However, addition of diphenyl
carbazide another artificial electron donor to photosystem II (9) restores electron transport. Typical pho-
tosystem-I-reactions are not affected by the antiserum.
From this it follows that the inhibition site is on the
oxygen-evolving side of the electron transport scheme
between the sites of electron donation of tetramethyl
benzidine and diphenyl carbazide (Fig. 6).

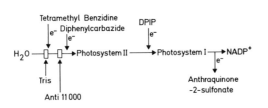

Fig. 6. Scheme to
illustrate the in-
hibition site of
the photosynthetic
electron transport
by the antiserum
to polypeptide
11000.

The inhibitory action of the antiserum on electron transport requires light. Despite the fact that agglutination of the chloroplasts occurs immediately upon addition of the antiserum, the inhibition of electron transport becomes only apparent in the course of the illumination period.

The localization of the inhibition site was also attempted by measuring fluorescence rise kinetics of chloroplasts in the presence of antiserum. In these measurements the influence of light on the inhibitory action of the antiserum is especially obvious. The assay system contained tetramethyl benzidine/ascorbate as the electron donor and anthraquinone-2-sulfonate as the acceptor. A dark adapted sample which has received no light prior to fluorescence excitation shows no influence of the antiserum on the steady state level of fluorescence (Fig. 7).

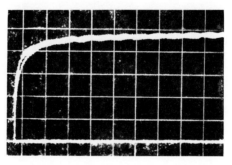

Fig. 7. Fluorescence rise in a chloroplast preparation in the assay system tetramethyl benzidine/ ascorbate→anthraquinone-2-sulfonate without illumination prior to the fluorescence assay. One assay with antiserum to polypeptide 11000, the other assay with control serum. Excitation wavelength 440 nm, excitation slit 40 nm; emission wavelength 685 nm, emission slit 34 nm. One grade of the abscissa represents 0.2 sec.

If, however, the reaction mixture has been illuminated with red light until the inhibitory effect was established otherwise, then the fluorescence level was lower in the presence of antiserum than in the presence of control serum (Fig. 8). This shows according to the literature that the inhibition site is on the water-splitting side of photosystem II (10). It is obvious that if the inhibition is introduced on the oxygen-evolving side of photosystem II that the fluorescence quencher Q is not or only slowly reduced. In this condition electrons are faster removed by photosystem I

than they are supplied by photosystem II which means
that the quencher Q is preponderantly in the oxidized
state in which the fluorescence is quenched and consequently low. If the drain of electrons through photosystem I is prevented by the addition of DCMU,fluores-
cence is immediately high because the quencher stays in
the reduced state (Fig. 8).

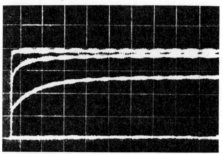

Fig. 8. Fluorescence rise curve of the same
assay as in Fig. 7. The assay mixture was shaken in the Warburg apparatus at 30.000 ergs·
$sec^{-1} \cdot cm^{-2}$ of red light 580 nm $<\lambda<$ 700 nm until
the inhibition caused by the antiserum was 93 %.
The assay was kept in the dark for 20 min prior
to the fluorescence measurement. Lower rise curve
in the presence of antiserum; middle curve in
the presence of control serum; upper curve in
the presence of control serum plus 10 /μmol DCMU.

The addition of DCMU to the sample with antiserum shows,
as DCMU entirely blocks the reoxidation of the reduced
Q and as the antiserum slows down the dark reduction of
photooxidized chlorophyll a_{II} that fluorescence rises
up again. The same result as with normal chloroplasts
is obtained with Tris-washed chloroplasts in the presence of an artificial electron donor system. From this
data it clearly follows that the antiserum inhibits
electron transport near the Tris-block on the oxygen-
evolving side of photosystem II.
    The use of chloroplasts from tobacco mutants which
differ with respect to electron transport properties
and photosynthetic unit sizes further confirms what has
been described above. The general electron transport
properties of the wild type tobacco and the two tobacco
mutants used are seen from the low temperature fluorescence spectra (Fig. 9).

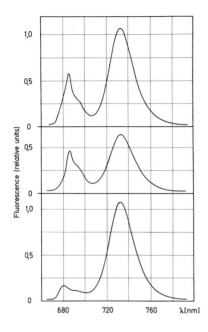

Fig. 9. Low temperature fluorescence emission at 77° K of tobacco chloroplasts. Excitation wavelength 440 nm; excitation and emission slit 10 nm; upper spectrum: Wild type tobacco N. tabacum var. John William's Broadleaf; middle spectrum: Tobacco aurea mutant Su/su^2; lower spectrum: Yellow leaf patch from variegated tobacco mutant NC95.

The green type shows the usual peak ratios at 685 nm versus 732 nm known from the literature (11). The aurea mutant Su/su^2 which has a smaller photosynthetic unit than the green type, namely 300 Chl/O_2 evolved/flash in comparison to 2400 Chl/O_2 evolved/flash with the wild type has a much higher ratio of the short wavelength emission bands versus the long wavelength emission, which indicates an increased photosystem II/photosystem I ratio. The yellow leaf patch from the variegated tobacco mutant NC95 described some years ago (12) shows a clear preponderance of the long wavelength emission band which is attributed in the literature to photosystem I (11). The effect of the antiserum on these three types of chloroplasts in the electron transport system $H_2O \rightarrow$ anthraquinone-2-sulfonate or tetramethyl benzidine $\rightarrow$ anthraquinone-2-sulfonate is described in figure 10. It is clearly seen that chloroplasts from the tobacco mutant NC95 are not inhibited by the antiserum. The functional difference of these chloroplasts in comparison to those of the wild type is the lack of the oxygen evolving capacity. As seen from figure 10 the chloroplasts support some tetramethylbenzidine-dependent electron transport to anthraquinone-2-sulfonate and exhibit normal rates of the photosystem I-dependent photoreduction of anthraquinone-2-sulfonate with 2,6-dichlorophenol indophenol/ascorbate as the electron donor.

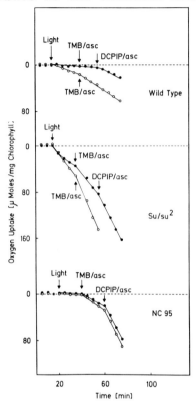

Fig. 10. Electron transport measured as oxygen uptake in an anthraquinone-2-sulfonate mediated Mehler reaction with various electron donors in tobacco chloroplasts. TMB/asc, tetramethyl benzidine/ascorbate; DCPiP/ascorbate, 2,6-dichlorophenol indophenol. O, assay in the presence of control serum; ●, assay in the presence of antiserum. Upper curve: Wild type tobacco; middle curve: tobacco aurea mutant Su/su^2; lower curve: chloroplasts from yellow leaf patch of the variegated tobacco mutant NC95. Red light 580 nm $< \lambda <$ 700 nm, $25°$ C.

The effect of the antiserum on the mutant chloroplasts from NC95 fits into the observations made with Tris-washed chloroplasts. Tris-washing of the chloroplasts from wild type tobacco abolishes the capactiy for oxygen evolution (13). Addition of the artificial electron donor couple tetramethyl benzidine/ascorbate restores electron transport to $NADP^+$ or to a suitable artificial electron acceptor such as anthraquinone-2-sulfonate. Also in the Tris-washed condition the antiserum exerts its inhibitory effect (Fig. 11). It appears that the relative degree of inhibition is lower than with unwashed chloroplasts. If instead of tetramethyl benzidine the artificial donor diphenyl carbazide is added then the antiserum does not influence the electron transport rate (Fig. 11). Again, it shows that the inhibition site is between the site of electron donation of tetramethyl benzidine and that of diphenyl carbazide.

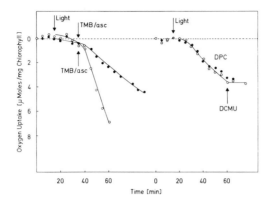

Fig. 11. Effect of the antiserum on electron
transport in the Tris-buffer washed chloroplasts
from tobacco. The electron transport system is
TMB/asc→ anthraquinone-2-sulfonate or diphenylcarbazide (DPC) → anthraquinone-2-sulfonate.
●, in the presence of antiserum, O, in the presence of control serum. Concentration of DCMU
after addition $2 \times 10^{-5} M$ in the DPC-assay.

Occasionally it was observed that the antiserum to polypeptide 11000 did not inhibit anymore the electron
transport rate in Tris-washed chloroplasts in the presence of the donor couple tetramethyl benzidine/ascor-
bate. This led us to the idea that Tris-buffer treatment might wash out the polypeptide 11000. This is not
the case. By means of our different antisera we have
found that amongst other proteins, coupling factor, reductase and plastocyanin were extracted, whereas the
antiserum to the polypeptide 11000 gave no precipita-
tion with the supernatant from the Tris-wash. The reason for the loss of inhibition might be that a too complete removal of the above listed proteins leads to an
altered molecular structure of the thylakoid membrane
which does not permit access of the antibodies to the
antigen. It might also be possible that the active component has lost its functionability. This leads to the
explanation that the site of inhibition is closely behind the Tris-block (see Fig. 6).
    The inhibition of electron transport is also ob-
served in the presence of uncouplers of photophosphorylation such as dicyclohexyl carbodiimide (DCCD), m-
chlorophenyl hydrazone (CCCP), gramicidin and valinomycin which means that the inhibitory action is directed

towards the electron transport chain (14).

Application of the repetitive flash spectroscopy to the electron transport systems described in figure 5 permits the direct measurement of the photoreaction of the reaction center chlorophyll a_{II} (15) and that of the primary electron acceptor X_{320} (16). It appears that the initial amplitude of the absorption change of $Chla_{II}$ at 690 nm and that of the primary electron acceptor X_{320} at 334 nm are both diminished by the antiserum (17). Comparison of the degree of inhibition on the amplitudes of the slow and fast components of the 690 nm absorption change with that of the manometrically measured degree of inhibition of electron transport shows that the reaction centers of photosystem II are affected in two ways by the antiserum. In green tobacco chloroplasts 50 - 60 per cent of the reaction centers are fully put out of function whereas the remaining 40 - 50 per cent change their way of functioning by mediating either a fast cyclic electron flow around photosystem II as proposed by Renger and Wolff for Tris-washed chloroplasts or a direct charge recombination (18). In Tris-washed chloroplasts Renger and Wolff had observed that the initial amplitude of the absorption change of chlorophyll a_{II} at 690 nm and that of the primary electron acceptor X_{320} were not altered whereas the overall electron transport to P_{700} was decreased (18). Therefore, from the fact that the initial amplitudes of the 320 nm and 690 nm absorption changes are decreased by the antiserum by more than 50 per cent it follows that the inhibition site of the antiserum must be closer towards the reaction center than the Tris-block (17) which fits into the observations described with figure 11.

The antiserum also affects the low temperature fluorescence emission at 77° K in a way which is opposite to Murata's effect of the Mg^{2+}-ion induced inhibition of energy spill-over from photosystem II to photosystem I (19). Under the condition shown in which the Hill reaction is inhibited, the antiserum lowered the 685 nm emission and enhanced the 732 nm emission which shows that the energy spill-over to photosystem I is enhanced (17). These experiments show that the antiserum affects the functionability of photosystem-II-reaction centers and that the energy spill-over to photosystem I is enhanced when photosystem-II-reaction centers are put out of function (17,20). From comparison of the low temperature fluorescence of Tris-washed chloroplasts, with that of normally active chloroplasts and the effect of the antiserum on these two types of chloroplasts it appears that the antiserum effect is similar

to Tris-treatment where the ratio F732/F686 is also increased (17).

We think to have presented the evidence that the antiserum to polypeptide 11000 inhibits photosynthetic
electron transport on the oxygen-evolving side of photosystem II. The data presented above means that the
polypeptide 11000 is located in the outer surface of
the thylakoid membrane. Possibly polypeptide 11000 is
part of a protein molecule which is composed of several
polypeptide chains. The polypeptide 11000 or the protein is associated with photosystem II and somehow par-
ticipates in a reaction which occurs on the water-
splitting side of photosystem II. Our result does certainly not permit any conclusion as to where the splitting of the water molecule itself occurs.

The data obtained by the low temperature fluorescence and absorption change measurements (17) show that
the antiserum affects the mode of functioning of the reaction center of photosystem II. In this context it
should be noted that Radunz et al. have shown in 1971
by means of antibodies to chlorophyll that part of the
reaction center chlorophyll of photosystem II was surface located (21). This observation gains some interest
in context with the results obtained with Chlorella by
P. Joliot and A. Joliot in which the authors reach the
conclusion that the reaction center chlorophyll Chl_{aII}
is under their conditions surface located (22).

A not easily explainable effect is the influence of
light on the onset of inhibition. The effect was originally observed in the DCPiP-Hill reaction (23) but is
best seen in the fluorescence measurements. A minimum
reaction time of 2 minutes is required. However, it
appears that a mere preillumination of the chloroplasts
in the presence of antiserum but in the absence of the
electron transport components of the assay system does
not shorten the induction time. It rather looks as if
in the light in the presence of the complete electron
transport system a membrane condition or redox state is
established in which the inhibition on electron transport by the antiserum becomes possible. The long induction time until the onset of inhibition might speak in
favour of a rough structural change followed by other
events. Light might open up partitions an event which
then or at the same time leads to a rearrangement of
the membrane or in the membrane. Here, an observation
by Giaquinta et al. should be mentioned. The author observed that diazonium benzene (^{35}S)-sulfonate was incorporated into the lamellar system when electron flux
through photosystem II and plastoquinone occurred (24).

Finally the bearing of our results on the current
concept on the topology of the photosystems in the thy-

lakoid membrane could be summarized and emphasized as
follows:
A protein to which a polypeptide with the apparent molecular weight 11000 belongs is surface exposed in the
thylakoid membrane or becomes surface exposed in the
light. This protein is associated with the reaction
center II and is somehow involved in reactions on the
oxygen-evolving side of photosystem II.

ACKNOWLEDGEMENT: G.H. Schmid thanks the Deutsche For-
schungsgemeinschaft and the Max-Planck-Gesellschaft
for a travel grant.

REFERENCES

(1)   W. Kreutz and W. Menke, Z.Naturforsch. 17b, 675
       (1962).
(2)   K. Mühlethaler, Z.Wissenschaftl.Mikroskopie u.
       mikroskopische Technik 64, 444 (1960).
(3   R. Berzborn, W. Menke, A. Trebst, F. Pistorius,
       Z.Naturforsch. 21b, 1057 (1966).
(4)   F. Koenig, W. Menke, H. Craubner, G.H. Schmid and
       A.Radunz, Z.Naturforsch. 27b, 1225 (1972).
(5)   W. Menke, F. Koenig, A. Radunz and G.H. Schmid,
       FEBS Letters 49, 372 (1975).
(6)   G.H. Schmid, A. Radunz and W. Menke, Z.Natur-
       forsch. 30c, 201 (1975).
(7)   W. Menke, A. Radunz, G.H. Schmid, F. Koenig and
       R.-D. Hirtz, Z.Naturforsch. 31c, 436 (1976).
(8)   E. Harth, W. Oettmeier and A. Trebst, FEBS Let-
       ters 43, 23 (1974).
(9)   L.P. Vernon and E.R. Shaw, Plant Physiol. 44,
       1645 (1969).
(10)   H.G. Aach, U.F. Franck and R. Bauer, Naturwiss.
       58, 525 (1971).
(11)   J.C. Goedheer, Ann. Rev. Plant Physiol. 23, 87
       (1972).
(12)   P.H. Homann and G.H. Schmid, Plant Physiol. 42,
       1619 (1967).
(13)   T. Yamashita and W. Butler, Plant Physiol. 44,
       435 (1969).
(14)   G.H. Schmid, W. Menke, F. Koenig and A. Radunz,
       Z.Naturforsch. 31c, 304 (1976).
(15)   G. Döring, G. Renger, J. Vater and H.T. Witt,
       Z.Naturforsch. 24b, 1139 (1969).
(16)   H.H. Stiehl and H.T. Witt, Z.Naturforsch. 23b,
       220 (1968).
(17)   G.H. Schmid, G. Renger, M. Gläser, F. Koenig,
       A. Radunz and W. Menke, Z.Naturforsch. 31c, 594
       (1976).

(18)   G. Renger and Ch. Wolff, Biochim. Biophys. Acta
       423, 610 (1976).
(19)   N. Murata, Biochim. Biophys. Acta 189, 171 (1969).
(20)   R. Satoh, R. Strasser and W.L. Butler, Biochim.
       Biophys. Acta 440, 337 (1976).
(21)   A. Radunz, G.H. Schmid and W. Menke, Z.Natur-
       forsch. 26b, 435 (1971).
(22)   P. Joliot et A. Joliot, Comptes Rendus Acad. Sc.
       Paris, D, 283, 393 (1976).
(23)   F. Koenig, G.H. Schmid, A. Radunz, B. Pineau and
       W. Menke, FEBS Letters 62, 342 (1976).
(24)   R.T. Giaquinta, D.R. Ort and R.A. Dilley,
       Biochemistry 14, 4392 (1975).

BIOLOGICAL CONVERSION OF LIGHT ENERGY INTO ELECTRO-CHEMICAL POTENTIAL

MITSUO NISHIMURA, YASUSI YAMAMOTO, UMEO TAKAHAMA,
MICHIKO SHIMIZU AND KATSUMI MATSUURA

Department of Biology, Faculty of Science,
Kyushu University 33, Fukuoka 812, Japan

The conversion of light energy into electrochemical potential in photosynthetic organelles such as chloroplasts, membrane systems of blue-green algae and photosynthetic bacteria, and purple membrane of Halobacterium, takes place within the framework of "boundary conditions" predicted by the hypothesis of Mitchell (4).

Gibbs free energy change accompanying the translocation of 1 g-ion (mole) of an ionic species from compartment 1 to compartment 2 against a membrane potential, $\Delta\psi$, would be

$$\Delta G = RT \ln\frac{a_2}{a_1} + ZF \Delta\psi ,$$

where a_1 and a_2 are activities of the ionic species in the compartment 1 and 2, respectively.

When we consider translocation of H^+ across the membrane of vesicles,

$$\Delta G = RT \ln\frac{a_{H^+ in}}{a_{H^+ out}} + F \Delta\psi_{in-out} .$$

At 300K (room temperature), the free nergy change becomes, in kcal and mV,

$$\Delta G \text{ (kcal)} = 1.37 \text{ (pH}_{out} - \text{pH}_{in}) + 0.023 \Delta\psi \text{(mV)}.$$

The formation of membrane potential and ionic gradient across the photosynthetic membrane are driven by the light-induced electron transfer. Some of them are directly or chemically coupled to the separation of charges and electron transfer. However, part of them may be indirectly driven by energetical coupling in the membranes.

ATP synthesis (involving reversal of the ATP hydrolysis by the coupling factor protein CF_1) is accompanied by the lowering of the energetical level of the
energy-transducing membranes, the dissipation of the
membrane potential and the hydrogen ion gradient.
Actual <u>chemical</u> mechanism of ATP synthesis is not
known; the mechanism possibly requires multiple binding of adenine nucleotide molecules to the coupling
factor protein, and transphosphorylating reactions may
be involved. In the present discussion we will concentrate on the "black box" of energetics and the relaxation characteristics of the energy transducing
processes.

<u>Sidedness of the proton pump and the electric field</u>
<u>formation in natural and reconstituted membrane</u>
<u>preparations</u>

The sidedness of the proton pump and the electric
field formation in the photosynthetic membranes was
studied in cells, spheroplasts, spheroplast membrane
vesicles (hypotonically treated spheroplasts), chromatophores and reconstituted proteoliposome vesicles of

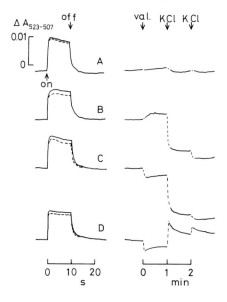

Fig. 1. Absorbance changes
of carotenoid induced by
light (left traces) and K^+
gradients (right traces).
Cells (A), spheroplasts
(B), spheroplast membrane
vesicles (C), and chromatophores (D) were suspended
in 5 mM $MgCl_2$–5 mM Na phosphate, pH 7.4. In addition
0.5 M sucrose was present
in the spheroplast sample
(B).

a purple bacterium <u>Rhodopseudomonas</u> <u>sphaeroides</u> (3).
Comparison of the electrochromic shift of the absorption spectrum of carotenoid, induced by the diffusion
potential after the K^+-valinomycin pulse (2), and the
light-induced absorption spectrum shift showed that;
(i) the same light-induced change of the "red shift"
type was observed in the membrane structures of either
sidedness; and (ii) the absorption changes by the K^+
addition in the presence of valinomycin in the right-
side-out membrane structures (cells, spheroplasts and
spheroplast membrane vesicles) were opposite to that
in the inverted vesicles (chromatophores), "blue shift"
in the former and "red shift" in the latter (Fig. 1).
Asymmetry of the membrane structure and function can
also be shown in the light-induced translocation of H^+.
Cells, spheroplasts and spheroplast membrane vesicles
showed the efflux of H^+ by illumination, and chromatophores showed the influx. It can be concluded that

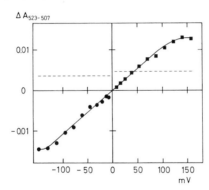

Fig. 2. Absorbance change of carotenoid as a function of calculated membrane
potential change. Spheroplast membrane
vesicles (circles) and chromatophores
(squares) were subjected to potassium
phosphate pulses in the presence of
90 nM valinomycin. Membrane potential
is expressed with a reference on the
cytoplasmic side. Dashed lines represent the levels of absorbance before the
addition of valinomycin.

the cytoplasmic-side-negative potential change induced
by either illumination or ionic gradient causes the
"red shift" type spectral change of carotenoid, and
the cytoplasmic-side-positive potential change causes
the "blue shift" type change (Fig. 2).
The light-induced absorption spectrum change of the

proteoliposomes, reconstituted from phospholipid and
pigment protein complexes of Rhodopseudomonas sphaeroides, was of the "red shift" type indicating that the
vectorial relationship between the local field and the
molecular orientation of carotenoid was kept in the
same manner as in the cells. The inside-positive
membrane potential induced by the K^+-valinomycin pulse
gave a "blue shift" type change showing that the sidedness of the reconstituted proteoliposomes was the same
as in cells or spheroplasts.

Dissipation of the membrane potential by the trans-location of H^+ in chloroplasts

    The parallelism between the decay of membrane poten-
tial and the formation of H^+ gradient (uptake of H^+)
after a short flash (1) has been studied at various
temperatures and in the presence of the field-decay-
or H^+-translocation-accelerating agents (7). In all
of the cases studied, a good agreement was observed
between the half-recovery time of the membrane potential (515-nm absorbance change) and the half-rise time
of the H^+ change (measured with the pH indicator
bromcresol purple).
    The primary photoact and the succeeding electron
transfer in the thylakoid membrane should cause vectorial displacement of charges across the membrane,

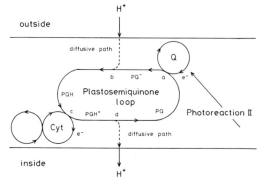

Fig. 3. A model for the electron transfer-
coupled H^+ translocation across the thylakoid membrane involving a plastosemiquinone
loop. PQ, plastoquinone; PQ^-, unprotonated
plastosemiquinone; PQH, protonated plastosemiquinone; PQH^+, protonated plastoquinone.
Q indicates the primary electron acceptor in
photosystem II.

making the inside of the membrane more positive. The rapid translocation (or binding) of H^+ in the direction apparently opposing the electric field suggests that H^+ is not driven electrophoretically but its uptake or release is chemically coupled to a mechanism such as one involving a plastosemiquinone loop or shuttle as shown in Fig. 3 (7). The disappearance of the membrane-potential-sensing absorbance change in parallel to the H^+ uptake suggests that the inside-positive electrostatic field is neutralized by the movement of H^+. The protonation and deprotonation steps are probably rate-limited by diffusion of H^+. This model nicely explains the characteristics of the changes of the physical parameters when chloroplasts are illuminated (7).

This chemically-coupled and diffusion-controlled H^+-translocating plastosemiquinone-plastoquinone shuttle may also be responsible for the temperature-jump-induced efflux of H^+ from chloroplasts (5, 6). The temperature-jump-induced H^+ efflux requires supply of H^+ from inside of the thylakoid vesicle. Pre-illumination increases the temperature-jump-induced H^+ efflux.

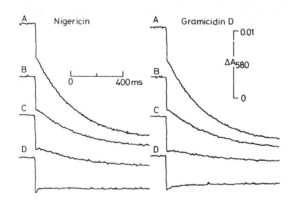

Fig. 4. Concentration dependences of effects of nigericin and gramicidin D on temperature-jump-induced H^+ efflux from chloroplasts. Nigericin: A, none; B, 0.9 nM; C, 2.25 nM; D, 6.8 nm. Gramicidin D: A, none; B, 0.13 uM; C, 0.67 uM; D, 2 uM. $\Delta A = 0.01$ corresponds to 14 nmol of OH^-.

The efflux of H^+ is lost when a H^+-gradient-dissipating agent such as nigericin or gramicidin D is added

(Fig. 4). The major part of the H^+ efflux is coupled to the reversed electron flow accompanying the establishment of a new redox steady state of electron transfer components.

References

(1) Grünhagen, H. H. and H. T. Witt: Umbelliferone as indicator for pH changes in one turn-over. Z. Naturforsch. 25b: 373-386 (1970).

(2) Jackson, J. B. and A. R. Crofts: The high energy state in chromatophores from Rhodopseudomonas spheroides. FEBS Lett. 4: 185-189 (1969).

(3) Matsuura, K. and M. Nishimura: Sidedness of membrane structures in Rhodopseudomonas sphaeroides. Electrochemical titration of the spectrum changes of carotenoid in spheroplasts, spheroplast membrane vesicles and chromatophores. Biochim. Biophys. Acta in press.

(4) Mitchell, P.: Chemiosmotic Coupling in Oxidative and Photosynthetic Phosphorylation. Glynn Research, Bodmin, Cornwall (1966).

(5) Takahama, U., M. Shimizu and M. Nishimura: Temperature-jump-induced release of hydrogen ions from chloroplasts and its relaxation characteristics in the presence of ionophores. Biochim. Biophys. Acta 440: 261-265 (1976).

(6) Takahama, U., M. Shimizu and M. Nishimura: Temperature-jump-induced release of H^+ from chloroplasts. A relationship between the release of H^+ and reverse electron transfer. Plant & Cell Physiol., Takamiya Memorial Issue in press.

(7) Yamamoto, Y. and M. Nishimura: Characteristics of light-induced 515-nm absorbance change in spinach chloroplast at lower temperatures. II. Relationship between 515-nm absorbance change and rapid H^+ uptake in chloroplast after a short flash illumination. Plant & Cell Physiol. in press.

SYNTHESIS OF ORGANIC COMPOUNDS
FROM CARBON DIOXIDE IN LAND PLANTS

James A. Bassham
Lawrence Berkeley Laboratory, University of California

Solar energy conversion by green plant photosynthesis supplies virtually all the energy for living cells on earth. Fossil fuels, on which we depend for a large part of our other energy requirements, are the products of photosynthetic solar energy conversion in past ages. Man has used wood and straw for energy and materials from the beginning, and it has been only about a decade since nuclear power surpassed wood combustion in importance as an energy source in the United States.

Man has modified the earth's surface to a much greater extent with agriculture than for any other purpose. It is economically practical and necessary to cover vast areas of land with selected plants for solar energy conversion when the product is to be used for food, as materials and chemicals, and even as fuel in some cases. There is no doubt about the importance of using plant photosynthesis for solar energy conversion. The only questions are how much plants can contribute to our energy needs that are now being filled by other exhaustible or environmentally unfavorable energy sources, and what is the best technology for increasing our use of plants for this purpose.

Three basic facts have to be faced in any scheme for using plants for industrial energy and chemical needs. Enormous amounts of energy are required, there is a relatively low energy conversion efficiency for the plant photosynthetic process, and energy as a crop has low current market value compared to conventional food and material crops. Solar energy farming shares with some other proposed energy technologies the problem that while it will eventually be feasible and needed, the short range economic prospects appear to be unfavorable. Nevertheless, ten to twenty years of research on plant selection, growth, and utilization for this purpose may be needed in order to be ready by that critical period around the year 2,000 when the supply of petroleum and gas have fallen seriously behind the world requirements for energy. There are indications that other alternative energy sources including coal, shale, and nuclear power will be sufficiently expensive, when the environmental and technological costs are paid, to allow energy farming to become economically competitive. In a shorter time span, photosynthetic solar energy conversion may provide us with portable fuels and chemicals at competitive prices. Some of the systems envisaged might produce basic foods such as protein and sugars at lower energy costs than conventional agriculture.

The total energy needs of a large industrial nation are so great that very large projects have to be considered. However, if the focus is on a specific need and area, such as electrical power generation in the U. S. Southwest, the impact of a system covering even 10,000 Km^2 can be very great.

Green plants can be used to cover large areas at relatively low cost, but the efficiency of solar energy conversion in plants is usually very low in conventional agriculture. The value of food (biological energy supply) is high enough to make such low efficiencies acceptable. For other energy uses--industrial, transportation, residential, electrical--such low effeciencies may be economically unacceptable. It follows that schemes for energy farms (purposeful growing of plants for abiological energy) should be very concerned with conversion efficiency, so that a reasonable yield of the product (energy) per unit area can be realized. Land cost and, to some extent, harvesting cost will be directly proportional to yield per unit area.

Energy farms are not likely to be able to compete with even inefficient food-producing agriculture for good land. Forests may be considered, but the expected efficiencies are low (less than 0.5%). Moreover, as with food production, competing uses as materials (fibre, chipboard, etc.) often have a higher economic value. For steep slopes (where much of the forests grow), ecological damage resulting from removing essentially all organic material during harvesting may rule out such harvesting even where selected timbering for lumber is permitted.

For any crop with low conversion efficiencies, collection costs make conversion to useful fuels or power economically unattractive. When collection of forest or agricultural wastes is feasible, higher economic use of the collected products as specialized materials will tend to rule out their use for energy.

We must therefore examine the factors limiting conversion efficiency in plants, determine what efficiency we might achieve, and define the conditions necessary for achieving it. At the same time, we should conceive of energy farms that could use land not suitable for conventional agriculture, yet could produce enough material per unit area to bring collection costs to reasonable levels. Finally, a scheme which could provide for an economic byproduct (food or chemical) with a value per acre equal to or greater than that of the energy produced might make the first, limited installations more attractive economically.

This discussion will be limited to land plants. There are other schemes involving fresh water plants (for example, water hyacinths) and marine plants (kelp). Probably it is worthwhile to explore all such approaches, but we should recognize that each faces severe economic considerations.

The maximum expected efficiency of solar energy conversion in green plants is directly predictable from our present day knowledge of the detailed mechanism of this process. Plant photosynthesis makes many organic products, but as a reasonable approximation we can consider the formation of the carbohydrates, starch and cellulose, which are composed of glucose subunits. For further simplification let us examine the formation of one sixth of a mole of a glucose subunit from CO_2 and water:

$$CO_2 + H_2O \xrightarrow{\text{Light}} (CH_2O) + O_2$$

1/6 glucose

The free energy stored by this reaction is about 114 Kcal per mole of CO_2 reduced to starch or cellulose. This overall reaction can be considered as the transfer of four electrons from the oxygen atoms of two water molecules to CO_2 resulting in the formation of a water molecule, oxygen, and carbon reduced to the level of carbohydrate.

From knowledge of the detailed mechanism of light absorption and electron transfer, we know that each electron must be transferred through a number of intermediate steps, two of which require light energy. Further examination of this process shows that for each electron transferred through a light reaction, one photon of light is used up. The theoretical quantum requirement is thus four (for four electrons) times two (for two light steps per electron) equals eight. Each mole of CO_2 reduced to sugar requires that eight moles of photons (eight einsteins) must be converted.

Green plants can use only light of wave lengths from 400 nm to 700 nm. This photosynthetically active radiation (P.A.R.) constitutes only about 0.43 of the total solar radiation at the earth's surface at a location such as the U. S. Southwest. All of this light is used as if it were 700 nm light, but since the photosynthetically active radiation (P.A.R.) includes all wavelengths from 400 nm to 700 nm, the energy input is equivalent to that of monochromatic light of about 575 nm wavelength.   An einstein of 575 nm light has an energy of 49.74 Kcal. Multiplying by 8, we get 398 Kcal required per mole of CO_2 reduced to glucose. Since this process stores 114 KCal as chemical potential, the maximum efficiency of photosynthesis is 114/398 = 0.286.

This is the efficiency of conversion of P.A.R. The efficiency based on total absorbed solar radiation is 0.286 multiplied by 0.43 (P.A.R./total radiation) or 0.123, a figure sometimes quoted as the maximum for aquatic plants (usually unicellular algae) where it is often assumed that there is total light absorption. For land plants with a well developed leaf canopy, an absorption of 0.80 of the P.A.R. is considered to be maximal. Using this value for energy absorption, we get an efficiency of 0.123 x 0.80 = 0.0984.

So far we have the maximum daylight efficiency in the green cells of leaves. However, plant cells also use up stored chemical energy when not photosynthesizing, and this introduces the fourth factor. At night, plants carry out respiration, which means they're burning glucose with oxygen. Also, the stems and roots respire during the day as well as night. The amount of such respiration varies greatly, depending on the weather, the temperature, the species of plant, and many other factors. Taking an overall figure which agronomists say is reasonable, we reduce the efficiency by a third, giving us a factor of 0.667.

When we multiply all these factors (0.43 x 80 x 0.286 x 0.67), we come out with about 6.6% overall maximum daily energy efficiency. From this value, we can calculate the absolute upper limit of stored chemical energy to be expected from land plants. The solar energy incident at the earth's surface, averaged over 24 h and 365 days is 3,930 Kcal/m^2 day for the United States (average) or 4,610 Kcal/m^2 day in the U.S. southwest (Table I). In the U.S. southwest during the summer the average is 6,775 Kcal/m^2 day. Of course, we should realize that the amount of energy changes during the day and with the weather, and there may be too much or too little at various times. This is one reason the actual energy storage under the most optimal conditions will be less than the theoretical.

TABLE I

SOLAR ENERGY AT EARTH'S SURFACE IN U.S.

	b.t.u./ft^2 day	cal/cm^2 day	Kcal/m^2 day	watts/m^2
Average (annual basis)	1,450	393	3,930	190
U.S. Southwest (annual basis)	1,700	461	4,610	223
U.S. Southwest (summer)	2,500	678	6,775	329

Using 4,610 Kcal/m^2 day and an efficiency of 0.066, we get a daily energy storage of 304.3 Kcal/m^2, enough to form (304.3/114)27 = 72 g/m^2 day. This is equivalent to 262.8 metric tons/hectare year (Table II). Other values of daily solar energy in Kcal/m^2 day may be converted to expected maximum dry weight stored in metric tons/hectare yr by multiplying by 0.057.

What we have done so far, of course, is to establish the upper (and doubtless unobtainable) limit, based on theoretical constraints. What are the actual rates measured? The figures in parentheses (Table II) are rates during the active growing season, not annual rates. For C-4 plants, these maximum rates range from 138 up to 190 metric tons per hectare per year. The maximum (190) is about half the calculated maximum. Similarly, the maximum reported annual yield, with sugar cane in Texas, is 112 metric tons per hectare--again about 1/2 the calculated maximum (263) for the U. S. Southwest. The energy storage efficiency for these maximum reported yields is about 3.3%. We can thus set a range of 3.3% to perhaps 5% as the best we can hope for with land plants in the future. One reason for going above the highest reported yields of total dry material (3.3%) is that we should be able to make some improvements if we can provide for year round growth and frequent harvesting of organic matter.

The term C-4 refers to certain plants such as sugar cane that evolved in semi-arid tropical or sub-tropical areas, and which have a special added metabolic pathway. Some of the intermediate compounds in this pathway are four-carbon acids, hence the term, "C-4." Those plants use some of their light energy to drive this extra path, but their overall energy efficiency in air and bright sunlight is higher than for other plants. This is because, by investing energy in the C-4 pathway, the C-4 plants avoid a wasteful process called photorespiration that occurs in other plants at high light intensities. Photorespiration results in the reoxidation of freshly formed sugar to carbon dioxide. The C-4 plants are more efficient at high light intensities and temperatures and low CO_2 pressures such as in air (0.03% CO_2) because they avoid photorespiration.

At higher levels of CO_2, photorespiration doesn't occur and some plants that are not C-4 plants become just as efficient. Even some non-C-4 plants, sugar beets, alfalfa and Chlorella in air (Table II) at certain times of the year produce at very respectable rates. On an annual basis, though, the yield drops down. This is in part because many of these plants are not grown year round. A plant such as sugar cane that grows year round can produce a very high annual yield. In general, therefore, the non-C-4 plants produce less than C-4 plants, but keep in mind that this is with air levels of CO_2 and low winter temperatures. Eucalyptus trees are considered by some as a good choice for energy farms because they grow rapidly. Sugar beets grow about as fast as alfalfa.

TABLE II

MAXIMUM PHOTOSYNTHETIC PRODUCTIVITY AND MEASURED MAXIMUM YIELDS

IN SELECTED PLANTS

	gmm^2/day	tons/ acre yr.	metric tons/ hectare yr.
Theoretical max. (Table II)			
U.S. Average annual	61	100	224
U.S. Southwest ave. ann.	72	117	263
U.S. Southwest, summer	106	172	387
Maximum Measured			
C-4 plants			
Sugar cane	38	(62)	(138)
Napier grass	39	(64)	(139)
Sudan grass (Sorgum)	51	(83)	(186)
Corn (Zea mays)	52	(85)	(190)
Non-C-4 plants			
Sugar beet	31	(51)	(113)
Alfalfa	23	(37)	(84)
Chlorella	28	(46)	(102)
Annual Yield C-4 plants			
Sugar cane	31	50	112
Sudan grass (Sorgum)	10	16	36
Corn (Zea mays)	4	6	13
Non C-4 plants			
Alfalfa	8	13	29
Eucalyptus	15	24	54
Sugar beet	9	15	33
Algae	24	39	87

Let us consider next the factor of CO_2 pressure (Table III).
At the level of CO_2 in air, corn and sugar cane grow faster than
the non-C-4 plants such as soybean and sugar beet. But when the

level of CO_2 in a greenhouse is raised by a factor of three or so, one observes higher rates with some of these temperate zone plants than with corn or sugar cane. This suggests that we should somehow enrich the atmosphere with CO_2. When CO_2 is released in the open, the wind blows it away. This leads to the idea of using covered agriculture, using inexpensive desert land, where it is necessary to save water.

In conventional covered agriculture we grow tomatoes or flowers in the winter, and they can be sold at a high price to justify this expensive installation. What I have in mind (Fig. 1) is a much less costly installation; namely, inflatable plastic covers such as are already used for temporary warehouses. Perhaps these can be coated in special ways to help control the flow of heat in and out. The greenhouse may have to have a floor under the soil--a plastic layer of some kind so the water isn't lost to the earth.

In these greenhouses we could grow some crops, such as alfalfa that can be harvested ten or twelve times a year. It's been found that five tons or more of protein per hectare can be grown in the form of leaf protein of alfalfa. This is possibly the highest amount of protein known to be produced per acre by any land plant.

To make the process more economical, we'll remove some of the protein from the leaves and sell this as an economic product. The scientists at the USDA Western Regional Laboratories found that they can remove protein from alfalfa leaves by presses. They can clean up this protein and deodorize it and take bad tastes out. It has very high nutritional value, better than soy protein, better than most cereal proteins, and is, in fact, as good as milk protein, according to nutritional studies with various animals. It doesn't have to be enriched with amino acids. Also, as prepared by the process developed at the U. S. Department of Agriculture, the purified protein is essentially free of the flatulence factors, stachyose and raffinose.

The supply of CO_2 for covered greenhouses could come from a powerhouse. As fuel for this powerhouse, we would burn all of the residue of the cells, after we've taken out about 15% of the dry weight of protein. The other 85% is mostly cellulose, sugars, lipids, and unextracted protein. The carbon taken from the system as protein could be made up as CO_2 from some fossil fuels that would also be burned in the powerhouse. All the CO_2 and the water vapor from combustion would go back into the greenhouse. The heat from combustion would be used to generate electricity which would be sold to the city.

If the CO_2 is enriched to a tenth of a percent or so, studies have shown that for nitrogen-fixing plants, such as alfalfa, the

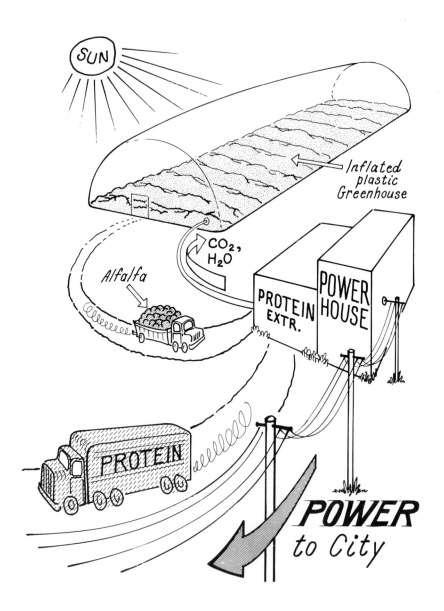

Figure 1. Alfalfa Energy Farm

fixation of nitrogen increases by a factor of five. Presumably this is because the photosynthesis rate has gone up and more of the photosynthate gets down to the roots to feed the bacteria that are living in root nodules and fixing N_2. This means that with CO_2 enrichment, we may not have to put in any fixed nitrogen made by the combustion of fossil fuels. Instead, the root nodules may be able to fix all the nitrogen required for this production of protein by using nitrogen from the atmosphere in the greenhouse.

TABLE III

RATE OF PHOTOSYNTHESIS AT AIR LEVELS AND ELEVATED LEVELS OF CO_2

(milligrams CO_2/dm^2 hour)

Plant	Air	Elevated CO_2
Corn, grain, sorghum,		
Sugar cane	60-75	100
Rice	40-75	135
Sunflower	50-65	130
Soybean, sugar beet	30-40	56
Cotton	40-50	100

At the moment, the market for plant protein (presently soy protein) is rather limited. Given the growing world population, as well as the development of new vegetable protein products in the U. S., and the escalating cost of animal protein, the market for plant protein for human nutrition should expand. Eventually, though, the energy generation, if successful, might grow to a point where there would be no possible market for all the protein produced.

Of course, there are some serious problems. If you put a transparent cover over fields, the system absorbs all of the solar energy and the plants are using only a small part of it for photosynthesis (3-5%). All the rest is converted to heat. Thus, there is a tremendous heating effect. This is a very serious problem, but it may not be insurmountable. Perhaps these covered energy farms could be placed in some of the high deserts, such as exist in Nevada and California at high elevations where the temperature at night drops down very low. The hope would be that by using a very large structure with a large volume of air to warm up, there would be enough heat capacity in that air to be able to absorb the input of heat during the course of one day without the temperature rising above a permissible level. Also, the transpiration of water from the plant leaves during the day would absorb a large amount of heat. Then at night the water vapor which has been transpired during the day would condense and rain back down on the plants.

Special coatings on the plastic to facilitate the flow of heat from inside to outside (since there will always be a temperature gradient) might help. Finally, solar-energy driven heat pumps could be employed, although this would be costly. A preliminary engineering calculation indicates that with certain assumptions made about meteorological conditions in the U.S. southwest and humidity in the greenhouse, a volume of air some 200 meters high would have to be enclosed in order that the temperature would not rise above an acceptable limit for plants during one daylight period. Perhaps an equally serious problem could be the failure to lose enough heat across the plastic barrier during the night, with radiation to the desert sky, to allow the temperature to return by morning to the starting level.

Another problem is the possibility of poisoning of the plants by gaseous contaminants from combustion of both the plant material and the fossil fuel that would be added to produce make-up CO_2 to compensate for carbon removal from the system as protein. Fortunately, research on effects of CO_2 and SO_2 (an expected contaminant) on greater leaf photosynthesis suggests that the deleterious effect of low levels of SO_2 are to some extent mitigated by elevated CO_2. This is due to the fact that SO_2 at low levels causes partial closure of the stomata through which the CO_2 enters the leaves. Higher levels of CO_2 can overcome this effect.

The choice of plants for such a system may require an extensive examination throughout the plant kingdom for plants with appropriate characteristics. The environment in the covered energy farm would be very different from that in which most temperate zone plants grow. The humidity and temperature could undergo large daily variations with very high levels in the late afternoons. Disease resistance would be most important, along with high growth rates at elevated CO_2 pressures, good leaf canopy characteristics, and high leaf protein content. This might suggest plants from certain tropical areas, but the selected plants would have to use high light intensities efficiently. Plant breeding or genetic manipulation of cells growing in tissue culture might be required.

The system should show some promise of becoming economically viable in the next 20 years or so. Conventional covered agriculture has usually been limited to frame and glass greenhouses which produce crops with values running to hundreds of thousands of dollars per hectare per year. An "energy crop," with the best efficiencies we can expect will bring in only a few percent of that amount. Thus we need to give serious consideration to possible valuable byproducts that might help in the economics.

Let us assume that we can achieve somewhat better than the best measured energy conversion efficiencies of about 3% but of course below the maximum of 6.6% I have calculated earlier. Allowing for expected improvements to come from CO_2 enrichment and continuous growth and harvesting, let us assume about 75% of the 0.0666 conversion efficiency calculated above, or a 0.05 conversion efficiency. This would result in the production of 200 metric tons dry matter per hectare year. The energy stored would be 0.05 x 2230 (kilowatts/hectare) x 24 x 365 = 9.77×10^5 kilowatt-hr/hectare year. If we extract 15% of the dry matter as protein, and assume a 30% conversion efficiency of the residue to electricity, each hectare would produce (9.77×10^5) x 0.85 x 0.3 = 2.49×10^5 Kw-hr per year. At the 1975 Southern California Edison price of $0.02 per kw-hr, this energy crop would be worth $5,000 per hectare. Obviously this is a very poor yield compared with even conventional agricultural crops, and much worse compared with present day covered agriculture crops. Let us therefore consider also the possible value of the protein.

The 15% of 200 metric tons would be 30,000 Kg. Green leaf protein, because of its higher nutritional value, should in time have a higher value than soy protein, so let us assume a value of $1.00 a Kg. We could then obtain $30,000 a hectare for the protein, which added to the $5,000 for the energy gives a $35,000 crop. Of course, the market for protein might not match the amount produced when we get into really large scale energy production, but perhaps by then the economics for pure energy production will have improved.

Let us next consider the impact of 10,000 Km^2 of such agriculture (an area 100 x 100 Km or 60 x 60 miles). At 2.49×10^5 Kw-hr/hectare year we would have 2.49×10^{11} Kw-hr/year. The total electrical sales in the state of California for the year 1975 was 1.39×10^{11} Kw-hr, and is projected to reach 2.67×10^{11} Kw-hr/year by the year 1995, according to a recent study by the California Energy Conservation and Development Commission.

What would it cost? Perhaps too much, if we accept figures like $20 per m^2 of plastic cover, plus all the costs of fabrication, inflating mechanism, cooling, farming, power plants, etc. With a huge market for plastic and other materials the cost should

come down. In time we might have to learn how to use some of the
leaf substance as a starting material for the plastic synthesis.

At this point, we should take the view that this scheme is as
worthy of further study as most of the other long range energy
farm proposals. We are entering a new era in which the economic
factors of the last century (very inexpensive energy and food) may
be poor signposts to the future.

At the present time, residual material after removal of
protein from alfalfa leaves has a higher market value as cattle
fodder than as fuel for a power plant. As the production of grain
and cotton for world markets continues to displace forage crops,
there will be a growing demand for some time for this cellulosic
residue from leaves. The current market price in California is
about $80 per metric ton, so that the 170 tons of dry matter left
after protein extraction is worth $13,600. This suggests that in
the early stages of development, it would be more economic to sell
all the product of covered agriculture as agricultural commodities. In this case, of course, the CO_2 for enrichment would have
to come entirely from fossil fuel combustion. Even on this basis,
the system is worth developing for the purpose of water, land and
energy conservation. In time, as fossil fuel becomes more
expensive, the system could change to one generating electricity
from biomass.

While covering of large areas of desert with plastic greenhouses may not be practical at this time, there is an intermediate
type of project that can act as a forerunner by providing information about the feasibility of solar energy conversion via covered
agriculture with CO_2 enrichment and water conservation. This is
to use waste gases and heat from fossil fuel combustion at plants
in the U.S. Southwest. Such programs already are being proposed
and would appear to be very deserving of support. The initial
focus would be on high value crops such as tomatoes in the winter
as well as algae which could be fed to other organisms, harvested
for their food value directly. Such projects might very well
prove to be economic in the near future. At the same time,
through such projects, information about plastic or other inexpensive structures, about heating and cooling, diseases in plants
under such artificial environments, productivity, etc. will be
obtained.

To summarize, covered agriculture for solar energy conversion leading to electrical energy production may be some years in
the future, but there are practical, forerunner projects that can
be started now. It will be imperative, given the economic
considerations, to maximize photosynthetic efficiency. What may
be ultimately achieved is not certain, but careful consideration
of plant physiology and biochemistry suggests that we can reason-
ably expect coversion efficiencies of 5%, given optimal

conditions$_2$ including CO_2 enrichment. Rather large projects (10,000 km^2) could have significant impacts on the electrical energy requirements in the U.S. Southwest.

One way to improve the economics of energy farms may be, as just suggested, to produce a valuable byproduct. Another approach is to use plants to produce an energy product of greater value than power from a stationary source. One such proposal, discussed by Calvin, is to raise plants as suppliers of hydrocarbons. An indirect method of doing this is to produce sugars and cellulose, convert the cellulose enzymatically to sugars, and ferment all of the sugars to alcohol. Since sugar cane and sugar beets are among the fastest growing plants, even the fermentation of sugar alone may be attractive in some situations. In Brazil, for example, sugar-containing residues left after crystallization of sugar as a commercial product, are being converted to alcohol which is in turn being used as an extender or additive to gasoline. Something like 5% of the total volume of gasoline engine fuel in Brazil is or soon will be ethanol. In general, though, sugar has a higher value as food than as a precursor for alcohol for fuel.

A more direct approach is to raise plants that produce hydrocarbons convertible to portable fuels. As Calvin points out, productivity of Hevea trees in Indonesia has been increased from 200 Kg per hectare in 1945 to 2000 Kg per hectare in 1965, and there are expectations, based on experimental plots, etc., that the yield could go as high as 8 metric tons per hectare. While latex would not be useful as fuel, other hydrocarbon producing plants might be found among the the thousands of species, that would yield large quantities of lower molecular weight polyisoprene compounds. As Calvin suggests "....there are genuine possibilities for harvesting economic amounts of crude-oil like hydrocarbons from land in dry, sunny regions such as Southern California or Southwest Texas which today cannot be easily used for food or fiber production." A search for such compounds and plants has been undertaken in Calvin's laboratory, and already species of Euphorbia have been found which produce sizable quantities of non-latex (lower molecular weight) "hydrocarbon" compounds. Some of this material has proved to be sterols, which, though derived from isoprene units, are perhaps not useful as potential fuels. Probably we are going to need to learn a lot more about the regulation of biosynthetic pathways leading to various end-products made from isopentenyl pyrophosphate, the biochemical form of isoprene.

While the yield per hectare of hydrocarbons from plants is not likely to be as great the yield of cellulose and other materials, we can again envisage a dual productivity, in which the hydrocarbons are extracted and sold as a valuable product, while the residue is burned in a power plant.

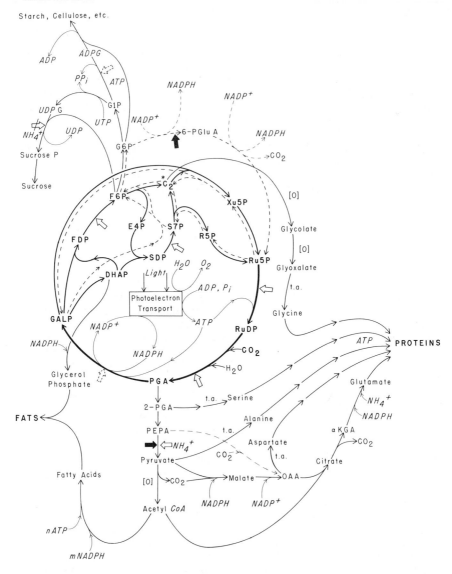

Fig. 2. Regulation of Photosynthetic Metabolism

XBL 706-5260

A knowledge of the regulation of the pathways of biosynthesis in plants from CO_2 to end products will be useful in selection of optimal physiological conditions, and selection and breeding of plants for energy farms. Clearly, if in some processes the byproduct is more valuable than the energy product, it will be helpful to optimize for the byproduct (protein, hydrocarbon, or other material of commercial value.)   At the same time, the preceding discussion illustrates the importance of maximizing overall efficiency. It is therefore necessary to understand the metabolic consequences of physiological variables such as CO_2 pressure.

Many of the early sites of metabolic regulation of CO_2 uptake by photosynthesis and conversion of photosynthate to secondary products have been identified, and some are shown in Fig. 2. There are important points of regulation at the carboxylation of ribulose 1,5-diphosphate and its formation, and at the conversion of hexose and heptose disphosphates to monophosphates. We have postulated that a balance between the rates of these conversions helps to regulate the pool sizes of intermediate compounds of the photosynthetic reductive pentose phosphate cycle. The enzymes at these sites also function to turn off the reductive cycle in the dark, when respiratory metabolism sets in. It is also necessary to control the flow of carbon away from the photosynthetic cycle to secondary products. One example of this is the regulation of the reaction converting phosphoenolpyruvate to pyruvate. As we examine the further conversion of intermediates to fats, amino acids, polyisoprenes, carbohydrates, etc., we will have to learn about many more sites of regulation. In short, a look into the future suggests that as we attempt more and more to use plants as factories to produce specific products in high yield, we will need more than ever to learn how the production is regulated in the plants and how it might be turned more to our advantage.

The preparation of this paper was sponsored by the U. S. Energy Research & Development Administration.

WAVELENGTH-EFFECTS OF INCIDENT LIGHT ON CARBON METABOLISM IN
CHLORELLA CELLS[1]

SHIGETOH MIYACHI,[2,3] AKIO KAMIYA[2] AND SHIZUKO MIYACHI[3]

Institute of Applied Microbiology[2] and Radioisotope Centre[3], University of Tokyo, Bunkyo-ku, Tokyo, Japan

Summary

1. When a colorless mutant of *Chlorella vulgaris* (Mutant #125) was starved in phosphate medium in darkness rates of respiration and dark CO_2 fixation were suppressed. Illumination with blue light to the starved mutant cells immediately enhanced CO_2 fixation. Respiration was also enhanced after a lag period which lasted for several minutes. The same enhancements in CO_2 fixation and respiration were induced by adding ammonium chloride to the starved mutant cells in the dark. The main initial $^{14}CO_2$ fixation product under blue light as well as in the presence of ammonium ion in the dark was aspartate.

2. Prolonged illumination with blue light induced *de novo* synthesis of phosphoenolpyruvate (PEP) carboxylase (EC 4.1.1.31) in colorless mutant of *C. vulgaris*. The same treatment brought about decreases in the activities of isocitrate lyase (EC 4.1.3.1) and malate synthase (EC 4.1.3.2), key enzymes of the glyoxylate cycle. Fractionation of crude homogenate of the colorless mutant *Chlorella* on sucrose density gradient revealed that these three enzymes are localized in the cytoplasm.

3. Analysis of photosynthetic $^{14}CO_2$ fixation products under illumination of monochromatic blue and red light to wild type cells of *Chlorella vulgaris* (#11h) showed that the rates of ^{14}C incorporation into sucrose or glucose-polymer fraction were higher under red light than under blue light, while blue light specifically enhanced ^{14}C incorporation into alanine, lipid-fraction, aspartate, malate and protein-fraction. Superimposition of blue

---

Abbreviations: pcv, packed cell volume; PEP, phosphoenolpyruvate; PGA, 3-phosphoglycerate; OAA, oxalacetic acid; TCA cycle, tricarboxylic acid cycle.

[1]The request for reprints should be addressed to S. Miyachi, Institute of Applied Microbiology, University of Tokyo, Bunkyo-Ku, Tokyo 113, Japan. This research was supported by a grant from the Japanese Ministry of Education given to S. Miyachi.

light to red light at saturating intensity showed that photosynthetic carbon metabolism is modulated by blue light at intensity
as low as 450 erg/cm^2·sec.

4. Based on the above mentioned results discussion was made as
to the regulatory effects of blue light on photosynthetic and non-
photosynthetic carbon metabolism in *Chlorella*.

Introduction

The rapid increase in world population requires the enhancement
of production of food crops. Increases in crop yields have been
achieved by increasing fertilizer and pesticide applications as
well as the use of high yielding cultivars. To further increase
the plant productivity we must develope the methods other than
mentioned above since not only these have been exploited extensively but also the fertilizer, pesticide and irrigation all require an extra energy input and we are faced with a global energy
shortage. To increase agricultural efficiency with minimum energy
input it is necessary to elucidate those factors which modulate
the metabolic conversion of photosynthetic intermediates to the
desired end products.

Environmental factors such as light quality are known to alter
the chemical composition of plant cells (10,26). This indicates
that photosynthetic carbon metabolism to various end products is
modulated by quality of illuminating light. With green unicellular algae *Chlorella ellipsoidea* and *Chlorella vulgaris* 11h as well
as colorless mutant of *C. vulgaris* (Mutant #125) we have been
studying the regulation of carbon metabolism by wavelengths. The
results with *C. ellipsoidea* were reported elsewhere (12,15-18).
The results thus far obtained with green and colorless *C. vulgaris*
·will be reported in this paper.

Results and Discussion

Wavelength effects on carbon metabolism in colorless Chlorella
mutant cells

*(1) Effects of ammonium chloride on CO$_2$ fixation and respiration
in the dark*

The material used was the colorless *Chlorella vulgaris* mutant
(Mutant #125) which contains neither chlorophyll nor carotenoid.
When the colorless mutant cells of *C. vulgaris* which had been
grown in the glucose medium were transferred into M/150 phosphate
buffer (pH 6.5) and shaken in the dark, the rates of dark respiration and CO$_2$ fixation decreased (6). When the algal cells which
had been starved in phosphate medium in the dark were illuminated
with blue light, both respiration and CO$_2$ fixation were significantly

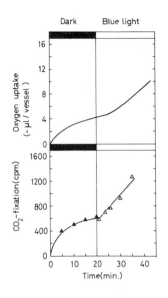

Fig. 1.   Effects of blue light on respiration and $^{14}CO_2$ fixation (C. vulgaris #125, 6).   Blue light, 456 nm, 1,400 erg/cm^2·sec. Cell density, 5 ml pcv/liter in M/150 phosphate buffer (pH 6.5). Temperature, 30° C.   Final concentration of NaH$^{14}CO_3$, 1.7 mM.

enhanced (Fig. 1). The figure shows that there was some **difference** in the kinetics of these two effects of blue light: Usually it took several minutes before respiration reached the accelerated steady rate under blue light, while the rate of $^{14}CO_2$ fixation increased immediately after the blue light was switched on.

Except the above-mentioned kinetics these two blue light effects were similar to each other. At 456 nm, 100 erg/cm^2·sec sufficed for half-saturation and 400-800 erg/cm^2·sec for complete saturation of the blue light effects of $^{14}CO_2$ fixation and respiration. Action spectra for these two enhancing effects were similar to each other; both showed peaks at 460 nm and 380 nm, which correspond to the absorption maxima of flavin. The light at wavelengths longer than 550 nm was without effect (6). Pronounced blue light effects were observed in the cells which had been starved in darkness. Respiration and CO_2 fixation in young growing cells of the mutant was not stimulated by blue light. These results indicate that the same mechanism underlies the effects of blue light on CO_2 fixation and respiration and the pigment responsible for both blue light-stimulations is a flavin.

The main $^{14}CO_2$ fixation product under blue light was aspartate, indicating that C_1-C_3 carboxylation reaction is increased by blue light which causes an enhancement in the formation of oxalacetic acid (OAA), which is converted to aspartate (6). We assumed that an enhancement in supply of OAA and other compounds will accelerate the tricarboxylic acid (TCA) cycle and consequently the oxygen uptake. The lag period observed in the blue light respiration (Fig. 1) presumably corresponds to the time required for the blue light-induced increase in the levels of intermediates of the TCA cycle.

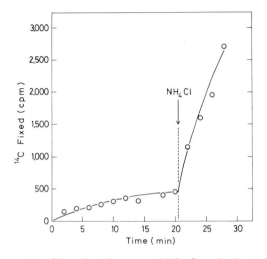

Fig. 2. Effect of ammonium chloride (5 x 10⁻³ M) on dark $^{14}CO_2$ fixation (C. vulgaris #125). The cells were grown in the glucose medium for 16 days and then shaken in phosphate buffer for 3 days. Cell density, temperature and concentration of $NaH^{14}CO_3$ were the same as shown in Fig. 1.

Schmid and Schwarze (20) found that the rate of oxygen consumption in the presence of FMN and glycine was significantly enhanced on illumination of the extract of a colorless mutant of *C. vulgaris*, indicating that the enzymatic glycine oxidation was stimulated by blue light. They therefore assumed that the blue light respiration is a result of an increase in the level of FMN-dependent amino acid oxidase induced by blue light. Their results are in accord with the action spectrum on the light enhanced respiration (6). However, they reported that there was no lag period in the blue light-induced stimulation of glycine oxidation. In contrast, there was a lag period in the blue light-induced respiration (Fig. 1). We therefore assumed that the blue light respiration is not directly correlated to the oxygen consumption due to the amino acid oxidation.

Fig. 2 shows that the rate of dark $^{14}CO_2$ fixation in the colorless *Chlorella* cells was greatly enhanced immediately after the addition of ammonium chloride. During the initial period of dark $^{14}CO_2$ fixation in the presence of ammonium chloride the highest radioactivity was found in aspartate, followed by malate, citrate, fumarate and glutamate in decreasing order (Table 1). The percent incorporations of radioactivities into aspartate kept decreasing during the experimental period. On the other hand, the percent incorporations of radioactivities in glutamate, glutamine and insoluble compounds which were very small or negligible during the initial period kept increasing throughout the experimental period. These results indicate that C_1-C_3 carboxylation reaction in the starved mutant *Chlorella* cells is greatly enhanced by the addition of ammonium ion and the most of the radioactivities is transferred to the intermediates of the TCA cycle and its closely related compounds. Fig. 3 shows that the dark respiration was stimulated several minutes after the addition of ammonium chloride. Thus the effects of ammonium ion on CO_2 fixation and respiration are very similar to those of blue light. These results indicate that the

Table 1. ^{14}C-incorporation into products during dark $^{14}CO_2$ fixation in the presence of ammonium chloride ($5 \times 10^{-3} M$, C. vulgaris #125)

Compound	Radioactivity (cpm/30 μl pcv) after $^{14}CO_2$ fixation for			
	20 sec	40 sec	2 min	5 min
Aspartate	165 (45.2)	335 (44.4)	900 (30.8)	1,350 (26.2)
Glutamate	5 (1.4)	55 (7.3)	290 (10.0)	575 (11.2)
Glutamine	0 (0.0)	5 (0.7)	220 (7.5)	1,230 (23.9)
Malate	100 (27.4)	210 (27.8)	730 (25.0)	955 (18.5)
Citrate	40 (11.0)	45 (6.0)	245 (8.4)	225 (4.4)
Fumarate	15 (4.1)	35 (4.6)	180 (6.2)	220 (4.3)
Insolubles	0 (0.0)	15 (2.0)	45 (1.5)	180 (3.5)
Others	40 (11.0)	55 (7.3)	313 (10.8)	415 (8.1)
Total	365 (100)	755 (100)	2,925 (100)	5,150 (100)

The figures in parenthesis show the percent incorporation of ^{14}C. The growth conditions of the alga used in this experiment were the same as described in Fig. 2.

role of FMN-dependent amino acid oxidase, if it is involved in the observed blue light effects, consists of a release of ammonia which, in turn, enhances carbon dioxide fixation into OAA. The enhancement of oxygen uptake due to the enhanced amino acid oxidation is assumed to be negligible immediately after the blue light was switched on. The rate of oxygen uptake will become significantly

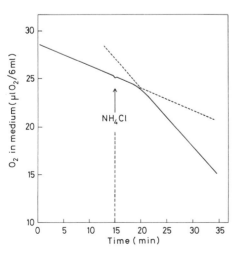

Fig. 3. Effect of ammonium chloride (5×10^{-3} M) on dark respiration (C. vulgaris #125). Cell density, 10 ml pcv/liter in M/150 phosphate buffer (pH 6.5). Other conditions were the same as described in Fig. 3.

large when the levels of intermediate compounds of TCA cycle increased after several minutes.

The experimental results with *Chlorella pyrenoidosa* by Hiller (3)
and Kanazawa *et al.* (8) indicated that pyruvate kinase activity
was stimulated by the addition of ammonium ion. Assuming that this
stimulation is occurring in the colorless mutant *Chlorella* cells,
one might expect that the intracellular level of pyruvate is increased by ammonium ion and this, in turn, brings about an enhancement of the formation of OAA (*via* pyruvate carboxylase (EC 6.4.1.1))
or malate (*via* NAD(P)-malic enzyme (EC 1.1.1.38(40))). However,
the former possibility is not in accord with our finding that no
pyruvate carboxylase activity was detected in the colorless mutant
cells (7). The latter possibility is also not in accord with our
experimental results that the main initial $^{14}CO_2$ fixation product
in the presence of ammonia was not malate but aspartate (Table 1).
The results indicate that the formation of OAA rather than malate
is stimulated by ammonia. Moreover, pyruvate kinase activity in
the eluates obtained from non-starved as well as starved mutant
cells was not affected by the addition of 5 mM ammonium chloride
(data not shown).

*(2) Effects of continuous illumination with blue light on PEP
carboxylase, isocitrate lyase and malate synthase in colorless*
Chlorella *mutant cells*

Colorless *Chlorella* mutant cells suspended in phosphate medium
were placed in a transparent round vessel. The control vessel was
kept in the dark while others were illuminated with red or blue
light all being continuously bubbled with sterilized air. At
intervals, portions were taken from each vessel and assayed for
their enzyme activities (7). We found that PEP carboxylase activity
was markedly increased by blue light while red light did not enhance its activity (See also Fig. 5). The activity attained its
maximum level after 2 days of illumination. On the other hand,
under both red and blue lights, aspartate amino transferase (EC
2.6.1.1) and NADP-malic enzyme activities were slightly lower than
the values obtained for cells which were kept in the dark. We
further found that the blue light effect on PEP carboxylase activity was completely suppressed by cycloheximide (5 μg/ml), indicating that blue light induces *de novo* synthesis of this enzyme
in the cytoplasm (7).

We have found that the rate of CO_2 fixation into aspartate in *C.
vulgaris* #125 cells increases immediately after blue light is
switched on (Fig. 1). This indicates that C_1-C_3 carboxylation reaction is enhanced immediately on illumination with blue light.
Since pyruvate carboxylase activity was not detected in the colorless mutant cells we concluded that PEP carboxylase activity is
enhanced by blue light. (Although data are not shown here, no
PEP carboxytransphosphorylase (EC 4.1.1.38) activity was detected
in the extract of *C. vulgaris* #125 cells, either.) We therefore

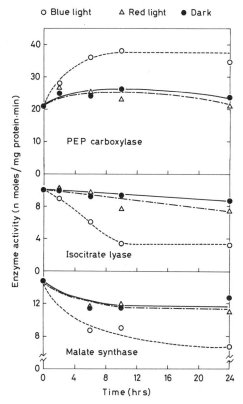

Fig. 4.   Effects of blue and red lights on the activities of PEP carboxylase, isocitrate lyase and malate synthase (C. vulgaris # 125). Mutant cells which had been shaken in the glucose medium in the dark for 113 hr were transferred to M/100 phosphate buffer (pH 7.0) and shaken again. After 20 hr of shaking in the dark, the cell suspension was divided into three transparent glass vessels and continuously bubbled with air under different illuminating conditions. Temperature, 28° C. Cell density, 14 ml pcv/liter. Open circles; blue light (462 nm, 800 erg/cm²·sec). Open triangles; red light (>550 nm (mainly 600-650 nm), 8,400 erg/cm²·sec). Closed circles; control in the dark. At intervals, portions were taken out and each suspension was disrupted in a French-pressure cell (7). Isocitrate lyase and malate synthase activities were determined by the same methods described elsewhere (14), except non addition of Triton X-100. PEP carboxylase activity was assayed according to Kamiya and Miyachi (7).

assume that blue light exerts a dual effect on PEP carboxylase in Chlorella cells; an immediate enhancement of the enzyme reaction and induction of the synthesis of the enzyme.

Fig. 4 shows that the decreases in isocitrate lyase and malate

Table 2. *Effects of dark starvation in activities of various enzymes in* Chlorella vulgaris *#125*

		Enzyme activity	
Expt.	Enzymes	non starved cells	cells starved for 2 days[A]
		(n moles/mg protein·min)	
1	PEP carboxylase	66.2	7.1
	Isocitrate lyase	32.4	24.3
	Malate synthase	76.6	74.2
2	PEP carboxylase	128.3	26.7
	PEP carboxykinase	27.3	0.0
		(μ moles/mg protein·min)	
	NAD-malic dehydrogenase	15.6	8.0

[A]*Cells were shaken in phosphate medium in the dark.*

synthase activities were significantly enhanced by blue light, while red light did not exert any effect on all the enzymes activities tested. After 24 hr, the both levels decreased less than half the respective values observed in the starting material. Table 2 shows that when the mutant cells were shaken in phosphate medium in the dark for 2 days PEP carboxylase activity decreased to about one tenth of its original level, while the activities of isocitrate lyase and malate synthase decreased only slightly. These data show that PEP carboxylase activity decreased during dark starvation and was increased by blue light, while the activities of malate synthase and isocitrate lyase did not show appreciable change during the dark starvation and was decreased upon blue light illumination. Similar blue light effects on the activities of PEP carboxylase and isocitrate lyase have been reported in *Chlorogonium elongatum* cells grown with acetate in the dark (23).

Ashworth and Kornberg (1) reported that PEP carboxylase defective mutant of *Escherichia coli* was unable to grow in the medium containing glucose as a sole carbon source but could grow at high rate on malate, aspartate and glutamate. Hsie and Rickenberg (4) reported that PEP carboxykinase (EC 4.1.1.32) defective mutant of *E. coli* failed to grow with any of the intermediates of the TCA cycle. These results indicate that the role played by PEP carboxylase is in the anaplerotic CO_2 fixation to supply OAA into TCA cycle (9), while that played by PEP carboxykinase is in the decarboxylation of OAA to PEP, which is required for the gluconeogenesis. Table 2 and 3 show that PEP carboxykinase activity decreased during dark starvation but was not affected by blue light illumination.

Table 3. *Effects of red and blue light on PEP carboxykinase activity in* Chlorella vulgaris #125

Activity (n moles/mg protein·min) after the cells were bubbled
      with air in phosphate medium for 24 hr
                        in

darkness	blue light	red light
12.7	13.1	11.0

Regulatory effects of blue light on carbon metabolism in the
colorless mutant cells of *Chlorella vulgaris* which was assumed
from the above results is shown in Fig. 5. When the algal cells
are starved in the dark in phosphate medium, PEP carboxylase activity, hence the activity of CO_2 fixation which supplies OAA from
PEP starts to decrease. (Correspondingly the respiratory activity
also starts to decrease.) Since isocitrate lyase and malate
synthase activities do not decrese there would be no change in the
rate of glyoxylate by-pass. On the other hand, PEP carboxykinase
activity decreases significantly during the starvation period.
Therefore in the starved cells, the glyoxylate by-pass is playing
a role in the anaplerotic supply of malate and succinate to TCA
cycle rather than in the gluconeogenesis (2). Thus the decrease
in the intermediates in the TCA cycle due to the decrease in PEP
carboxylase activity would be compensated at least to some extent
by the action of the glyoxylate by-pass. When the starved cells
are illuminated by blue light, anaplerotic supply of OAA from
endogenous glucose-polymers *via* PEP carboxylase starts to increase.

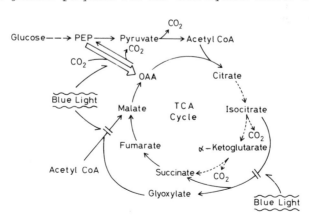

Fig. 5. Regulatory effects of blue light on carbon metabolism in
Chlorella vulgaris #125. *Open arrow and the arrows crossed by
parallel lines indicate the blue light-enhanced and the blue light-
suppressed reaction, respectively.*

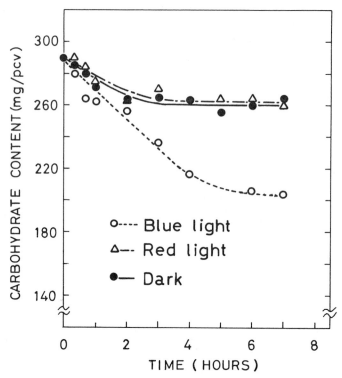

Fig. 6. Effects of blue and red lights on carbohydrate content
(C. vulgaris #125). Mutant cells which had been shaken in the
glucose medium in the dark for 4 days were transferred to M/100
phosphate buffer (pH 7.0) and shaken again. After 20 hr of shaking
in the dark, the suspension was divided into three transparent
glass vessels and continuously bubbled with air under different
illuminating conditions (See Fig. 4 for details and symbols).
Portions were taken out at intervals and the total amount of carbo-
hydrate was determined by the anthron method (22).

At the same time, due to the decrease in isocitrate lyase and
malate synthase activities, the supply of malate and succinate
via glyoxylate by-pass is suppressed.

Kowallik and Ruyters (11) found that pyruvate kinase in a Chloro-
phyll-free, carotenoid containing mutant of *Chlorella vulgaris*
(Mutant #211) was enhanced either by blue light illumination or
by the addition of glucose in the dark. These results led them to
the conclusion that "blue light may induce the synthesis of pyruvate
kinase by supplying free sugars at the site of enzyme synthesis".
Similar effects of glucose on the synthesis of pyruvate kinase in
Euglena gracilis was reported by Ohmann (19). Teraoka *et al.* (24)
showed that PEP carboxylase in *E. coli* was nutritionally induced

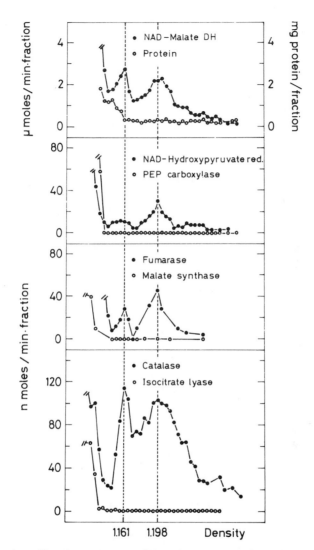

Fig. 7. Distribution of protein and enzymes from the crude homogenate of Chlorella vulgaris #125 *in a sucrose density gradient.*

by glucose. Both PEP carboxylase and pyruvate kinase are known to catalyze catabolic conversion of PEP. It is therefore possible that blue light induced-formation of PEP carboxylase (7, see also Fig. 3) is also due to the increased supply of free sugars induced by blue light. In this connection we found that decrease of the total carbohydrate in colorless *Chlorella* mutant cells during the starvation in phosphate medium was specifically enhanced by blue light illumination (Fig. 6). However, our preliminary experiment

showed that, irrespective of the illumination conditions, there
was practically no change in the acid soluble carbohydrate content
(ca. 1.1-1.4 glucose eq./pcv). At the moment, therefore, it is
not possible to conclude whether the products of blue light-enhanced
carbohydrate breakdown induces the synthesis of PEP carboxylase
or blue light first induces PEP carboxylase synthesis and the
resulting enhancement in the conversion of PEP to OAA eventually
brings about the decrease in glucose-polymer.

Recently we fractionated crude homogenate of the colorless *Chlo-rella* mutant on sucrose density gradients (14). Fig. 7 shows that
catalase (EC 1.11.1.6), which is known as the marker of microbodies
(25), showed two peaks at densities 1.161 and 1.198 g/cm^3. The
density values indicate that the peaks at 1.161 and 1.198 g/cm^3
correspond to the mitochondrial and the microbody fraction, re-spectively. It has been shown that catalase, isocitrate lyase
and NAD-hydroxypyruvate reductase (EC 1.1.1.29) were readily solu-bilized when the microbodies isolated from castor bean endosperm
were ground vigorously or osmotically shocked, while malate synthase
was rather tightly bound to microbody membranes (5). Recoveries
of catalase and NAD-hydroxypyruvate reductase in particulate
fractions were about 60 and 20% of the total. This indicates that
about 20% of the microbodies was recovered in intact form. Since
in this gradient, no activity of isocitrate lyase, malate synthase
and PEP carboxylase was detected in any particulate fraction we
concluded that all these emzymes, of which functions are regulated
by blue light, are localized in the cytoplasm.

*Wavelength effects on photosynthetic carbon metabolism in green
alga* Chlorella vulgaris *11h*

Distribution of radioactivities among photosynthetic $^{14}CO_2$ fixation

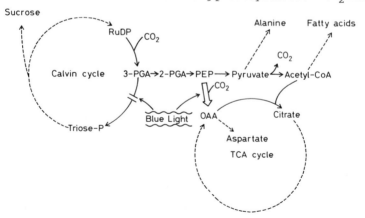

*Fig. 8. Regulatory effects of blue light on photosynthetic carbon
metabolism in* Chlorella vulgaris *11h. Open arrow and the arrow
crossed by parallel lines indicate the blue light-enhanced and the
blue light-suppressed reaction, respectively.*

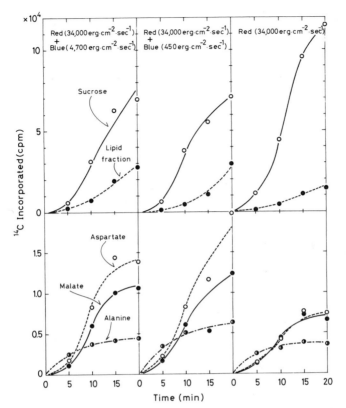

Fig. 9. Effects of blue light added to saturating red light on photosynthetic $^{14}CO_2$ fixation (C. vulgaris 11h). Cells were suspended in M/400 phosphate buffer (pH 7.6) at density of 10 ml pcv/liter. Wavelength: red light, >500 nm (mainly 600-650 nm); blue light, 456 nm. Temperature, 25° C. NaH$^{14}CO_3$ concentration, 1 mM.

products under monochromatic red and blue light was studied with wild type cells of *Chlorella vulgaris* (# 11h). Light intensities of red (5,600 erg/cm^2·sec at 660 nm) and blue light (10,500 erg/cm^2·sec at 456 nm) were adjusted so that the rates of ^{14}C fixation were equal under both wavelengths. $^{14}CO_2$ fixation was started by adding NaH$^{14}CO_3$. At intervals, portions were taken out to the test tubes containing methanol (final concentration, 80%). The methanol suspension was paperchromatographed two dimentionally and the radioactivity of each radioactive spot was determined. We found that the rates of ^{14}C incorporation into sucrose and glucose-polymer were greater under red light than under blue light. In contrast, red light specifically increased ^{14}C incorporation into alanine, lipid-fraction, aspartate, malate and protein-fraction (13). The above observation is most simply explained if we assume that quality of light affects the fate of photosynthetic CO_2 fixation product, 3-phosphoglyceric acid (PGA): Under red light, much more PGA is reduced to triose phosphate than under blue light, while blue light enhanced the conversion of PGA to acetyl CoA through the glycolytic pathway (Fig. 8). Enhancement in the production of triose phosphate under red light will enhance the formation of sucrose and other carbohydrates, whereas fatty acids (lipid fraction), alanine, and aspartate and malate will be converted from acetyl CoA, pyruvate and PEP, respectively.

The right hand side of Fig. 9 shows the time courses of photosynthetic $^{14}CO_2$ fixation into various products which was carried out under red light of saturating intensitiy (34,000 erg/cm^2•sec). When blue light of very low intensity (450 erg/cm^2•sec) was superposed to the saturating red light, the rates of ^{14}C incorporation into sucrose decreased, while the same rates into lipid-fraction, aspartate, malate and alanine were enhanced (Middle of Fig. 9). Practically the same pattern of ^{14}C incorporation was observed when blue light with 10 times higher intensity (4,700 erg/cm^2•sec) was superposed to the saturating red light (Left side of Fig. 9). The results show that photosynthetic carbon metabolism is also regulated by blue light of very low intensity. One of the possible explanations for the above results is that the reduction of PGA to triose phosphate is inhibited by blue light of low intensity so that much more PGA is converted to acetyl CoA *via* PEP and pyruvate than under red light (Fig. 8). The observation that $NADP^+$-glyceraldehyde-3-phosphate dehydrogenase activity in *Lemna gibba* was inhibited by superposition of monochromatic blue light (476 nm) to red light (669 nm, 21) supports this inference.

Literature cited

1. Ashworth,J.M. and H.L.Kornberg. 1966. The anaplerotic fixation of carbon dioxide by *Escherichia coli*. Proc.Roy.Soc. 165: 179-188.
2. Beevers,H. 1969. Glyoxysomes of castor bean endosperm and their relation to gluconeogenesis. Ann.N.Y.Acad.Sci. 168: 313-324.
3. Hiller,R.G. 1970. Transients in the photosynthetic carbon reduction cycle produced by iodoacetic acid and ammonium chloride. J.Expt.Bot. 21: 628-638.
4. Hsie,A.W. and H.V.Rickenberg. 1966. A mutant of *Escherichia coli* deficient in phosphoenolpyruvate carboxykinase activity. Biochem.Biophys.Res.Comm. 25: 676-683.
5. Huang,A.H.C. and H.Beevers. 1973. Localization of enzymes within microbodies. J.Cell Biol. 58: 379-389.
6. Kamiya,A. and S.Miyachi. 1974. Effects of blue light on respiration and carbon dioxide fixation in colorless *Chlorella* mutant cells. Plant & Cell Physiol. 15: 927-937.
7. Kamiya,A. and S.Miyachi. 1975. Blue light-induced formation

of phosphoenolpyruvate carboxylase in colorless *Chlorella* mutant cells. Plant & Cell Physiol. 16: 729-736.

8. Kanazawa,T., K.Kanazawa, M.R.Kirk and J.A.Bassham. 1972. Regulatory effects of ammonia on carbon metabolism in *Chlorella pyrenoidosa* during photosynthesis and respiration. Biochim. Biophys.Acta 256: 656-669.

9. Kornberg,H.L. 1965. Anaplerotic sequences in microbial metabolism (1). Angew.Chem.internat.Edit. 4: 558-565.

10. Kowallik,W. 1970. Light effects on carbohydrate and protein metabolism in algae. *In:* P.Halldal, ed., Photobiology of Microorganisms. Wiley-Interscience, London. pp. 165-185.

11. Kowallik,W. und G.Ruyters. 1976. Über Aktivitätssteigerungen der Pyruvatkinase durch Blaulicht oder Glucose bei einer chlorophyllfreien *Chlorella*-Mutante. Planta 128: 11-14.

12. Miyachi,S. and D.Hogetsu. 1970. Effects of preillumination with light of different wavelengths on subsequent dark CO_2-fixation in *Chlorella* cells. Can.J.Bot. (Krotkov-Nelson Memorial Issue). 48: 1203-1207.

13. Miyachi,S. 1975. Abstracts, XII International Botanical Congress, Leningrad. Vol. 2. pp. 480.

14. Miyachi,S., A.Kamiya and S.Muto. Isolation of peroxisomes from colorless *Chlorella* mutant cells and intracellular localization of isocitrate lyase, malate synthase and phosphoenolpyruvate carboxylase. *Structure & Function of Photosynthetic Organelles.* Eds., S.Miyachi *et al.* (Special Issue of Plant & Cell Physiol.) in press.

15. Ogasawara,N. and S.Miyachi. 1969. Effect of wavelength on $^{14}CO_2$-fixation in *Chlorella* cells. *In:* H.Metzner,ed., Progress in Photosynthesis Res. Vol. 3. Verlag C. Lichtenstern,München. pp. 1653-1661.

16. Ogasawara,N. and S.Miyachi. 1970. Regulation of CO_2-fixation in *Chlorella* by light of varied wavelengths and intensities. Plant & Cell Physiol. 11: 1-14.

17. Ogasawara,N. and S.Miyachi. 1970. Effects of disalicylidene-propandiamine and near far-red light on $^{14}CO_2$-fixation in *Chlorella* cells. Plant & Cell Physiol. 11: 411-416.

18. Ogasawara,N. and S.Miyachi. 1971. Effects of dark preincubation and chloramphenicol on blue light-induced CO_2-incorporation in *Chlorella* cells. Plant & Cell Physiol. 12: 675-682.

19. Ohmann,E. 1969. Die Regulation der Pyruvat-Kinase in *Euglena gracilis*. Arch.Mikrobiol. 67: 273-292.

20. Schmid,G.H. and P.Schwarze. 1969. Blue light enhanced respiration in a colorless *Chlorella* mutant. Hoppe-Seyler's Z.f. Physiol.Chemie 350: 1513-1520.

21. Schmidt-Clausen,H.J. and I.Ziegler. 1969. The influence of light quality on the activation of $NADP^+$-dependent glyceraldehyde-3-phosphate dehydrogenase. *In:* H.Metzner,ed., Progress in Photosynthesis Res. Vol. 3. Verlag C. Lichtenstern,München. pp. 1646-1652.

22. Scott,Jr.,T.A. and E.H.Melvin. 1953. Determination of dextran with anthrone. Anal.Chem. 25: 1656-1661.
23. Stabenau,H. 1972. Aktivitätsänderungen von Enzymen bei *Chlorogonium elongatum* unter dem Einfluss von rotem und blauem Licht. Z.Pflanzenphysiol. 67: 105-112.
24. Teraoka,H., T.Nishikido, K.Izui and H.Katsuki. 1970. Control of the synthesis of phosphoenolpyruvate carboxylase and phosphoenolpyruvate carboxykinase in *Escherichia coli*. J.Biochem. 67: 567-575.
25. Tolbert,N.E., A.Oeser, T.Kisaki, R.H.Hageman and R.K.Yamazaki. 1968. Peroxisomes from spinach leaves containing enzymes related to glycolate metabolism. J.Biol.Chem. 243: 5179-5184.
26. Voskresenskaya,N.P. 1972. Blue light and carbon metabolism. Ann.Rev.Plant Physiol. 23: 219-234.

TWO PHASES OF CO_2 ABSORPTION ON LEAVES

TERUO OGAWA AND KAZUO SHIBATA

Laboratory of Plant Physiology
Institute of Physical and Chemical Research
(Rikagaku Kenkyusho), Wako-shi, Saitama, Japan

Summary

Two phases were found in CO_2 uptake and transpiration on leaves
of *Taraxacum officinale* L.(dandelion), *Triticum aestivum* L.(wheat)
and several other species of plants.   The gas exchange in the
earlier phase denoted as phase 1 was not controlled by stomata and
the exchange in the later phase denoted as phase 2 was controlled
by stomata.   The rate of phase-1 CO_2 uptake reached saturation at
low intensity, while the rate of phase-2 CO_2 uptake at high intensity, so that the shears of these two phases of CO_2 uptake were
greatly dependent on the light intensity.   The importance of
phase-1 CO_2 uptake in photosynthesis in shaded areas or on cloudy
days was stressed from these data.   On the contrary, the contribution of phase-2 CO_2 uptake is high at high intensities.
Very low O_2 concentrations below 1% suppressed both phase-2 CO_2
uptake and phase-2 transpiration, which suggested the importance
of respiration in stomatal opening.   High O_2 concentrations also
suppressed phase-2 CO_2 uptake and transpiration, but this type of
suppression occurred only in the presence of CO_2.   The suppression at high O_2 concentrations on the leaves of C_3 plants was due
to both photorespiration and stomatal closure, but the suppression
on leaves of C_4 plants was solely due to stomatal closure.

Introduction

It has been assumed that stomata on higher plant leaves control
most of the gas exchange between the external atomosphere and the
internal space in leaves.   However, there has been reported some
data suggesting an appreciable exchange which is not controlled by
stomata.   Transpiration through cuticle is known for various
plants (5,10,12) and CO_2 uptake through the upper epidermis of
hypostomatous plants has been reported by Freeland (3) and Dugger
(2).   Zelitch and Waggoner (15) observed appreciable rates of
transpiration and CO_2 uptake on the leaves which had been sprayed
with phenylmercuric acetate, a reagent for closing stomata artificially in the light.   In 1908, Lloyd (7) observed complete

closure of stomata only for a few species of plants out of many other species examined, and Loftfield (8) reported in 1921 that stomata of many plants do not close completely even after a long period of incubation in the dark.   The CO_2 uptake through cuticle or through such unclosed or incompletely closed stomata, both being not controlled by stomata, may shear a certain part of photosynthetic production in leaves, which should be evaluated from modern approaches.

It was found in a previous study that the rates of CO_2 uptake and transpiration increase in two steps when a leaf kept in dark- ness was exposed to light (11).   The rates first go up to a stationary level in 1 to 2 minutes and then rise again to a higher stationary level. CO_2 uptake and transpiration on this second stationary level, which we call phase-2 photosynthesis and trans- piration, respectively, in this paper, was completely suppressed by addition of abscisic acid in a very low concentration to the water in which the petiol of the sample leaf is dipped, whereas those on the first stationary level, phase-1 photosynthesis and transpiration, were not affected at all by abscisic acid.   These two phases of gas exchange were found for the lower epidermis of hypostomatous plant leaves but only phase-1 exchange for their upper epidermis.   These previous experiments opened up the way to distinguish the two types of gas exchange, one controlled and the other not controlled by stomata.

The present paper describes these two types of gas exchange measured at various light intensities and at various O_2 concent- rations.   The two types of CO_2 uptake as a function of light intensity clarified the contribution of each type to the produc- tion of organic matters at different places, at different climates and at different times of day or seasons, and those at various O_2 concentrations revealed inhibitory effects of oxygen at low and high concentrations on stomatal opening.

Materials and Methods

Leaves of *Taraxacum officinale* L.(dandelion), *Raphanus stivus* L., *Spinacia oleracea* L., *Citrus Natsudaidai* Hayata, *Camellia sasanqua* Thunb. and *Camelia communis* L. were harvested on our campus.   Seeds of *Triticum aestivum* L.(wheat) were germinated on wet paper soaked with distilled water, and those of *Zea mays*(corn) on a vermiculite bed with water.   These plants were grown at 25°C for 1 week under continuous illumination with white light from fluorescent lamps.   Sample leaves for measurements were excised and incubated in the dark with their petioles dipped in water.

The gas exchanges of excised leaves were measured at 25°C in normal air or in a mixture of nitrogen and oxygen containing 0.04% CO_2 with the open gas-analysis system described previously (11), which directly records the exchange rates of CO_2 and H_2O against time.   The air in our laboratory or the gaseous mixture with CO_2 was led to an acrylic sample chamber(diameter = 53mm, height =

88 mm, volume = 120 ml), and the exchanged gas from the chamber
was led to a hygrometer with an aluminum oxide sensor(developed
by Dr. Furuichi in our institute), then to $MgClO_4$ powder for
drying, and finally to an infra-red CO_2 analyser(Jasco PAX-10).
The relative humidity in the air or in the mixture of nitrogen and
oxygen with CO_2 led to the chamber was 50% at 25°C, and the flow
rate was adjusted with a needle valve to be 250 ml/min throughout
the experiments.

The sample leaf in the chamber was illuminated from one side
with white light from a 300W projector lamp through a fan-cooled
heat-absorbing filter and 0.5% $CuSO_4$ solution(7 cm in thickness).
The light intensity was varied with Toshiba neutral filters placed
between the projector and the sample chamber.   The light intensity was measured with a Kipp and Zonen E2 thermopile standardized
with a standard lamp(National Bureau of Standards U.S.A.).   The
light intensity without filter on the sample surface was 50 mW/cm^2
in the above condition through 0.5% $CuSO_4$ solution and 380 mW/cm^2
without this $CuSO_4$ solution and 100 mW/cm^2 through water in 7 cm
in thickness.

Abscisic acid was a gift from Hoffman La Roche & Co., Basel,
Switzerland.   This reagent was dissolved at a concentration of
10 mM in 10 mM NaOH solution, and 0.1 ml of this solution was
added to 10 ml of water in which the petiol was dipped.

Results

Two phases of gas exchange

A typical example of the two phases of gas exchange found for
a dandelion leaf at a light intensity of 50 mW/cm^2 is shown in
Fig. 1, which clearly indicates two steps of rate increase both in
CO_2 uptake and in transpiration (11).   The rates of phase-1 and
phase-2 photosynthesis were estimated from the height of the first
stationary level and from the difference in height between the two
stationary levels, and were designated in this paper as P_1 and P_2,
respectively, as shown in the figure.   These rates were corrected
for respiration in the dark.   The rates of phase-1 and phase-2
transpiration were similarly estimated and designated as T_1 and
T_2, respectively.

In the case of wheat leaves, phase-1 gas exchange proceeded
much more rapidly as seen from the curves in Fig. 2.   At a high
intensity of 50 mW/cm^2, phase-2 photosynthesis and transpiration
proceeded but, at a low intensity of 5 mW/cm^2, only phase-1 exchanges were observed.   Application of abscisic acid on the
higher stationary level in the light decreased the rates to lower
levels which are approximately equal to the levels of P_1 and T_1,
respectively, and turning the light off on these levels reduced
the rates to zero(the curves on the left side of Fig. 3).   The
leaves preincubated with abscisic acid for 5 hrs showed only
phase-1 gas exchange(the curves on the right side).   These

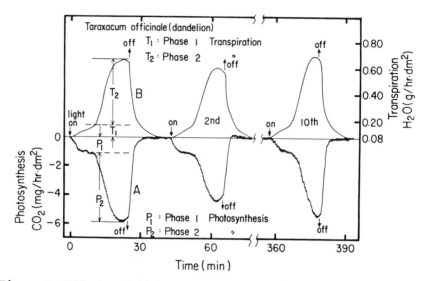

Figure 1. Effects of light on CO_2 uptake and transpiration on a dandelion leaf. Strong white light(50 mW/cm²) was turned on and off, and the rates of CO_2 uptake(curve A) and transpiration(curve B) were recorded against time.

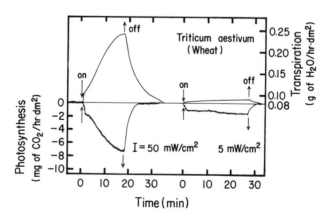

Figure 2. Effects of light on CO_2 uptake and transpiration on wheat leaves. Leaves were illuminated with strong(50 mW/cm²) and weak(5 mW/cm²) light.

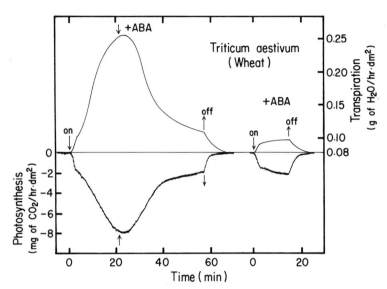

Figure 3. Effects of abscisic acid on CO_2 uptake and transpiration on wheat leaves in the light (the curves on the left side, light intensity = 50 mW/cm^2) and in the dark (the curves on the right side). These curves on the right side were observed for the leaves incubated with abscisic acid for 5 hrs in darkness.

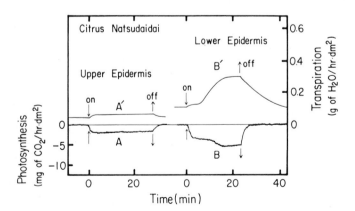

Figure 4. Effects of light on CO_2 uptake and transpiration on a Citrus leaf. Grease was pasted on the lower (curves A and A') and upper (curves B and B') epidermises. The leaf was illuminated with strong light (50 mW/cm^2).

results bear evidence that phase-2 gas exchange is controlled by
stomata, since abscisic acid induces stomatal closure (6,9,13).
    This view is supported by the observation on a hypostomatous
leaf of *Citrus Natsudaidai* Hayata shown in Fig. 4.   Only phase-1
gas exchange was found for its upper epidermis, while the two
phases were found for the lower epidermis.   In this experiment,
grease was pasted on the upper epidermis when the gas exchange on
the lower epidermis was to be observed, and *vice versa*.   It may
be worth to be noted that the phase-1 CO_2 uptake and transpiration
on the lower epidermis thus observed are higher than those on the
upper epidermis.   This difference may be ascribed to the contribution of unclosed stomata present only on the lower epidermis of
this type of plant.

Effects of light intensity

    The effects of light intensity described above on the two
phases of gas exchange was studied more precisely with wheat and
corn leaves.   Curves P_1 and P_2 in Fig. 5 obtained for wheat
leaves show the rates of P_1 and P_2 plotted against light intensity, which indicates different responses of the two phases of
photosynthesis.   The rate of phase-1 CO_2 uptake reached saturation below 10 mW/cm^2, while the rate of phase-2 CO_2 uptake
reached saturation at a higher intensity of 25 mW/cm^2.   The ratio
(P_1/P_t) of phase-1 CO_2 uptake to the total uptake($P_t = P_1 + P_2$)
estimated from these curves was 0.22 at a high intensity of 50 mW/
cm^2 but was much higher, 0.68, at a low intensity of 5 mW/cm^2.
A similar experiment was conducted for a corn leaf of C_4 plant,
and the result shown in Fig. 6 indicated the same trend in the
effect of light intensity, although the P_2 rate as well as the P_t
rate did not show saturation as is well known for C_4 plants.   The
much lower rates of CO_2 uptake observed on these corn leaves were
mostly due to the low light intensity in growing the sample corn
leaves.
    Seven species of leaves were examined for the rates of phase-1
and phase-2 CO_2 uptake at the low and high intensities of 5 and
50 mW/cm^2, and the results are summarized in Table 1.   The ratios
of P_1/P_t determined for the six species excluding a Commelia leaf
was as high as 0.57-1.00 at 5 mW/cm^2 but as low as 0.15-0.31 at
50 mW/cm^2.   This indicates a great contribution of phase-1 photosynthesis at low intensity.   The rate of phase-1 CO_2 uptake
through a Commelia leaf was practically zero.

Effects of O_2 concentration

    The rates of CO_2 uptake and transpiration greatly depended on
O_2 concentration.   The curves in Fig. 7 show the rate changes of
these gas exchanges observed on a dandelion leaf in nitrogen
containing various concentrations of O_2 and 0.04% CO_2.   Only
phase-1 gas exchange was observed at 0% of O_2 under illumination
with strong white light(50 mW/cm^2).   The absence of phase-2 gas

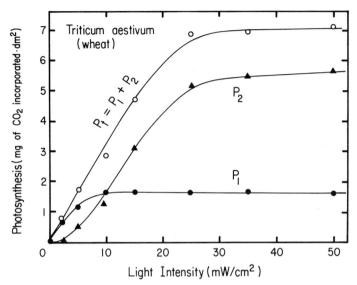

Figure 5. Effects of light intensity on the rates(P_1, P_2 and P_t) of phase-1, phase-2 and total CO_2 uptake on wheat leaves.

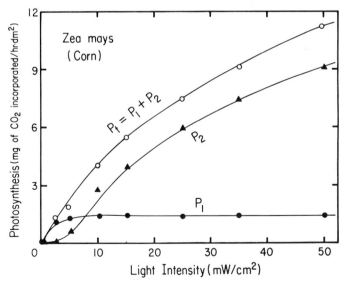

Figure 6. Effects of light intensity on the rates(P_1, P_2 and P_t) of phase-1, phase-2 and total CO_2 uptake on corn leaves.

Table 1. *Two phases of CO_2 uptake ($mg/h \cdot dm^2$), P_1 and P_2, by seven species of leaves at light intensities of 5 and 50 mW/cm^2.*

Leaves	Light intensity	P_1	P_2	$\dfrac{P_1}{P_1 + P_2}$
Spinacia oleracea L.	5	2.8	1.2	0.58
	50	2.8	11.4	0.20
Taraxacum officinale Weber	5	0.8	0.3	0.73
	50	0.8	4.3	0.14
Raphanus stivus L.	5	2.7	0.0	1.00
	50	3.0	13.9	0.18
Citrus Natsudaidai Hayata	5	2.3	0.0	1.00
	50	2.7	5.9	0.31
Camellia sasanqua Thunb.	5	1.3	0.7	0.65
	50	1.3	3.8	0.26
Commelia communis L.	5	0.0	0.5	0.00
	50	0.0	3.3	0.00
Zea mays	5	1.4	0.4	0.78
	50	1.5	8.3	0.15

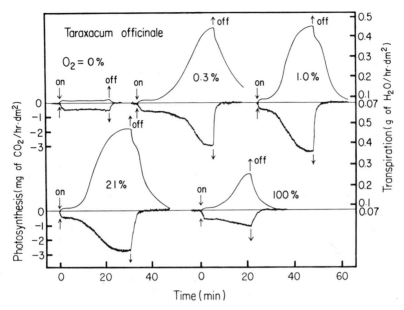

Figure 7. Effects of light on CO_2 uptake and transpiration on a dandelion leaf at various O_2 concentrations.   The leaf was illuminated with strong light(50 mW/cm^2).

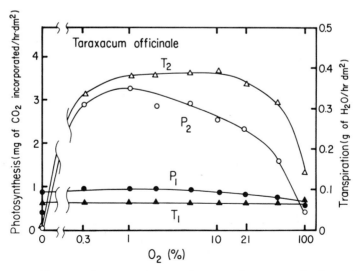

Figure 8. Effects of O_2 concentration on phase-1 and phase-2 CO_2 uptake(P_1 and P_2) and transpiration(T_1 and T_2) on a dandelion leaf.

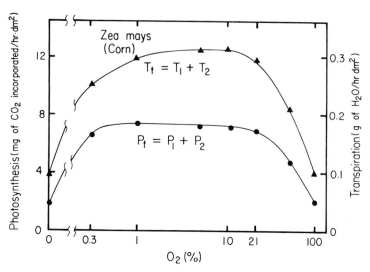

Figure 9. Effects of O_2 concentration on total CO_2 uptake($P_t = P_1 + P_2$) and transpiration($T_t = T_1 + T_2$) on corn leaves.

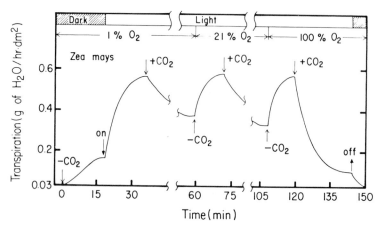

Figure 10. Effects of O_2 and CO_2 concentrations on transpiration on corn leaves.

exchanges indicates suppression of stomatal opening by the O_2-free nitrogen, as previously reported by Heath and Orchard (4) and Walker and Zelitch (14).   The importance of respiration in stomatal opening may be suggested from this fact.   Phase-2 gas exchanges were observed with 0.3% O_2 but it took about 10 min, for the P_2 and T_2 rates to start to increase from the phase-1 levels. With 1% and 21% O_2, this induction period for phase-2 gas exchanges was shorter and the rates increased to higher levels with 21% O_2.   With 100% O_2, however, the P_2 and T_2 rates decreased remarkably.

The curves in Fig. 8 show the rates of P_1, P_2, T_1 and T_2 thus measured for a dandelion leaf as a function of O_2 concentration. The concentration did not affect the P_1 and T_1 rates of phase 1, but the phase-2 rates decreased remarkably at low and high O_2 concentrations.   It is also evident from this figure that the inhibitory effect of high O_2 concentration on the P_2 rate starts slightly above 1% and progressively increases at higher O_2 concentrations whereas the inhibitory effect on the T_2 rate starts at 21% of O_2 and increases abruptly above this concentration.   The P_2 rate in normal air(21% O_2) was 72% of the rate with 1% O_2 and the rate with 100% O_2 was as low as 12% of the rate with 1% O_2. Experiments for corn leaves showed the same trend as for the dandelion leaf in the dependency of transpiration on O_2 concentration but the P_t rate stayed constant between 1% and 21% O_2 (Fig. 9).   The corn leaves used in this experiment showed poor separation of the two phases so that the total rates of P_t and T_t are plotted in this figure.

The transpiration rate with 0% O_2 was very low, regardless of the presence or absence of CO_2, but the rate in the presence of O_2 was greatly dependent on CO_2 concentration.   This is shown in the experiment of Fig. 10 on corn leaves.   When the sample corn leaves incubated in darkness in nitrogen containing 1% O_2 and 0.04% CO_2 was transferred into CO_2-free N_2 gas with 1% O_2 in the dark, the transpiration rate increased and reached a plateau during 20 min, showing stomatal opening in the dark by the absence of CO_2.   Illumination of these leaves at the plateau caused an increase of rate to a higher plateau, and addition of 0.04% CO_2 at this stage decreased the rate to a level which is 60% of the higher level attained in the absence of CO_2.   The rate increased again to the same higher level when exposed to CO_2-free air containing 21% O_2 in the light but addition of CO_2 decreased the rate to 50% of the original level.   The same experiment with 100% O_2 decreased the rate to 10% after addition of CO_2.   These data indicate that the suppression of stomatal opening at high O_2 concentrations was dependent on CO_2 concentration, and the suppression did not occur in the absence of CO_2.

Discussion

The presence of two phases was demonstrated in the present

study in CO_2 uptake and in transpiration for leaves of wheat and several other species of plants.   The earlier phase denoted as phase 1 was not controlled by stomata, and the later phase called phase 2 was controlled by stomata.   The rate of phase-1 CO_2 uptake reached saturation at low intensity while the rate of phase-2 uptake at high intensity.   This implies different shears of these two phases of CO_2 uptake at different intensities.   For example, the phase-1 CO_2 uptake of wheat leaves was 22% of the total rate at a high intensity of 50 mW/cm^2 but was much higher, 68%, at 5 mW/cm^2.   Experiments with several other species of leaves showed the same trend as shown in Table 1.   These data stressed the importance of phase-1 CO_2 uptake in photosynthesis at low intensities.   The light intensity in open area on a sunny day was 100-150 mW/cm^2 and the same measurements through water and through 0.5% $CuSO_4$ solution both in 7 cm in thickness were 60-90 and 35-53 mW/cm^2.   The intensity measured through the $CuSO_4$ solution was 2.0-2.5 mW/cm^2 in a shaded place on a sunny day and was 1.5-2.0 mW/cm^2 in open area on a cloudy day.   It seems therefore that the contribution of phase-1 CO_2 uptake is considerably high under these conditions of low intensity and most of the CO_2 uptake is not controlled by stomata.   Further studies on photosynthesis in field condition are awaited to estimate the contribution of phase-1 CO_2 uptake.

The rate of phase-2 CO_2 uptake on a dandelion leaf in normal air (21% O_2) was 72% of the rate with 1% O_2 and the rate decreased gradually to 17% with 100% O_2.   Similar suppression of CO_2 uptake by O_2 has been found by Björkman (1) for leaves of several species of C_3 plants.   The gradual decrease of rate between 1% and 21% is interpreted as due to photorespiration, since such decrease was not found for corn leaves of C_4 plant (Fig. 9).   The rates of both CO_2 uptake and transpiration on corn leaves were abruptly suppressed above 21% of O_2, which indicates that the suppression is due to the suppression of stomatal opening.

The suppression of stomatal opening at high O_2 concentrations occurred in the presence of CO_2 but did not occur in its absence. Stomata were completely open in the light even with 100% O_2 in the absence of CO_2.   This suggests that a substance produced from CO_2 in the light at high concentrations of O_2 may be responsible for this suppression of stomatal opening.   Some way of artificial control of CO_2 uptake through stomata in mass production of organic matters may be deduced from the results obtained in the present study.

Acknowledgements

The present study was supported by a grant for the study of Life Sciences at The Institute of Physical and Chemical Research (Rikagaku Kenkyusho).

Literature Cited

1. Björkman, O. 1967. Photosynthetic inhibition by oxygen in higher plants. *Carnegie Instn. of Wash. Year Book 65: 446-454.*
2. Dugger, W. M. 1952. The permeability of non-stomate leaf epidermis to carbon dioxide. *Plant Physiol. 27: 489-499.*
3. Freeland, R. O. 1948. Photosynthesis in relation to stomatal freequency and distribution. *Ibid. 23: 595-600.*
4. Heath, O. V. S. and B. Orchard. 1956. Effects of anerobic conditions upon stomatal movements - a test of Williams' hypothesis of the stomatal mechanism. *J. Exp. Bot. 7: 313-325.*
5. Jarvis, P. G. and M. S. Jarvis. 1963. The water relation of tree seedlings. IV. Some aspects of the tissue water relations and drought resistance. *Physiol. Plant. 16: 501-516.*
6. Little, C. H. A. and D. C. Eidt. 1968. Effect of abscisic acid on bud break and transpiration in wood species. *Nature 220: 498-495.*
7. Lloyd, F. E. 1908. The physiology og stomata. *Publ. Carnegie Instn. Wash. 82: 1-142.*
8. Loftfield, J. V. G. 1921. The behavier of stomata. *Ibid. 314: 1-104.*
9. Mittelheuser, C. J. and R. F. M. Van Steveninck. 1969. Stomatal closure and inhibition of transpiration induced by (RS)-abscisic acid. *Nature 221: 281-282.*
10. Monsi, M. 1944. Untersuchungen über die pflanzliche Transpiration, mit besonderer Berücksichtigung der stomatären und inneren Regulation. *Jpn. J. Bot. 13: 367-433.*
11. Ogawa, T. 1975. Two steps of gas exchange in leaf photosynthesis. *Physiol. Plant. 35: 91-95.*
12. Oppenheimer, H. R. 1960. Adaptation to drought xerophytism. - *In Plant-water Relationships in Arid. Zone Research 15, pp. 105-137. Unesco Paris.*
13. Uehara, Y., T. Ogawa and K. Shibata. 1975. Effects of abscisic acid and its derivatives on stomatal closing. *Plant Cell Physiol. 16: 543-546.*
14. Walker, D. A. and I. Zelitch. 1963. Some effects of metabolic inhibitors, temperature, & anerobic conditions on stomatal movement. *Plant Physiol. 38: 390-396.*
15. Zelitch, I. and P. Waggoner. 1962. Effect of chemical control of stomata on transpiration and photosynthesis. *Proc. Natl. Acad. Sci. 48: 1101-1108.*

STARCH DEGRADATION IN ISOLATED SPINACH CHLOROPLASTS[1]

Dwight G. Peavey, Martin Steup[2], and Martin Gibbs
Institute for Photobiology of Cells and Organelles
Brandeis University, Waltham, Mass. 02154

In 1862 Sachs (8) demonstrated that carbon assimilated by
chloroplasts in the presence of light and chlorophyll accumulated
as starch. The synthesis of starch from the photosynthetic intermediates of the carbon reduction cycle is well established. In
our recent publication (9) we demonstrated the regulatory role of
orthophosphate in the synthesis and degradation of starch in
isolated intact spinach chloroplasts. An initial investigation
of Levi and Gibbs (6) suggested that starch was catabolized via
a glycolytic pathway. Using chloroplasts preloaded with ^{14}C starch,
we have investigated starch breakdown, the resultant products and
the regulation of the degradation. Our results have shown that
the majority of starch was phosphorolytically degraded. Products
of starch are mobilized via a glycolytic pathway to the level of
PGA and triose-phosphates.

MATERIALS AND METHODS

Spinach (Spinacea oleracea L. var. Long Standing Bloomsdale)
was grown according to Peavey and Gibbs (7). Intact chloroplasts
were isolated by the method of Gibbs and Robinson (4). Deribbed
leaves (10-15 gm) were homogenized for 2 sec in 50 ml of chilled
isolation medium (50mM HEPES pH 6.8, 330 mM sorbitol, 10mM EDTA,
2mM PP, 1mM MgCl2 and 1mM DTT). The chloroplasts were pelleted
at 750 xg for 1 min. The photosynthetic reaction medium contained
50mM HEPES pH 7.8, 330mM sorbitol, 2mM EDTA, 2mM MgCl2, 1mM DTT
and 3.5-8.85mM NaHCO3 (7-15mCi/mmol). The photosynthetic conditions were: 10-15 min, 22.5°C, 400 W/M[2] and 50-150 µg·chloro-
phyll per ml. After photosynthesis 10 ml of chloroplast suspension was washed with 30 ml chilled buffer (50mM HEPES pH 7.8, 330
mM sorbitol, 2mM EDTA, 2mM MgCl2 and 5mM P(except where noted).
The chloroplasts were pelleted at 750 xg for 1 min. The pelleted
chloroplasts were resuspended in the same buffer, and starch
degradation was monitored for 30-60 min in the dark.

During starch degradation and photosynthesis, 0.5ml aliquots
were removed at intervals, and the reaction was stopped with 50
µl of concentrated formic acid. Total ^{14}C-incorporation into acid
stable intermediates was determined on a Nuclear Chicago gas flow

[1]This research was supported by NSF (BMS-71-00978) and by ERDA
(ET(11-1)-3231).
[2]Present address: Pflanzenphysiologisches Institut d. Universitat,
Untere Karspuele 2, D-3400 Goettingen, Germany (BRD)
[3]Abbreviations: DTT, dithiothreitol; IAA, Iodoactic acid; P, orthophosphate; PP, pyrophosphate, PGA, glycerate-3-phosphate;
triose-P, mostly dihydroxyacetone phosphate.

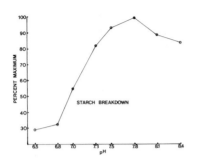
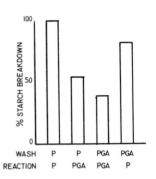

Figure 1.   (Left)   Effect of Varying pH on Starch Degradation
Figure 2.   (Right)   Effect of Phosphate and PGA on Starch Degrada-
            tion

end window detector system. ^{14}C-labeled starch was determined in
two different ways.   The sample was filtered through a MF-Milli-
pore membrane filter (0.45 μm pore size) and washed several times
with ethanol.   The filter retained only polyglucans.   In a second
procedure the ^{14}C labeled starch together with the ^{14}C intermediates was separated by one dimensional paper chromatography in
the GW3 solvent (10).   Hydrolysis (in HCL at 100° C) of the starch
fraction obtained either by filtration or by chromatography
yielded only labeled glucose.   The results obtained by both methods
did not differ more than 5%.   The identity of maltose was confirmed by co-chromatography, hydrolysis (1 M HCL at 100° C) to glucose and glucosidasing to glucose.   The ^{14}C labeled intermediates
were identified by co-chromatography with ^{14}C labeled standards.
The phosphorylated intermediates were phosphatased and rechroma-
tographed against known free sugars in GW3 and "semi-stench" (2).

RESULTS

As outlined in our previous report (9), intact spinach
chloroplasts were loaded with ^{14}C starch during 10-15 minutes of
photosynthesis in the presence of low phosphate (0.25mM).   These
plastids, containing 50-80% of ^{14}C in starch, were washed,
pelleted and resuspended with no significant breakage (<10%) as
checked by $^{14}CO_2$ fixation.   With a constant velocity for one
hour, starch breakdown in the dark occured at a rate of 1-2 μg-
Atom c/mg chlorophyll per hour.   In contrast to starch synthesis,
optimal conditions for starch breakdown required high phosphate
between 25-50 mM.   Osmotically shocked or phosphate-free chloroplast preparation showed no significant starch breakdown (<5%).
As shown in Figure 1, a pH profile for starch degradation had an
optimum at pH 7.8.   Starch breakdown was greatly inhibited at
lower pH's.   At these lower pH's, the chloroplasts intactness

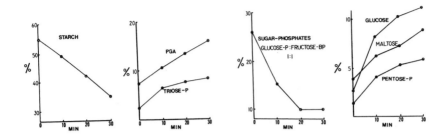

Figure 3.   Distribution of ^{14}C Intermediates of Starch Breakdown.

altered and various glycolytic enzymes were inhibited.

In contrast to starch synthesis which was inhibited by phosphate and activated by PGA, starch breakdown required phosphate
and was inhibited by PGA. As shown in Figure 2, maximal inhibition
of starch breakdown by PGA was demonstrated with chloroplast that
are not pre-loaded phosphate during the wash. Addition of phosphate overcame PGA inhibition. Likewise the enhanced degradation
of starch by washing the plastids with phosphate is reversed upon
the addition of PGA to the reaction medium. Under in vitro conditions, the chloroplast in the dark would have high concentrations of phosphate (3).

The degradation products of starch breakdown during the dark
are shown in Figure 3. The major portion of the ^{14}C labeled
compounds derived from the ^{14}C starch are phosphorylated intermediates. As the starch decreased, PGA and triose-phosphates
increased proportionately. The rate of PGA to triose-phosphates
was maintained at 2:1 throughout starch breakdown. The sugar
phosphates, of which a large percentage were photosynthetically
derived decrease rapidly and contributed to the flow of ^{14}C into
PGA, Triose-P and especially the pentose-phosphates. The pool of
sugar-phosphates plateaued at 10% due to glucose phosphates and
fructose 1-6, bisphosphate derived from starch breakdown. Under
conditions which do not favor starch breakdown, these sugar phosphates fall much lower (5). During starch breakdown, the ratio
of glucose-phosphates to fructose bisphosphate remained at 1:1.
In contrast to the phosphorylated intermediates, between 5-10%
of the ^{14}C derived from starch was found in maltose and glucose
indicating an anylolytic cleavage of starch. This was not surprising in light of the presence of α and β-amylase known to be
found in spinach leaves.

To determine whether the intermediates of starch breakdown
were mobilized via the pentose phosphate pathway, the preparation
was checked for the liberation of $^{14}CO_2$. No $^{14}CO_2$ was trapped
by ethanolamine or hyamine hydroxide. It should be noted that
no reducing agents were used at any stages of the experiments.

	0 min		15 min		30 min	
	control	DTT	control	DTT	control	DTT
	%^{14}C					
Starch	69.4	67.6	56.9	59.7	45.9	44.4
Maltose	6.3	6.4	8.3	8.1	9.0	9.3
Glucose	1.4	1.4	5.3	3.3	7.4	4.2
Glucose-P	7.3	9.1	5.4	11.6	5.2	12.6
Fructose-bP	5.1	4.3	2.9	3.0	3.8	3.6
Fructose-P	2.1	2.4	1.4	2.6	1.8	2.8
Pentose-P	1.3	1.4	5.3	3.0	7.4	4.1
Triose-P	2.4	1.4	5.9	1.3	7.5	1.6
PGA	4.7	6.0	8.5	13.3	11.7	17.4

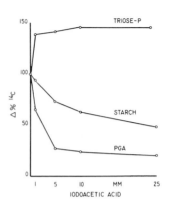

Table I.   Effect of 10mM Dithiothreitol on Starch Breakdown.
Figure 4. Effect of Iodoacetic Acid on Starch Breakdown.

Likewise, dithiothreitol, a known inhibitor of glucose-6-phosphate dehydrogenase, was tested as seen in Table I.   Starch breakdown was not inhibited.   The distribution of ^{14}C between triose-phosphate and PGA was shifted in favor of PGA.   Dithiothreitol is known to enhance glyceraldehyde-3-phosphate dehydrogenase activity. The glucose-phosphate pool was doubled at the expense of glucose and pentose phosphate.   At no time was 6-phosphogluconate, an intermediate of the pentose phosphate cycle, detected among the products of starch breakdown.   It seems very unlikely that any starch was degraded via the pentose phosphate shunt.

Since the conversion of Triose-phosphate to PGA generates ATP necessary for the glycolytic breakdown of starch, Iodoacetic acid (1), a known inhibitor of glyceraldehyde-3-phosphate dehydrogenase was tested.   As shown in Figure 4, starch breakdown was inhibited fifty percent.   The addition of IAA blocked the conversion of Triose-P to PGA shifting the ratio from 2:1 to 1:1. Except for the increase in triose phosphate and the sharp decrease in PGA, the ^{14}C distribution among the breakdown intermediates was unaffected.   The addition of ATP overcame the inhibition of starch breakdown by iodoacetic acid (Data not shown).

SUMMARY

Starch degradation in the dark in isolated intact spinach chloroplast requiring high phosphate indicated a phosphorolytic cleavage.   The distribution of intermediates derived from ^{14}C labeled starch confirmed the glycolytic pathway of degradation. The presence of maltose and glucose suggested a concomitant amylolytic breakdown.   The inhibition of starch breakdown by iodoacetic acid demonstrated a source of ATP necessary for the fructose-6-phosphate kinase reaction.   The phosphorolysis of starch in the chloroplast represented a highly favored conservation of energy versus the amylolytic degradation of starch in other plant tissues. The mobilization of carbon from starch via the glycolytic pathway produced PGA and Triose-phosphate which are readily exported from the chloroplast. The pertinent question is the mechanism of NAD regeneration necessary for the oxidation of triose phosphate.

LITERATURE CITED

1. Calo, N. and M. Gibbs. 1960. Z. Naturforschung 15b:287-291.
2. Growley, G.T., V. Moses and J. Ullrich. 1963. J. Chromatogr. 12:219-228.
3. Heber, V.W. and K.A. Santarius. 1965. Biochim. Biophys.Acta. 109:390-408.
4. Gibbs, M. and M.J. Robinson. 1974. in: Experimental Plant Physiology ed. A. San Pietro, pp. 13-20.
5. Latzko, E. and M. Gibbs. 1969. Plant Physiol. 44:396-402.
6. Levi, C. and M. Gibbs. 1976. Plant Physiol. 57:933-935.
7. Peavey, D.G. and M. Gibbs. 1975. Plant Physiol. 55:799-802.
8. Sachs, J. 1862. Bot. Z. 20:365-373.
9. Steup, M., D.G. Peavey and M. Gibbs. 1976. Biochem. Biophys. Res.Commun. 72:1554-1561.
10. Wood, T. 1961. J. Chromatogr. 8:142-154.

CONVERSION OF GLUCOSE-POLYMER TO SUCROSE IN *CHLORELLA* INDUCED BY HIGH TEMPERATURE

YASUNORI NAKAMURA AND SHIGETOH MIYACHI

Radioisotope Centre, University of Tokyo, Bunkyo-ku, Tokyo, Japan
and
Institute of Applied Microbiology, University of Tokyo, Bunkyo-ku, Tokyo, Japan

Summary

Analysis of photosynthetic products under various temperatures as well as the transitional changes in the products induced by changing temperature during the course of the experiment revealed that the conversion of glucose-polymer to sucrose in *Chlorella vulgaris* 11h cells was greatly accelerated at temperatures higher than 30°C. This temperature-induced degradation of glucose-polymer occurred even in the presence of DCMU or in the dark.

Introduction

Longer than a decade ago Oullet studied the relationship between the experimental temperatures and the distribution of labeled compounds after 2 min of photosynthetic $^{14}CO_2$ fixation in green alga *Scenedesmus*. He showed that as the temperature was raised from 25 to 37°C the amount of radioactivity incorporated into sucrose remarkably increased, while that into phosphate esters gradually decreased. These results led him to the inference that the formation of sucrose from hexose phosphates is greatly accelerated at higher temperatures (6). Recently, Sawada and Miyachi reported that photosynthetic capacity and carbon metabolism in various higher plants were strongly dependent on the temperature during growth (7-9). Also, Döhler studied the effects of growth and experimental temperatures as well as the CO_2 concentration during growth on the carbon metabolism during the induction period of photosynthesis in *Chlorella vulgaris* (3). However, among various environmental factors which affect the photosynthetic carbon metabolism, relatively little attention has been paid to the effects of temperatures.

Abbreviation used: pcv, packed cell volume.

The present study with *Chlorella vulgaris* 11h cells confirmed
that ^{14}C incorporation into sucrose was accelerated as the temperature during photosynthetic $^{14}CO_2$ fixation was raised. At the
same time our results indicated that the degradation of glucose-
polymer was also enhanced at higher temperatures. In this paper
the results will be described which confirmed that the increased
^{14}C incorporation into sucrose is due to the enhancement of the
degradation of glucose-polymer at high temperatures.

Materials and Methods

Algal culture

Chlorella vulgaris 11h (Algensammlung des Pflanzenphysiolo-
gischen Institut der Universität Göttingen) was grown in a flat
oblong vessel with a light-dark rhythm of 16 and 8 hr. The inorganic culture medium contained 5 g KNO_3, 1.25 g KH_2PO_4, 0.1 g
K_2HPO_4, 2.5 g $MgSO_4 \cdot 7H_2O$, 2.8 mg $FeSO_4 \cdot 7H_2O$ and 1 ml of Arnon's
"A5" solution per liter (pH 5.7) (5). Air enriched with 2-3% CO_2
by volume was constantly bubbled through the algal suspension for
7 to 10 days at 23°C. During the light period, the vessel was
illuminated by an incandescent lamp. The light intensity was increased stepwise from 0.5 to 10 klux according to the increase in
the algal density. Cells were harvested 3 hr after the start of
the last light period and washed once with and suspended in the
culture medium which had been diluted with 9 volumes of deionized
water.

Photosynthetic $^{14}CO_2$ fixation

Photosynthetic $^{14}CO_2$ fixation was carried out in a water jacket-
ed glass vessel "lollipop" which was connected to the disphragm
pump with vinyl tubes. Solution of 10 mM Na-phosphate buffer (pH
6.9) dissolved in a diluted culture medium (x 1/20) was placed in
the lollipop and it was illuminated by metal halide lamp from both
sides. The light intensity was 2 x 20,000 lux at the surface of
the lollipop. The temperature was controlled by water running
through the water-jacket. The solution was continuously bubbled
with air circulating through the pump and the lollipop, and ^{14}C
labeled $NaHCO_3$ was added. Ten minutes after adding $NaH^{14}CO_3$,
photosynthetic $^{14}CO_2$ fixation was initiated by the addition of
Chlorella suspension (final density, 2-5 ml pcv/1). At intervals,
portions of algal suspension were quickly transferred to the test
tubes containing methanol (final concentration, 80%). A portion
taken from each sample was placed on a planchet and dried under an
infra-red lamp to determine the total ^{14}C fixed.

Analysis of $^{14}CO_2$ fixation products

The rest of the sample was centrifuged. The precipitate was

extracted again with 80% methanol and centrifuged. The precipitate thus obtained will be referred to "insoluble fraction" and the combined supernatant to "soluble fraction".

The amount of soluble fraction was reduced in vacuo and chromatographed two dimensionally on Whatman No. 1 filter paper, first with phenol-acetic acid-1 M EDTA-water (740:10:1:260, by volume) and then with n-butanol-propionic acid-water (140:71:100, by volume). After chromatography the autoradiogram was prepared and the radioactivity of each spot was measured. The individual compounds were identified with paper co-chromatography and/or paper co-electrophoresis with the authentic compounds.

The radioactive sucrose thus separated was purified by one dimensional paperchromatography with n-butanol-acetic acid-water (12:3:5, by volume) and then eluted from the paper with water. The eluate (1 ml) was treated with invertase (50 U) for 4 hr at 28°C and then the sample was chromatographed with n-butanol-acetic acid-water.

The insoluble fraction was heated at 96°C for 30 min in 1 N HCl solution. After this treatment the solution was neutralized with NaOH and chromatographed one dimensionally with the same solvent as above. The preliminary experiment with authentic ^{14}C-starch revealed that glucose-polymer was completely hydrolyzed to glucose by this treatment.

Results and Discussion

Time courses of photosynthetic $^{14}CO_2$ fixation under differnt temperatures

The rates of photosynthetic $^{14}CO_2$ incorporation increased with the increase in temperatures from 6 through 40°C (Upper part of Fig. 1). Analysis of $^{14}CO_2$ fixation products under varied temperatures revealed the following facts:

1) Under all temperatures tested, the active ^{14}C-incorporation into phosphate esters lasted for 3 to 5 min, followed by a slow incorporation which continued during the experimental period (30 min). The radioactivities incorporated into phosphate esters after $^{14}CO_2$ fixation for 30 min increased in parallel with the rise of temperature up to 30°C. At 40°C, however, the same radioactivity was far smaller than that observed at 30°C.

2) The rates of ^{14}C-incorporation into lipid fraction were sluggish at temperatures lower than 20°C. The incorporation was enhanced significantly at 30°C, but the rise of temperature to 40°C did not bring about the further increase.

3) Incorporation of ^{14}C into sucrose increased gradually as the temperature was raised from 6 to 21°C. The increase was remarkably accelerated by the rise of temperature to 40°C.

4) The initial rate of ^{14}C-incorporation into the insoluble fraction was enhanced in parallel with the rise of experimental temperatures. [The rates (cpm/2 min) at 6, 12, 21, 30 and 40°C

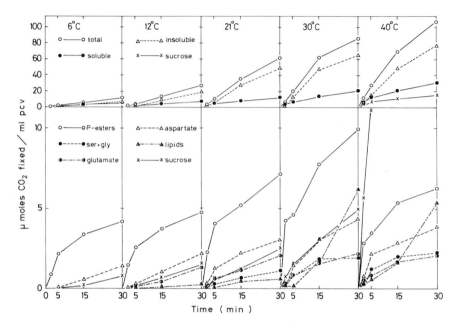

Fig. 1.   Time courses of photosynthetic $^{14}CO_2$ fixation by Chlorella vulgaris *cells under various temperatures.   Symbols: P-esters,*
phosphate esters; ser + gly, serine + glycine.

were 117, 633, 2513, 4868 and 9150.]   At 6 and 12°C, these rates
remained constant during the experimental period, while at 21
through 40°C, the rates were slowed down after 5 to 10 min.   As a
result, the radioactivities incorporated in this fraction after
30 min of photosynthetic $^{14}CO_2$ fixation increased as the temperature was raised from 6 to 21°C and then kept relatively constant
level at higher temperatures.   In some cases (*e.g.* Expt.III in
Fig. 2) the level at 40°C was   even lower than that at 20°C.   As
shown in Fig. 2, the percent incorporation of radioactivity into
insoluble fraction increased as the temperature was raised from
6 to 20°C and started to decrease when the temperature was further raised.   The inverse relationship was observed in those incorporated into soluble and sucrose.

    5) Acid hydrolysis of the insoluble fractions obtained under
various temperatures revealed that more than 90% of the radioactivities recovered in glucose.

    Based on the above observations we assumed that the rate of
photosynthetic $^{14}CO_2$ incorporation into glucose-polymer (insoluble) fraction was increased in parallel with the rise of temperatures.   On the other hand, when the temperature was raised
higher than 20°C, the breakdown of the glucose-polymer to form
sucrose was greatly stimulated.   Consequently, the percent incorporation of radioactivity into the insoluble fraction

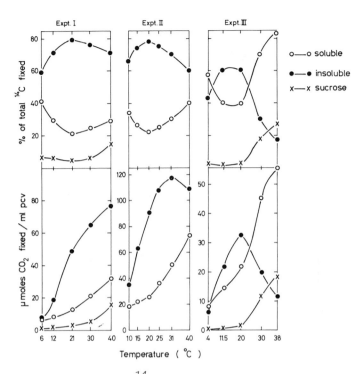

Fig. 2. Incorporation of ^{14}C into soluble, insoluble and sucrose fractions during photosynthetic $^{14}CO_2$ fixation under varied temperatures. Durations of $^{14}CO_2$ fixation in Expts. I, II and III were 30 min, 30 min and 10 min, respectively. The data shown in Expt. I were calculated from the results shown in Fig. 1.

decreased while that into sucrose increased at temperatures higher than 20°C. Following experiments were carried out to test this inference.

The levels of radioactivities in the products as influenced by the changes in the experimental temperature during photosynthetic $^{14}CO_2$ fixation

The rate of $^{14}CO_2$ incorporation was enhanced when the temperature of *Chlorella* suspension was quickly raised from 20 to 40°C during the course of photosynthetic $^{14}CO_2$ fixation (Fig. 3). Also, the radioactivities in the insoluble fraction started to decrease, while those in sucrose started to increase. These results indicate that, when the temperature is raised, the conversion of glucose-polymer to sucrose is accelerated in *Chlorella* cells. The figure further shows that the level of phosphate esters also dropped as a result of the temperature transition. Essentially the same results were obtained when the temperature was raised

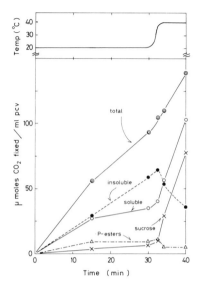

Fig. 3. Changes in the radioactivities in the products when the temperature was raised from 20 to 40°C during the course of photosynthetic $^{14}CO_2$ fixation.

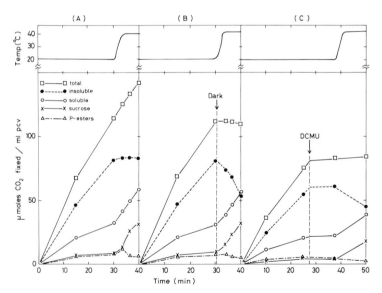

Fig. 4. Effects of light and DCMU on the temperature-induced changes in the photosynthetic products. (A); control. (B); light was turned off at 30.5 min. (C); DCMU (final concentration, 2 uM) was added to the reaction mixture at 27.5 min.

during the photosynthesis which was carried out by the provision of 0.2 mM $NaH^{14}CO_3$ (data not shown).

Effects of DCMU and darkness on the temperature-induced changes in photosynthetic products

To see whether photosynthetic electron transport and photophosphorylation is responsible to the above mentioned changes, the light was turned off or DCMU was added under continued illumination during the course of photosynthetic $^{14}CO_2$ fixation at 20°C, and then the temperature was quickly raised to 40°C. Fig. 4 shows that the temperature-induced changes occurred in the dark as well as in the presence of DCMU. In both cases, the decrease in radioactivities in insoluble (glucose-polymer) fraction was approximately accounted for by the increase in those in sucrose. We therefore assumed that the photosynthetic electron transport system is not involved in the mechanism of the temperature-induced changes in the photosynthetic products.

Fig. 5 shows that when the experimental temperature was lowered from 40 to 20°C simultaneously with turning off the light, there was practically no change in the radioactivities in the insoluble fraction and sucrose.

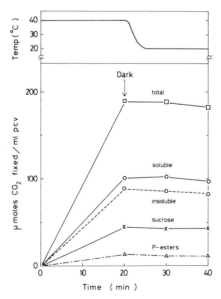

Fig. 5. Effects of temperature transition from 40 to 20°C in the dark after photosynthetic $^{14}CO_2$ fixation at 40°C. Light was turned off 20 min after photosynthetic $^{14}CO_2$ fixation had been carried out at 40°C, and then the temperature was quickly lowered to 20°C.

Temperature dependency of the conversion of glucose-polymer to sucrose

Following experiments were carried out to confirm that the conversion of glucose-polymer to sucrose in *Chlorella* cells is accelerated by the rise of the experimental temperatures. After photosynthetic $^{14}CO_2$ fixation for 30 min at 20°C, the suspension of *Chlorella* was quickly centrifuged and washed once with 10 mM Na-phosphate buffer (pH 7.0) at 0°C to remove unreacted $^{14}CO_2$. The cells were resuspended in the reaction medium except for the omission of sodium bicarbonate and exposed to various temperatures in the dark. Fig. 6 shows that, 20 min after the exposure to 40°C, the radioactivities in glucose-polymer decreased to about half, and this decrease was accounted for by the increase in those in sucrose. The changes were less pronounced at 30°C and practically no change was observed at temperatures lower than 20°C. The similar results were obtained when the experiments were conducted in the light (data no shown). We therefore concluded that, in *Chlorella* cells, the conversion of glucose-polymer to sucrose is significantly accelerated at temperatures higher than 30°C.

Hydrolysis of sucrose by invertase showed that, irrespective of the experimental conditions, the radioactivities in sucrose obtained from the cells which had been kept at 40°C were equally distributed in glucose and fructose (Table 1). This fact indicates

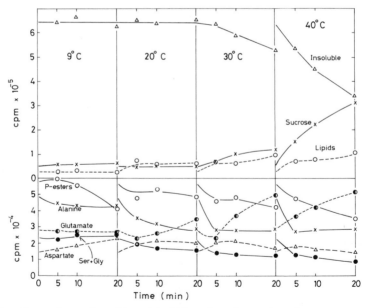

Fig. 6. Transfer of radioactivities in glucose-polymer to sucrose under various temperatures. For details see the text.

Table 1. *Radioactivities in glucose and fructose moieties obtained after the hydrolysis of radioactive sucrose by invertase*

Expt.	^{14}C-sucrose obtained after	Radioactivity (cpm) in		
		Glucose	Fructose	Sucrose
I.	$^{14}CO_2$ fixation for 30 min,40°C,L	753	746	51
II.	$^{14}CO_2$ fixation for 30 min,20°C,L and 6 min,40°C,L	961	939	64
III.	$^{14}CO_2$ fixation for 30.5 min,20°C,L and 6 min,40°C,D	635	641	30
IV.	DCMU added after $^{14}CO_2$ fixation for 27.5 min,20°C,L, then for 12 min,40°C,L	132	149	trace
V.	$^{14}CO_2$ was removed from the suspension after $^{14}CO_2$ fixation for 26 min, 20°C,L, then for 10 min,40°C,D	431	419	trace

The radioactive sucroses used in Expts.I, II, III and IV were obtained after the experiments shown in Figs. 1, 4-A, 4-B and 4-C, respectively.   Procedure of Expt. V is the same as that in Fig.6. L, light; D, dark.

that both glucose and fructose moieties in a sucrose molecule were formed from the glucose-polymer in *Chlorella* cells.

It has been shown that polyglucan phosphorylase and α-amylase are in *Chlorella* cells (10). It is unlikely, however, that α-amylase is involved in the temperature-induced conversion of glucose-polymer to sucrose, since no free glucose was detected during this process.   To study whether polyglucan phosphorylase is enhanced at higher temperatures, *Chlorella* cells which had fixed $^{14}CO_2$ at 20°C for 30 min in the light were disrupted with French pressure cell.   The homogenate thus obtained was kept at 40°C.   However, no temperature-induced breakdown of glucose-polymer was observed with this homogenate (data not shown).

In sharp contrast with the present results, it is well established that the degradation of starch in higher plant tissues is enhanced under low temperatures (0-2°C) (1,2,4). The problem whether the conversion of glucose-polymer to sucrose which is accelerated at higher temperatures is unique to algae is being studied in our laboratory.

Literature cited

1. Arreguin-Lozano,B. and J.Bonner. 1949. Experiments on sucrose
   formation by potato tubers as influenced by temperature.
   Plant Physiol. 24: 720-738.
2. Barker,J. 1968. Studies in the respiratory and carbohydrate
   metabolism of plant tissues XXIV. The influence of a decrease
   in temperature on the contents of certain phosphate esters in
   plant tissues. New Phytol. 67: 487-493.
3. Döhler,G. 1974. Einfluss der Kulturbedingungun auf die photo-
   synthetischen Carboxylierungsreaktionen von *Chlorella vulgaris*. Z.Pflanzenphysiol. 71: 144-153.
4. Isherwood,F.A. 1973. Starch-sugar interconversion in *Solanum
   tubersum*. Phytochemistry. 12: 2579-2591.
5. Ogasawara,N. and S.Miyachi. 1970. Effects of disalicylideno-
   propanediamine and near far-red light on $^{14}CO_2$-fixation in
   Chlorella cells. Plant & Cell Physiol. 11: 411-416.
6. Oullet,C. 1951. The path of carbon in photosynthesis XII. Some
   temperature effects. J.Exp.Bot. 2: 316-320.
7. Sawada,S. and S.Miyachi. 1972. Photosynthetic carbon metabolism in green plants acclimated to differnt temperatures. *In:*
   G.Forti, M.Avron and A.Melandri,eds.,Proceedings of the IInd
   International Congress on Photosynthesis Research,Vol.3.
   Dr.W.Junk N.V.Publishers,The Hague. pp.2035-2049.
8. Sawada,S. and S.Miyachi. 1974. Effects of growth temperature
   on photosynthetic carbon metabolism in green plants II. Photosynthetic $^{14}CO_2$-incorporation in plants acclimatized to
   varied temperatures. Plant & Cell Physiol. 15: 225-238.
9. Sawada,S. and S.Miyachi. 1974. Effects of growth temperature
   on photosynthetic carbon metabolism in green plants III.
   Differnces in structure, photosynthetic activities and activities of ribulose diphosphate carboxylase and glycolate oxidase in leaves of wheat grown under varied temperatures. Plant
   & Cell Physiol. 15: 239-248.
10. Wanka,F., M.M.J.Joppen and Ch.M.A.Kuyper. 1970. Starch-degra-
    ting enzymes in synchronous cultures of *Chlorella*. Z. Pflan-
    zenphysiol. 62: 146-157.

PHOTOSYNTHESIS AND RESPIRATION IN
RELATION TO PRODUCTIVITY OF CROPS

AKIRA TANAKA

*Faculty of Agriculture,
Hokkaido University
Sapporo, Japan 060*

Agriculture is the only device human beings have, at least at present, to produce food by converting solar energy through photosynthesis. Although solar energy is a key factor, its importance in agriculture was not seriously considered until recently.

In primitive agriculture the yield of crops was low because plant nutrients, especailly nitrogen, or water, were frequently the limiting factor, and losses caused by pests or diseases were large. In such situations the significance of solar energy in determining the yield was concealed.

With the advance of agricultural technology, fertilizers became more abundant, facilities for irrigation or drainage were improved, pests and diseases were properly controlled, and yields were pushed up. At this point, solar energy became one of the important yield-limiting factors. Because there is no adequate agricultural method of supplementing solar energy, improvement of the efficiency of crops of converting solar energy to foods became an important issue.

The yield of crops (economic yield, Ye) is the function of total dry matter production (biological yield, Yb) and harvest index (h): $Ye = h \cdot Yb$. Dry matter production(W) is controlled by photosynthesis(P) and respiration(R). The relation between photosynthesis and dry matter production has been studied intensively, and in these studies respiration was frequently considered as a loss of primary photosynthetic products: $W = P-R$, when P and R are expressed in terms of glucose, and the carbon content of dry matter is considered to be the same as glucose.

However, respiration is indispensable for dry matter production because the production of any substance requires energy and this energy is generated by respiration. In this sense, out of a given quantity of primary photosynthetic products(P) a certain amount(R) is bound to be respired to generate the energy by which the rest($P-R = W$) is converted to stable plant constituents. To express this relationship numerically: if the amount of dry

213

matter which is produced from a unit amount of primary photo-synthetic products is called growth efficiency(GE), then: GE = W/P = W/(W+R) (13). If the growth efficiency is constant, improvements in dry matter production could be based on the study of photosynthesis alone. Whereas, if it changes considerably with conditions, studies on respiration are also important.

The harvest index is the ratio of the amount of dry matter of the organ to be harvested to the amount of dry matter produced during a whole cycle of the growth of the crop. In grain crops the seeds are the organ to be harvested, and the seeds are gener-ally incapable of (or very weak in) photosynthesis and start to grow after flowering by using substances mostly produced during ripening. Thus, the harvest index is controlled by (a) the source; the capacity of vegetative organs to send substances to the seeds during ripening, more specifically the capacity of leaves to produce primary photosynthetic products to be sent to the seeds, and (b) the sink; the capacity of growing seeds to ac-cept the substances which are sent from the source.

The potential capacities of the source and the sink are deter-mined before flowering, and the realization of these capacities during ripening is strongly influenced by the source-sink inter-action, although it is also affected by the environmental condi-tions during ripening. Because of this, the total dry matter production and the harvest index are influenced by each other, and the equation $Ye = h \cdot Yb$ is not so simple as it appears.

LIGHT RESPONSE CURVE OF PHOTOSYNTHETIC RATE AND PLANT TYPE

Classic studies have demonstrated that there is a photo-saturation point in the light response curve of the photosyn-thetic rate of single leaves. The saturation point of many crop species was found to be at 30-40 Klux which is far below full sunlight. On the basis of this evidence, the concept prevailed that solar energy was not the limiting factor of the yield, and this concept did not conflict with agronomic experiences. For example, under a given climatic condition the yield of rice increased with an increase of nitrogen application, and it was considered that nitrogen was the limiting factor, and not solar energy.

However, it was gradually recognized that the optimum nitrogen level for a maximum yield shifts with changes of solar energy: The optimum nitrogen level as well as maximum yield become higher when solar energy is more abundant. On the basis of this obser-vation, the idea was established that when sufficient nitrogen is applied, solar energy will take over as the yield-limiting factor.

This idea seemed to conflict with the nature of the light re-sponse curve of photosynthetic rate. However, this complication was explained in the following manner: Leaves of plants, which constitute a population, shade each other when the leaf area

index is above a certain value. Under such a condition, even if
the light intensity at the top of canopy is at the saturation
point of single leaves, many leaves in the population receive
light below the saturation point due to the mutual shading. Thus,
an increase of light intensity at the top of the canopy beyond
the saturation point improves the photosynthetic rate of leaves
in the population (2)(3)(6). On the basis of this idea, intensive studies of the geometry of leaves were made to improve the
distribution of light in the plant population, and the concept of
plant type was established (1)(10).

For example, the yield of rice in Hokkaido, Japan has been
improved by both varietal improvement and an increase in the use
of nitrogen fertilizers (Fig. 1, left). The varietal improvement
in this case was to increase the efficiency of rice plants at
converting solar energy by increasing the leaf area index and
decreasing the extinction coefficient (Fig. 1, right) (16).

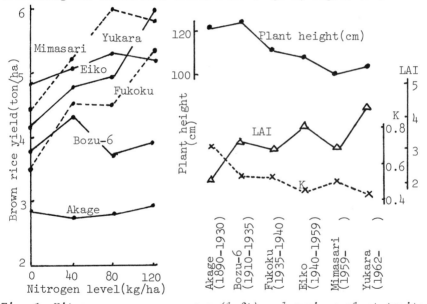

*Fig. 1. Nitrogen response curves (left) and various plant traits
(right) of old and new rice varieties in Hokkaido.
LAI: Leaf area index, K: Extinction coefficient.
Numbers in parentheses beside variety names in right-hand figure
indicate years when respective varieties were popular among farmers.*

The improvement of rice varieties at IRRI was successfully accomplished within a short period by following the plant type concept. Breeding programs of other crops are also following a
similar approach and producing promising results. However, it
can be argued that efforts through the plant type concept are

approaching a ceiling.

The discovery of C₄ plants opened a new avenue of research. Ambitious workers are trying to accomplish a remarkable jump in the photosynthetic rate by introducing the device of C₄ plants into C₃ plants.

It is not, however, certain whether the new avenue is real or fantasy. It is all very well for some scientists to proceed through the new avenue, but others should also consider where the avenue is going. I feel that the following questions should be answered before initiating a big breeding program based on the above-mentioned idea: Do leaves with a higher photosynthetic rate make a variety a high yielder? How can a varietal comparison of the photosynthetic rate of single leaves be made which takes into account the fact (15) that the rate differs among leaves depending upon their position in a plant, and furthermore that the rate fluctuates with the age of the leaf as shown in Fig. 2? And so on.

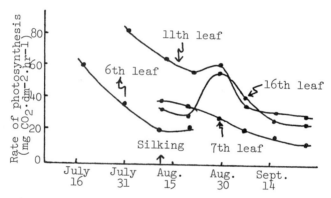

Fig. 2. Fluctuation of rates of photosynthesis of the leaves at various positions of a maize plant during growth (Fukko No. 8, 1969).

RESPIRATION OF VARIOUS ORGANS

The photosynthetic rate of the leaves of various crop species has been measured repeatedly by many scientists, and along with these measurements the respiratory rate of leaves has also been measured frequently. By using these data some scientists treat the dry matter production of plants as the balance between the rates of photosynthesis and respiration of the leaves. However, such treatments have extremely limited significance because leaves are not so active in respiration as some other organs.

The leaf which is photosynthesizing actively is not the organ which is growing actively. The photosynthetic products of actively photosynthesizing leaves are translocated to growing

organs and are partially consumed there by respiration to generate the energy needed to synthesize substances by which the
growing organs are composed. In this sense it is important to
have data for the respiratory rate of various organs, especially
of growing organs.

Figure 3 illustrates the rates of photosynthesis and respiration at stratified heights of a maize population during ripening
(15). The photosynthetic and respiratory rates are most active

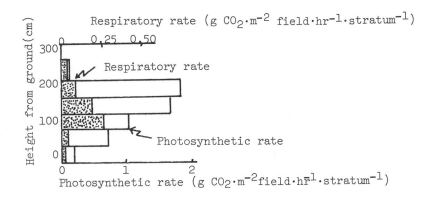

*Fig. 3. Photosynthetic rate and respiratory rate at
various strata of a maize population during grain-
filling (Fukko No.8, 1969).*

at the top of the canopy and at the middle height, respectively.
This means that active photosynthesis and respiration are occurring at different positions: Photosynthesis and respiration are
active at the leaves which are receiving strong light and at the
rapidly growing seeds, respectively.

Figure 4 illustrates the respiration of various organs of a
plant as proportions of the total respiration at successive
stages of growth (20). The sheath-and-culm in this figure includes young growing leaves and panicles which are not yet ex-
serted from the leaf-sheath at lower positions. The leaf-blades
are not the organs which respire most abundantly. In the early
stages of growth the respiration of the roots occupies quite a
large portion. During this stage the respiratory rate of the
roots per unit dry weight is high because the roots are growing
actively and also absorbing nutrients rapidly. After the initi-
ation of panicle premordia, the respiration of the sheath-and-
culm occupies a large portion due to rapid growth of the panicle
primordia and the culm. After flowering, the respiration of the
panicles occupies a large portion because the seeds are the organs growing most actively.

The rates of apparent photosynthesis and respiration and the
CO_2 compensation point (CCP) of various organs were measured on

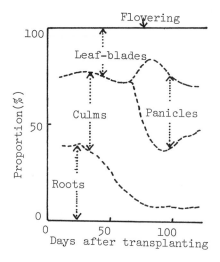

Fig. 4. Respiration of various organs of rice plants as proportions of total respiration at successive stages of growth.

a maize plant at 50 days after sowing (Table 1) (12). In the leaf-blade of the 11th leaf which had completed elongation, photosynthesis as well as respiration were active and the CCP was 10 ppm CO_2. On the other hand, in the culm the apparent photosynthetic rate was negative and there was no CCP (the CO_2 concentration of the ambient air of a closed system increased beyond 300 ppm CO_2). However, the CCP of the plant was as low as 10 ppm CO_2, if it is measured by enclosing the whole plant in a closed system. This means that in the C_4 plants the CCP of a whole plant is very low even if active respiration is going on at the organs which are growing rapidly.

TABLE 1. *Rates of apparent photosynthesis and respiration and CCP of various organs of a 50-day-old maize plant.*

Organs	Apparent[a] photosynthesis	Dark respiration	CCP[a] (ppm CO_2)
	$(mgCO_2 \cdot g^{-1} \cdot hr^{-1})$		
Leaf-blade of 16th leaf[b]	33.7	5.3	25
Leaf-blade of 11th leaf[c]	80.7	6.5	10
Leaf-sheath of 12th leaf[b]	−0.3	3.3	above 300
Leaf-sheath of 8th leaf[c]	2.8	3.1	150
Culm	−1.6	2.9	above 300

a) Measured at 30 Klux, b) elongating rapidly, c) completed elongation.

The heterogeneity of the organs of a plant should be recognized. A plant is composed of organs which have differing

characteristics, and most active respiration goes on at the actively growing organs where plant materials are rapidly synthesized from primary photosynthetic products which are sent from
actively photosynthesizing organs. From the agronomic standpoint, the photosynthesis or respiration of a whole plant is more
important than for a single leaf.

Another point to be added here is that even an organ which is
not growing, and which is not performing any positive function
for the growth of the plant, is respiring. In other words, there
is maintenance respiration. It is difficult to measure the maintenance respiration of plants. However, to get an estimate of
the value, a maize plant growing under a normal condition was
shifted to a dark room, and the elongation of the newly-coming-
out leaves and the respiratory rate of the plant were measured
every day. The elongation of new leaves slowed down gradually
and stopped 5 days after shifting the plant to the dark room.
The respiratory rate also decreased gradually, reached about 1
mg $CO_2 \cdot g$ dry matter$^{-1} \cdot$hr^{-1} when the elongation stopped, and this
value was maintained for several days. From these data it was
estimated that the maintenance respiration was about 1.5% of dry
matter per day (7). This method of estimation of the maintenance
respiration is not at all logical, but the value thus obtained is
similar to a reported value estimated by a completely different
method (5).

In any case maintenance respiration exists. Idle organs which
make no contribution to production, such as excessively elongated
internodes at the base of a culm or lower leaves which do not
photosynthesize due to the shade produced by the upper leaves,
etc., consume photosynthetic products by maintenance respiration.
Thus, these organs are parasitic.

GROWTH EFFICIENCY

Seedlings growing in the dark: The growth of seedlings germinating in the dark is supported by the substances in the seeds.
This means that the amount of seedling grown(W) is produced from
the amount of seed substances which are consumed during the
growth of the seedling(S). As defined earlier, the GE can be
calculated as W/S.

When rice seeds were germinated at temperatures between 20°C
and 30°C, the growth rate of the seedlings was higher at higher
temperatures than at lower temperatures, but the GE remained at
about 0.60-0.65 when the seedlings were growing actively by rapidly consuming substances in the seeds. The GE decreased rapidly, however, when the growth became slow due to extinction of the
substances in the seeds (Fig. 5) (7). It was also demonstrated
that the GE was lower than 0.60 when germination temperature was
at 15°C or 40°C even if there were sufficient substances in the
seeds. Repeated measurements showed that the GE's of seedlings

of soybeans, maize and rice germinating in the dark at temperatures within a normal range were about 0.73, 0.64 and 0.62, respectively (14).

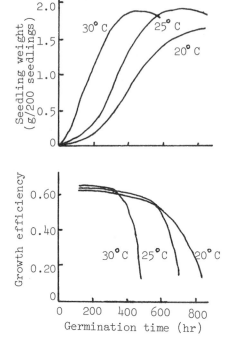

Fig. 5. Growth and growth efficiency of rice seedlings grown in the dark at various temperatures.

This means that with a given composition of substances for growth the GE remains constant when the temperature is within a normal range, although the growth rate is noticeably influenced by temperature within this range. On the other hand, the GE is influenced by the composition of the seeds: In seeds high in lipids, such as soybeans, the GE is higher than in seeds high in carbohydrates, such as rice.

Plants growing in the light: The growth of plants in the light is supported by primary photosynthetic products. Thus, $GE = W/P = W/(W+R)$. From this equation, the GE was estimated by measuring the rates of dry matter production and respiration.

Figure 6 is an example of the GE of maize plants at successive stages of growth (17). The data on rice (20) or soybean (18) followed a similar pattern. From these data it can be said that the GE during vegetative growth is 0.60-0.65 for all crops tested.

When maize plants were grown under graded amounts of solar energy, the growth rate decreased with a decrease of the solar energy, but the GE remained almost constant between 300 and 170 $cal \cdot cm^{-2} \cdot day^{-1}$ and decreased significantly when the solar energy fell below 100 cal·cm-2·day-1 (Fig. 7) (19).

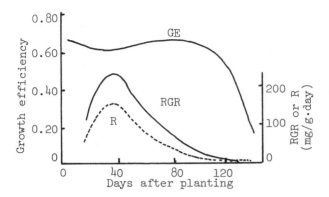

Fig. 6. Relative growth rate(RGR), respiratory rate(R) and growth efficiency(GE) of maize plants at successive stages of growth (respiratory rate is expressed as CH_2O).

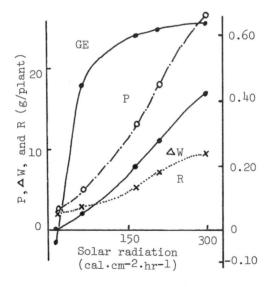

Fig. 7. Amounts of photosynthesis(P), respiration(R) and dry matter production ($\triangle W$) and growth efficiency of maize plant at graded amounts of solar energy.

Thus, it can be concluded that in plants growing rapidly by their own photosynthetic products, out of the carbon originally taken into primary photosynthetic products (such as glucose) 35-40% is respired, and the rest is converted into stable constituents by using the energy generated by the respiration, at least during vegetative growth. Because of the rigid coupling between growth, respiration and photosynthesis, the GE is maintained at a constant level if the growth conditions are within a normal range. The rates of growth, respiration and photosynthesis are influenced by the temperature and light intensity, but the GE is kept constant by the coupling. However, if these environmental factors go outside a normal range, the GE becomes smaller because the coupling is impaired and the uncoupled respiration or maintenance respiration occupies a large portion of the total respiration.

At later stages of growth, however, the GE frequently goes below 0.6 due to (a) the retranslocation of stored substances from the vegetative organs to the seeds during ripening, (b) the maintenance respiration of excessively elongated internodes and mutually shaded lower leaves, etc.

Ripening seeds: During ripening, the growth and respiration of the ripening seeds are active, and these are supported by substances which mostly come from other organs. By measuring the rates of growth and respiration, the GE of the ripening seeds of rice, maize and soybean were estimated. As it is difficult to measure the respiratory rate of seeds alone, measurement were made, in practice, on panicles, ears or pods. As shown in Fig.8,

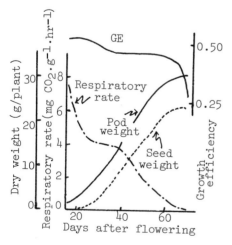

Fig. 8. Weight, respiratory rate, and growth efficiency of soybean seeds (pods) at successive stages of ripening.

the GE fluctuated to some extent (18), but it was possible to estimate the value during the whole ripening period.

From repeated measurements the values for rice, maize and soybean were estimated to be within the range of 0.63-0.73, 0.69-0.72, and 0.43-0.45, respectively. The GE is apparently lower in soybean than in rice or maize probably because the seeds are higher in lipids and proteins in soybean than in the others.

If it is assumed that (a) the seeds consist of carbohydrates, lipids, proteins and ash, (b) glucose and glutamate are the only organic substances which are translocated into the ripening seeds, (c) carbohydrates and lipids are derived from glucose, and proteins from glutamate and glucose, and (d) glutamate is synthesized from NH_4^+ and glucose by the mechanism given in the standard textbooks in vegetative organs, it is possible to calculate the conversion efficiency of each component of the seeds from data of the GE of ripening seeds and their composition. The conversion efficiency means here the weight of products derived from a unit weight of raw materials, which is a similar expression to the growth efficiency in the case of dry matter. The production value which was defined biochemically by Penning de Vries (14) is the equivalent value of the conversion efficiency.

By using the data obtained from many measurements on rice, maize and soybean, the conversion efficiencies of each component were computed by the method of least squares. The values for carbohydrates, lipids and proteins turned out to be 0.84, 0.31 and 0.38, respectively (18). If the conversion efficiencies are expressed in terms of calories instead of weight, the values are 0.94, 0.77 and 0.57, respectively.

From the above discussion it can be said that a comparison of yields among crops in terms of dry weight is not fair when the chemical compositon is very different from crop to crop. By using values for the chemical compositon of products and the conversion efficiency of each component, it is possible to work out the equivalent yield of various crops with different compositions at the same efficiency of solar energy utilization. In Table 2 examples of this computation are given, taking rice as a standard. As expected soybean has a lower value because it is higher in proteins and lipids than other cereal crops.

TABLE 2. *Equivalent yield and world average yield of various grain crops.*

Crop species		Rice	Wheat	Maize	Field beans	Soybean
Equivalent yield[a]		1.00	0.97	0.96	0.86	0.62
World average yield	(ton/ha)[b]	2.39	1.70	2.81	0.49	1.42
	(Relative value)	(100)	(71)	(117)	(21)	(59)

a) Relative yield at same efficiency of energy utilization taking rice as the standard (See: text).
b) From FAO Production Year Book.

RATES OF PHOTOSYNTHESIS AND RESPIRATION AS
AFFECTED BY THE SOURCE-SINK INTERACTION

During the ripening of grain crops the source is the leaves
and the sink is the seeds. The source-sink interaction is complicated and plays an important role in determining the yield.

Maize plants were grown, and three treatments were differentiated at the silk stage; viz. (a) control, (b) ears removed and
(c) top five leaves kept intact and lower leaves removed. The
net assimilation rates(NAR) during ripening were 4.77, 1.54 and
5.52 $g.m^{-2}.day^{-1}$ for (a), (b) and (c) treatments, respectively
(9). This means that the NAR decreases when the sink is reduced,
and increases when the source is reduced.

Figure 2 indicates that the rate of photosynthesis of the
leaves which were sending photosynthetic products to the ear
(11th and 16th leaves) became active when the seeds were growing
rapidly. This means that photosynthesis is accelerated when the
sink is active.

Such phenomena have been reported repeatedly and are frequently explained as follows: The photosynthetic rate is controlled
by the source-sink interaction. When the capacity of the sink is
large, the photosynthetic rate of the source is accelerated, and
when the capacity of the source exceeds the capacity of the sink,
it is retarded, because the photosynthetic rate is controlled by
the rate of removal of photosynthetic products from the site.

However, it should be remembered that the NAR is controlled
not only by the rate of photosynthesis, but also by the rate of
respiration: A decrease in the NAR may result either from a
decrease in the rate of photosynthesis or from an increase in
the rate of uncoupled respiration.

Field bean plants (a determinate type) were grown, and two
treatments were differentiated at full flowering; viz. (a) control and (b) all pods were removed and all buds which developed
later due to the removal of the pods were also removed. The
plants with these two treatments had almost identical leaf areas
during ripening because no leaves developed after the treatments
were differentiated. In the control treatment the photosynthetic
rate of the leaves was high when ripening was active and decreased rapidly to the end of growth, and the respiratory rate
of the leaves decreased gradually with growth (Fig. 9) (11). In
the pod removal treatment the photosynthetic rate was lower than
the control for about 15 days after the treatment was given due
most probably to a decrease of the sink. However, during later
growth stages the photosynthetic rate remained higher than the
control due to a higher nitrogen content of the leaves in this
treatment than in the control (Table 3). In the control the
nitrogen content of the leaves decreased because nitrogen in the
leaves translocated to the seeds. On the other hand, in the pod
removal treatment the nitrogen content remained higher because
there was no translocation of nitrogen. The respiratory rate of

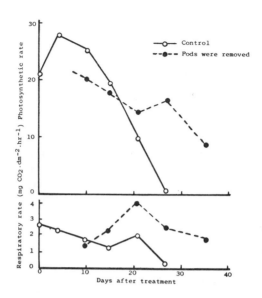

Fig. 9. Effect of removal of pods on rates of photosynthesis and respiration of the 3rd leaf.

TABLE 3. *Effect of removal of pods on dry weight and contents of nitrogen, sugars and starch of various organs at harvest in the field bean (Kintoki).*

Treatment	Control	Pods removed
Dry weight (g per plant)		
Pods and seeds	19.8	0
Leaves and stem	8.0	17.4
Roots	2.3	5.8
Total	30.1	23.2
N %		
Leaves and stem	1.11	2.47
Roots	2.18	2.12
Sugars %		
Leaves and stem	2.26	5.81
Roots	2.38	3.79
Starch %		
Leaves and stem	1.79	13.8
Roots	1.96	10.9

the leaves in this treatment was far higher than the control. This may be due to a higher content of nitrogen and sugars.   At harvest the total plant weight was far larger in the control than in the pod removal treatment because the weight of seeds was very large, although the weights of leaves-and-stem and roots were smaller. The dry matter production during ripening was 18.2 and 12.5 g in the control and the pod removal treatments, respectively, and the leaf areas of these treatments were almost identical. Thus, the NAR of the pod removal treatment was about 70% of the control. As the photosynthetic rate was more active at least in later growth stages, the smaller NAR value of the pod removal treatment was due to accelerated respiration caused by excessive accumulation of sugars.

From these phenomena it can be speculated that there is an uncoupled respiration which is not geared to growth when the source is in excess of the capacity of the sink. When uncoupled respiration constitutes a large portion of the total respiration, the growth efficiency becomes small.

PRACTICAL IMPLICATION OF FOREGOING DISCUSSIONS

In discussing agricultural problems the following statements are frequently cited in one way or another: The yield of rice is very high in Spain or Australia and is very low in tropical countries, the yield of soybean is lower than that of cereal crops like maize, and so on. As these statements are based on statistics compiled by reliable organizations, such as FAO, people are generally easily convinced by them. Although I have no doubt about the statistical basis for these data, it is difficult for me to accept these statements of their face value.

The average rice yields in Spain and Japan are 6.27 ton/ha and 6.02 ton/ha, respectively. Thus, one might conclude that technology of rice cultivation is slightly better in Spain than in Japan. However, Fig. 10, in which the seasonal changes of solar radiation at Tokyo and Seville are shown, indicates that a more satisfactory explanation is that the higher yields are more easily obtained when the solar energy flux is greater. Because of this, the productivity of crops should be expressed in terms of the amount of product per unit amount of solar energy flux rather than per unit of land area. In this expression the yield of rice in Japan is more than 1.5 times that of Spain.

The world average yields of rice and soybean are 2.39 ton/ha and 1.42 ton/ha, respectively. Thus, it might be argued that we should make more efforts to improve the productivity of soybean than of rice. However, Table 2 indicates that at the same efficiency of solar energy utilization the yield of soybean is 0.62 times of that of rice. On this basis, the productivities of these two crops are almost the same, and it is difficult to decide where the priority for research efforts lies.

In Table 2 the world average yields of various crops are given. From these values, along with the equivalent yield in

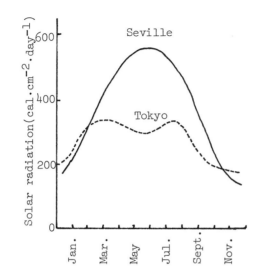

Fig. 10. Seasonal fluctuation of solar radiation at Tokyo and Seville.

the same table, it can be said that the present average yield is higher in maize than in the other crops, rice and soybean are at the same level, wheat is behind the other crops, and field beans are out of the running. The higher yield of maize could be due to a high efficiency of solar energy conversion based on a high rate of photosynthesis characteristic of a C_4 plant.

In discussing the productivity of agricultural land, it should be remembered that more than one crop can be grown on a piece of land during a year, if conditions allow. One crop of 10 ton/ha is inferior to three crops of 4 ton/ha. Therefore, the productivity of land should be expressed in terms of the amount of products per annum rather than per crop. With this expression the combination of crops to be planted in a field during the year becomes a very complicated problem, because the duration of the growth cycle differs among species or varieties, and there is seasonal fluctuation of solar radiation flux. Thus, the level of agricultural technology should be expressed in terms of the amount of energy utilized for the production of yield per unit amount of solar energy flux on a per annum basis.

What I intend to emphasize here is that a more rational expression of productivity is necessary for the scientific discussion of agricultural problems. Only by having such an expression it is possible to compare the levels of agricultural technology among countries with different climatic conditions, and the same is true of the level of improvements in breeding or agronomic technology among crops with different characteristics.

228   A. TANAKA

Instead of going into agricultural problems too deeply, let us now consider the improvement of the solar energy conversion of a crop. From the foregoing discussion it is clear that the source-sink interaction is extremely complicated and has a strong influence on the yield.

Generally speaking, when we are asked to improve the yield of a crop under given conditions, the first thing to do is to identify the yield limiting factor under those conditions. To improve the solar energy conversion it is first necessary, though very difficult, to identify which is the limiting factor, the source or the sink. I would like to say that as the productivity of a crop increases, the limiting factor swings in a zigzag between source and sink at each step (8). In the case of rice in southeast Asia in the early 1960's, the source was the yield limiting factor, and improvements in the photosynthetic facility through modifications of the plant type pushed up the potential yield considerably. But, an improvement in the photosynthetic facility does not necessarily result in an increased yield, because the source is not always the yield limiting factor. For example, it appears that in maize the sink is the limiting factor at least with presently available genetic materials (15).

REFERENCES

(1) JENNINGS, P. R. (1964): Plant type as a rice breeding objective. *Crop Sci.* 4:13-15.
(2) MONSI, M. and T. SAEKI (1953): Über dem Lichtfaktor in den Pflanzengensellshaften und seine Bedeutung für die Staffproduktion. *Jap. J. Bot.* 14:22-52.
(3) MURATA, Y. (1961): Studies on the photosynthesis of rice and its culture significance [in Japanese, English summary]. *Bull. Nat. Inst. Agr. Sci. (Jap.) Ser. D.* 9:1-169.
(4) PENNING de VRIES, F. W. T. (1972): Respiration and growth. *Crop Processes in Controlled Environments. etd.* A. R. REES et al. *Acad. Press.* London. 327-347.
(5) ROBSON, M. J. (1973): The growth and development of simulated swards of perennial ryegrass. II. Carbon assimilation and respiration in seedling sward. *Ann. Bot.* 37:501-518.
(6) TAKEDA, T. (1961): Studies on the photosynthesis and production of dry matter in the community of rice plants. *Jap. J. Bot.* 17:403-437.
(7) TANAKA, A. (1972): Efficiency of respiration. *Rice Breeding, Int. Rice Res. Inst.* 483-493.
(8) TANAKA, A. (1972): The relative importance of the source and the sink as the yield-limiting factors of rice. *Food & Fertilizer Tech. Center Tech. Bull. No. 6*, 18 p.
(9) TANAKA, A. and K. FUJITA (1971): Studies on the nutriophysiology of the corn plant. Part 7. Analysis of dry matter production from the source-sink concept [in Japanese]. *J. Sci. Soil and Manure, Japan* 42:152-156.

(10) TANAKA, A., K. KAWANO and J. YAMAGUCHI (1966): Photosynthe-
      sis, respiration, and plant type of the tropical rice
      plant. *Int. Rice Res. Inst. Tech. Bull. 7*, 46 p.
(11) TANAKA, A. and K. KIKUCHI (1976): Nutrio-physiological stu-
      dies on the field beans (*Phaseolus vulgaris L.*). Part IV.
      Effects of the removal of pods or leaves [in Japanese].
      J. Sci. Soil and Manure, Japan
(12) TANAKA, A., K. NAKABAYASHI and O. NAKAYAMA (1974): Nutrio-
      physiological studies on the photosynthetic rate of the
      leaf. Part 5. Photosynthetic rate and carbon-dioxide com-
      pensation point of rice and corn leaves [in Japanese]. *J.
      Sci. Soil and Manure, Japan 45:*513–516.
(13) TANAKA, A. and J. YAMAGUCHI (1968): The growth efficiency
      in relation to the growth of the rice plant. *Soil Sci.
      Plant Nutr. 14:*110–116.
(14) TANAKA, A. and J. YAMAGUCHI (1969): Studies on the growth
      efficiency of crop plants. Part 1. The growth efficiency
      during germinations in the dark [in Japanese]. *J. Sci. Soil
      and Manure, Japan. 40:*38–42.
(15) TANAKA, A. and J. YAMAGUCHI (1972): Dry matter production,
      yield components and grain yield of the maize plant. *J.
      Faculty Agr. Hokkaido Univ. 57:*71–132.
(16) TANAKA, A., J. YAMAGUCHI, Y. SHIMAZAKI and K. SHIBATA
      (1968): Historical changes viewed from plant type in rice
      varieties in Hokkaido [in Japsnese]. *J. Sci. Soil and
      Manure, Japan. 39:*526–534.
(17) YAMAGUCHI, J., T. HARA and A. TANAKA (1970): Studies on the
      growth efficiency of crop plants. Part 2. The growth effi-
      ciency of corn plants at successive stages of growth [in
      Japanese]. *J. Sci. Soil and Manure, Japan 41:*73–77.
(18) YAMAGUCHI, J., K. KAWACHI and A. TANAKA (1975): Studies on
      the growth efficiency of crop plants. Part 5. Growth ef-
      ficiency at successive growth stages and grain productivity
      of soybean in comparison with rice and maize [in Japanese].
      *J. Sci. Soil and Manure, Japan 46:*120–125.
(19) YAMAGUCHI, J. and A. TANAKA (1970): Studies on the growth
      efficiency of crop plants. Part 3. The growth efficiency of
      the corn plants as affected by growing condition [in Japa-
      nese]. *J. Sci. Soil and Manure, Japan. 41:*509–513.
(20) YAMAGUCHI, J., K. WATANABE and A. TANAKA (1975): Studies on
      the growth efficiency of crop plants. Part 4. Respiratory
      rate and growth efficiency of various organs of rice and
      maize [in Japanese]. *J. Sci. Soil and Manure, Japan. 46:*
      113–119.

INCREASING PHOTOSYNTHETIC CARBON DIOXIDE FIXATION BY THE BIOCHEMICAL AND GENETIC REGULATION OF PHOTORESPIRATION

ISRAEL ZELITCH, DAVID J. OLIVER, AND MARY B. BERLYN

Department of Biochemistry, The Connecticut Agricultural
Experiment Station, New Haven, Conn. 06504

An obvious method of increasing the efficiency of solar energy conversion by green plants would be to increase the efficiency of net CO_2 assimilation. The release of photo-respiratory CO_2 in C_3 species is so rapid under environmental conditions commonly encountered for photosynthesis that at least 50% of the net CO_2 assimilated is lost by photorespiration (1). The synthesis of glycolic acid and its further oxidation by reactions associated with the glycolate pathway of carbohydrate metabolism (Fig. 1) undoubtedly account for most of these losses (2).

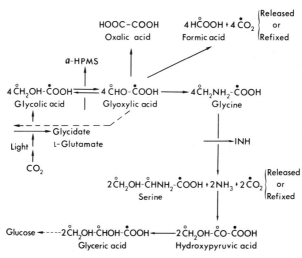

Fig. 1. Schematic diagram of the glycolate pathway of carbohydrate synthesis showing some regulators of the synthesis of glycolic acid and the likely sources of photorespiratory CO_2. (•) from C-1 of glycolate; (o) from C-2 of glycolate. α-Hydroxy-2-pyridinemethanesulfonic acid (α-HPMS) is shown blocking the glycolate oxidase reaction. This inhibitor causes glycolate to accumulate in leaves. Isonicotinic acid hydrazide (INH) inhibits the conversion of glycine to serine. Glycidate blocks the synthesis of glycolic acid. Evidence is presented here

that increasing the concentration of some common metabolites
such as glutamate or glyoxylate also inhibits glycolic acid
biosynthesis.

SOME PLANTS HAVE LOW PHOTORESPIRATION RATES

Several biochemical reactions for the biosynthesis of
glycolic acid are known to occur in green leaves (3,4), but the
relative contribution of each reaction to the total pool of
glycolic acid is unknown. However, it is well established that
photorespiration is slow and difficult to detect in C_4 species
(2), and that in maize leaves the rate of synthesis of glycolic
acid is only 10% as fast as in tobacco or sunflower (5). The
slow photorespiration and rapid photosynthesis in C_4 species
probably result from an inhibition *in vivo* of the enzymes that
catalyze glycolic acid synthesis and metabolism in the bundle
sheath cells (6,7).

Other examples of slower rates of photorespiration have
been found in nature besides those in the C_4 species. Kennedy
and Laetsch (8) found that leaves of carpetweed, *Mollugo*
verticillata, had morphological characteristics and photorespiration rates intermediate between C_3 and C_4 species. Brown
and Brown (9) similarly found the grass *Panicum miloides* has
decreased photorespiration. Zelitch and Day (10) showed that a
slow growing chlorophyll mutant of tobacco had faster photorespiration and lower rates of net photosynthesis in normal air
than its fast growing green sibling. By carrying out pedigree
selections on normal appearing tobacco plants, they obtained
about 25% of a population expressing slow photorespiration and
faster net photosynthesis (11). Thus it seems clear that
decreasing photorespiration need not have an adverse effect upon
net photosynthesis, and, on the contrary, the above examples
suggest the beneficial effect on photosynthetic capacity of
decreasing photorespiration.

PHOTOSYNTHESIS CAN BE INCREASED BY BLOCKING GLYCOLATE METABOLISM

There is also evidence from laboratory experiments that the
regulation of glycolic acid synthesis and oxidation can decrease
photorespiration and increase net photosynthesis. In order to be
metabolized, glycolate must first be oxidized to glyoxylate by a
reaction catalyzed by glycolate oxidase (Fig. 1). When an
effective α-hydroxysulfonate, an inhibitor of glycolate oxidase,
is added to illuminated tobacco leaf discs, glycolate oxidase is
blocked and glycolic acid accumulates at initial rates
sufficiently rapid to account for photorespiration (5). Under
suitable conditions these inhibitors also block photorespiration
in leaf discs and bring about large increases in net photosynthetic CO_2 uptake (12).

PHOTOSYNTHESIS CAN BE INCREASED BY BLOCKING GLYCOLATE
SYNTHESIS WITH GLYCIDATE

The turnover of glycolic acid in leaves is very rapid and its pool size is correspondingly small. To search for potential inhibitors of the synthesis of glycolic acid, illuminated tobacco leaf discs were floated for a preliminary period of one hour on solutions containing potential inhibitors. The solutions were removed, the discs washed with water, and then an effective α-hydroxysulfonate was added to block glycolate oxidase. The rate of glycolic acid accumulation in the next 3 minutes was then determined and compared with the rate for discs floated on water throughout the preliminary period (13). An inhibitor of glycolic acid synthesis would be expected to decrease the rate of glycolic acid accumulation in such an assay.

Since any compound that slows net photosynthesis, such as inhibitors of photosynthetic electron transport, would also block the synthesis of glycolic acid, three additional criteria were established to assure the biochemical specificity of an inhibitor. First, the inhibitory substance must block photo-respiration as determined by the assay of light-dependent $^{14}CO_2$ evolution in CO_2-free air (14). Second, it should not inhibit glycolate oxidase or cause glycolic acid to accumulate in the tissue. Third, and most important, under conditions similar to those used to inhibit glycolic acid synthesis, net photo-synthetic CO_2 uptake should be *increased* in the leaf tissue.

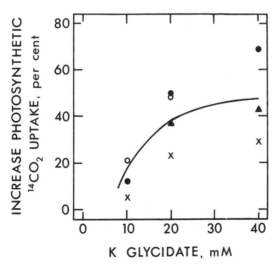

Fig. 2. *The effect of potassium glycidate concentration on the stimulation of* $^{14}CO_2$ *fixation by illuminated tobacco leaf discs* [*from (13)*].

Epoxide analogues of substrates were shown in several examples to react with carboxyl groups at the active sites of isolated enzymes (15, 16). Therefore epoxides similar in structure to glycolic acid were examined for their ability to block glycolic acid synthesis in tobacco leaf discs. When an inhibition was observed, the three additional criteria mentioned above were examined. By this procedure glycidic acid, 2,3-epoxy-propionic acid, was identified as a highly specific biochemical inhibitor of glycolic acid synthesis (13).

Floating leaf discs on 20 mM potassium glycidate caused a 40 to 50% inhibition of glycolic acid synthesis and a 40 to 50% increase in net photosynthetic CO_2 uptake (Fig. 2). Leaf discs previously treated with glycidate and then allowed to fix $^{14}CO_2$ showed large changes in the percentage distribution of ^{14}C in different fractions or compounds (Table I). Glycine and serine, products derived from glycolate metabolism (Fig. 1), decreased greatly in total radioactivity in response to glycidate treatment. This result is consistent with the observed inhibition of glycolic acid synthesis. Aspartate and glutamate increased two- to three-fold in the presence of glycidate.

Table I. Effect of Glycidate on the Percentage Distribution and $^{14}CO_2$ Fixation in Illuminated Tobacco Leaf Discs [from (13)]

Fraction or Compound	4 min in $^{14}CO_2$	
	Leaf Discs in Water	Leaf Discs in Glycidate
Neutral compounds	6.0	11.5
Aspartate	3.1	5.4
Glutamate	0.18	0.68
Phosphoglycerate	17.5	17.6
Phosphoglycolate	2.3	2.9
Fructose diphosphate, phospho-enolpyruvate, ribulose di-phosphate fraction	2.0	3.0
Glycolate	0.28	0.34
Glycerate	3.1	5.2
Glycine	15.3	7.6
Serine	22.8	6.8
Photosynthetic $^{14}CO_2$ fixation, μmoles per g fresh wt·hr	57.1	81.3 (+42%)

Leaf discs were floated on water or 20 mM K glycidate solution at 28 C and 2,000 ft-c for 60 min in air before $^{14}CO_2$ was supplied for 4 min.

GLYCIDATE DOES NOT DIRECTLY AFFECT RIBULOSE
DIPHOSPHATE CARBOXYLASE

We have found that glycidate does not affect the activity
of isolated ribulose diphosphate carboxylase/oxygenase (17, 18),
a reaction that may be responsible for some of the glycolic acid
produced during photosynthesis (19, 20). Similarly glycidate
did not affect the ribulose diphosphate carboxylase/oxygenase
activity in extracts from leaf discs previously treated with
glycidate, although glycolic acid synthesis in the discs was
inhibited over 50% (21). When $[1-^{14}C]$glycidate was supplied to
leaf discs, the radioactive glycidate predominantly labeled
proteins with molecular weights smaller than ribulose diphosphate carboxylase. Thus the synthesis of glycolic acid in
leaves could be greatly inhibited without directly affecting
ribulose diphosphate carboxylase activity or the Warburg effect
associated with the isolated enzyme.

Table II. Effect of Glycidate Treatment on the Activity of Some
Enzymes Extracted from Tobacco Leaf Discs

Enzyme Activity	Leaf Discs in Water	Leaf Discs in Glycidate
	μmoles per mg protein·hr	
Phosphoglycolate phosphatase	9.8	9.1 (-7.1%)
NADH-Glyoxylate reductase	158	164 (+3.8%)
NADPH-Glyoxylate reductase	15.9	13.2 (-17%)

Illuminated leaf discs were floated on water or 20 mM K
glycidate at 30 C for 2.5 hr. The leaf tissue was extracted and
the proteins collected for assay by means of column chromatography with Sephadex G-25.

The effect of treatment of leaf discs with glycidate on the
activities of phosphoglycolate phosphatase, NADH-glyoxylate
reductase, and NADPH-glyoxylate reductase was also investigated
(Table II). Only the NADPH-glyoxylate reductase actvitiy was
consistently decreased by glycidate additions. A time- and
glycidate concentration-dependent inhibition of about 30% could
be demonstrated with isolated NADPH-glyoxylate reductase. Thus
of these enzymes associated with glycolic acid synthesis, the
only direct inhibitory action of glycidate that could be shown
was a partial effect on the NADPH-glyoxylate reductase. These
results raised the question of whether the regulation of glycolic
acid synthesis exerted by glycidate was an indirect one.

GLYCOLIC ACID SYNTHESIS CAN BE INHIBITED
BY INCREASING THE CONCENTRATION OF NORMAL LEAF METABOLITES

It seems a remarkable omission in view of its importance
that to date little work has been carried out on the regulatory
aspects of the glycolate pathway in leaf tissue. There are perhaps two main reasons for this omission. First, a number of the
details such as the sites of CO_2 release and the mechanism of
glycolic acid biosynthesis are not fully resolved. Second, most
of the investigations have concentrated on the regulation of
isolated ribulose diphosphate carboxylase/oxygenase (20, 22, 23,
24), thus excluding from consideration other possible regulatory
sites in the pathway.

Investigators who assume that glycolic acid arises solely
from the oxygenase activity of ribulose diphosphate carboxylase
often assume that the oxygenase activity is a mandatory consequence of the enzyme's structure; hence a decrease of the
oxygenase activity is not possible without inhibiting the carboxylase activity (20, 22, 25). The assumption that the ratio of
carboxylase to oxygenase activity is fixed by the protein
chemistry of the enzyme suggests that glycolic acid synthesis is
not regulated by the rate of carbon flow through the rest of the
glycolate pathway. We believe instead that the glycolate pathway, like all important metabolic pathways, is probably subject
to feedback or other types of regulation. The elucidation of
control points in the glycolate pathway may enable us to obtain
biochemical mutants of higher plants with slower rates of
glycolic acid synthesis.

Table III. Effect of Prior Treatment of Leaf Discs with Various
Compounds on Inhibition of Glycolic Acid Synthesis

Compound Added (30 mM)	Effect on Rate of Glycolate Accumulation in Presence of the Sulfonate	
	Experiment 1	Experiment 2
	% change compared with discs on water	
L-Glutamate	-44	-35
L-Aspartate	-22	-18
Phosphoenolpyruvate		-24
Glyoxylate (20 mM)		-21
Glycine	-4	
α-Ketoglutarate	0	
L-Malate	0	
Pyruvate		0
3-Phosphoglycerate		0

Illuminated leaf discs were floated on solutions of the compounds shown (adjusted to pH 5.0 with KOH) for one hour at 30 C.
The fluid was then removed and replaced with water. α-Hydroxy-2-
pyridinemethanesulfonic acid solution (10 mM) was then added to

block glycolate oxidase, and the discs were killed after 3 min exposure to the sulfonate. Control discs floated on water accumulated glycolate at rates of (in μmoles per g fresh wt·hr) 36.9 in Experiment 1 and 71.2 in Experiment 2. Many other compounds tested had no effect.

We therefore investigated the effects of some common metabolites on the synthesis of glycolic acid. We used the conditions of the leaf disc assay in which the inhibitory effect of glycidate was discovered. Since the pool sizes of glutamate and aspartate increased in the presence of glycidate (Table I), these were among the first metabolites examined. Illuminated leaf discs were first floated on 30 mM solutions of the test metabolite, and glycolic acid accumulation was observed on subsequent treatment with the sulfonate. As shown in Table III, L-glutamate, L-aspartate, phosphoenolpyruvate, or glyoxylate inhibited the accumulation of glycolic acid.

Thus far the effect of added glutamate has been studied in greatest detail. Its effect as an inhibitor of glycolic acid synthesis in leaf discs is concentration- and time-dependent. Glutamate at 30 mM also inhibited photorespiration in the ^{14}C-assay (14), had no effect on glycolate oxidase activity, and stimulated photosynthetic CO_2 fixation 18 to 25% (Fig. 3). Thus glutamate fulfilled all the criteria expected of a specific inhibitor of glycolic acid synthesis as described previously.

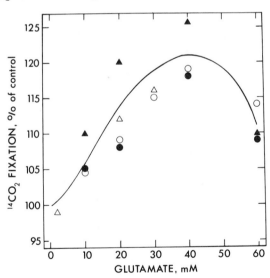

Fig. 3. The effect of L-glutamate concentration on the stimulation of photosynthetic $^{14}CO_2$ fixation by tobacco leaf discs. Leaf discs (1.6 cm in diameter) were floated on water or K glycidate, pH 5.0, at the concentrations indicated at 30 C in light for one hour. The fixation of $^{14}CO_2$ (600 μl/l passed con-

tinuously over the discs) was allowed to proceed for 5 min. The control rates (discs in water throughout) for the different experiments averaged 52 μmoles CO_2 per g fresh wt·hr.

After leaf discs were floated on 30 mM glutamate for one hour, the glutamate solution could be removed and the inhibition of glycolic acid synthesis (40%) remained unchanged for four further hours even though the total glutamic acid concentration in the tissue had returned to normal levels. Thus a catabolite of glutamate, rather than the amino acid itself, may be responsible for the regulation of glycolic acid synthesis.

Glutamate and aspartate are biochemically linked through the rapid transamination reaction involving oxaloacetate. Oxaloacetate can also be produced from phosphoenolpyruvate and CO_2 in the presence of phosphoenolpyruvate carboxylase. It is conceivable that the inhibition of glycolic acid synthesis and photorespiration that normally occurs in the bundle sheath cells of C_4 species is brought about in part by a regulatory mechanism similar to the one described here as a result of the higher concentration of C_4 acids found in these cells (26).

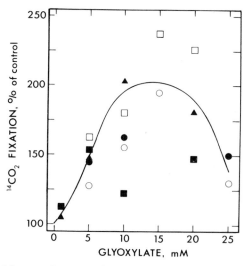

Fig. 4. The effect of potassium glyoxylate (pH 4.5) on the stimulation of photosynthetic $^{14}CO_2$ fixation by tobacco leaf discs. The discs were treated as described in Fig. 3. The control rates (discs in water throughout) for the different experiments averaged 46 μmoles per g fresh wt·hr.

Glyoxylate, however, may not equilibrate readily with the above substrates and probably regulates glycolic acid synthesis at a different site. We have observed that when tobacco leaf discs are floated on solutions of glyoxylate for one hour at 30 C, photorespiration in the ^{14}C-assay (14) is inhibited, and there is a concentration-dependent increase in photosynthetic CO_2 fixation (Fig. 4). Glyoxylate at 15 mM increased net photosynthesis about 100%. Thus we have demonstrated that photorespiration can be controlled and net photosynthesis increased by changing the concentration of some common metabolites. If stable changes in the intracellular concentration of such metabolites can be incorporated into higher plants, phenotypes showing large increases in net photosynthesis could be expected in species with rapid rates of photorespiration.

GENETIC ALTERATIONS IN CELL CULTURES SHOULD PROVIDE SLOW PHOTORESPIRATION PLANTS

Recent advances in somatic cell genetics in higher plants (27, 28, 29) show that procedures are available for obtaining biochemical mutants by screening large populations of mutagenized haploid plant cells under selective conditions that allow only cells with the desired phenotype to survive. Higher plants exhibiting the same biochemical phenotypes can then be obtained from such cells. The clarification of the control mechanisms involved in the photorespiratory pathway could therefore be exploited by devising rational means for selecting mutations which may block or bypass specific parts of the glycolate pathway (Fig. 1). The evidence presented above suggests that mutations affecting the synthesis, utilization, or accumulation of glutamate, aspartate, or phosphoenolpyruvate as well as glyoxylate and other compounds of the glycolate pathway would be important. We will discuss several examples of our work in progress utilizing inhibitors which may affect pool sizes of these compounds.

Haploid cell cultures of *Nicotiana tabacum* were grown in a standard liquid medium (30) with sucrose as a carbon source, treated with ultraviolet light as a mutagen, and after a short period of further growth were transferred to petri plates of the same medium with the addition of specific inhibitors. Resistant colonies which grew on these plates were transferred to nonselective medium (lacking inhibitor) and then rechallenged on selective medium. Pending genetic analysis of plants derived from these resistant cultures we shall refer to them as variant cell lines rather than mutants.

One group of these variants is resistant to glycine hydroxamate (a glycine analogue) at a concentration (2 mM) that is lethal to wild type cells. Experiments comparing the metabolism of [1-^{14}C]glycine and [1-^{14}C]glycolate in wild type and resistant cells have shown that glycine hydroxamate blocks the conversion of glycine to serine. This is confirmed by the finding in our laboratory that glycine hydroxamate is an effective competitive inhibitor of the condensation and decarboxylation of glycine in mitochondria isolated from tobacco plant cells (A. Lawyer, unpublished). Table IV shows that cells of one variant resistant to glycine hydroxamate (UGH 20) contain a much smaller pool of ^{14}C-glycine when the cells metabolize [1-^{14}C]glycolate. They also show differences in the ^{14}C-glycine/^{14}C-serine ratio compared to wild type cells. These results suggest that the metabolism of the variant cells has been altered with respect to the glycine-serine interconversion (Fig. 1).

Table IV. Metabolism of [1-^{14}C]Glycolate by Wild Type and Glycine Hydroxamate-Resistant Tobacco Callus Cells

Pool Size	Wild Type Cells		Glycine Hydroxamate-Resistant Cells	
	No GH	With GH	No GH	With GH
^{14}C-Glycine/ ^{14}C-Metabolized	0.11	0.15	0.02	0.09
^{14}C-Serine/ ^{14}C-Metabolized	0.18	0.20	0.07	0.04

Wild type and glycine hydroxamate-resistant cells (UGH 20) metabolized 10 mM [1-^{14}C]glycolate for one hour at 28 C in the absence or presence of glycine hydroxamate (GH) at 5 mM.

We have selected cell line variants resistant to isonico-tinic acid hydrazide, an inhibitor which also affects the glycine to serine conversion (31) and glycolic acid synthesis (5). In addition we have obtained variant cells resistant to an analogue of aspartate, α-methylaspartate (32). The variant cell lines are at different stages of testing, biochemical analysis, and regeneration into fertile plants.

The availability of photoautotrophic cells permits the selection of other types of photosynthetic mutants. We have previously shown that haploid tobacco cells can readily be grown photoautotrophically in a shallow liquid medium lacking sucrose in air enriched with 1% CO_2 (33). Based on the observation that tobacco seedlings with higher than usual rates of photorespira-tion are selectively killed in a 60% oxygen atmosphere(34), we are isolating variants selected from autotrophically grown cells that are resistant to prolonged exposure to 60% oxygen in the atmosphere.

In order to further our understanding of photosynthesis in cell cultures and to establish procedures for analyzing variants, we are examining net photosynthetic CO_2 fixation, photo-respiration, glycolate synthesis in the presence of sulfonate, and oxygen effects on photoautotrophically grown callus tissue. Autotrophic cells used for these studies were obtained from cultures grown on petri plates, where the previously described sucrose-free medium had been solidified with 1% agar (33). The cells were illuminated and incubated in transparent acrylic cabinets while continuously gassed with 1% CO_2 in air. Dry weight increases during successive passages were somewhat variable, ranging from two- to six-fold per three week growth period. Growth of one cell line is shown as an example in Table V. During the passages, the cells remained healthy looking, had rapid photosynthetic rates for callus cells (33), and grew well upon subsequent subculture. This small alteration in culture setup, making use of petri plates and solid medium, allows convenient maintenance of many individual autotrophic cultures and should facilitate the selection of variants from such cultures.

Table V. Photoautotrophic Growth of Tobacco Callus on Sucrose-Free Agar Medium in Successive Passages

Passage Number	Factor of Increase in Dry Wt per 3-Week Passage	
	Elevated CO_2 in Air	Air
2	2.0	1.3
4	4.2	0.8
5	5.5	
6	4.8	
6 (3% CO_2)	5.4	

One percent CO_2 in air was used in the first five passages. For the sixth passage, cultures were grown in 3% CO_2 in air as well as in 1% CO_2 in air.

Acknowledgements- This work was supported in part by a National Science Foundation Energy-Related postdoctoral fellowship to David J. Oliver and by a grant from the Rockefeller Foundation. Technical assistance by Pamela Beaudette and Cindy Spoor is gratefully acknowledged.

REFERENCES

1. Zelitch, I. (1975) Annu. Rev. Biochem., 44, 123-145.
2. Zelitch, I. (1971) *Photosynthesis, Photorespiration, and Plant Productivity.* Academic Press, New York.
3. Zelitch, I. (1975) Science, 188, 626-633.
4. Kelly, G. J., Latzko, E., and Gibbs, M. (1976) Annu. Rev. Plant Physiol., 27, 181-205.
5. Zelitch, I. (1973) Plant Physiol., 51, 299-305.
6. Black, C.C., Jr. (1973) Annu. Rev. Plant Physiol., 24, 253-286.
7. Chollet, R. (1974) Arch. Biochem. Biophys., 163, 521-529.
8. Kennedy, R.A. and Laetsch, W.M. (1974) Science, 184, 1087-1089.
9. Brown, R.H. and Brown, W.V. (1975) Crop Sci., 15, 681-685.
10. Zelitch, I. and Day, P.R. (1968) Plant Physiol., 43, 1838-1844.

11. Zelitch, I. and Day, P.R.   (1973) Plant Physiol., 52, 33-37.
12. Zelitch, I.   (1966) Plant Physiol., 41, 1623-1631.
13. Zelitch, I.   (1974) Arch. Biochem. Biophys., 163, 367-377.
14. Zelitch, I.   (1968) Plant Physiol., 43, 1829-1837.
15. Tang, J.   (1971) J. Biol. Chem., 246, 4510-4517.
16. Schray, K.J., O'Connell, E.L., and Rose, I.A.   (1973) J. Biol. Chem., 248, 2214-2218.
17. Bowes, G.W., Ogren, W.L., and Hageman, R.H.   (1971) Biochem. Biophys. Res. Commun., 45, 716-722.
18. Andrews, T.J., Lorimer, G.H., and Tolbert, N.E.   (1971) Biochemistry, 12, 11-18.
19. Bassham, J.A. and Kirk, M.   (1973) Plant Physiol., 52, 407-411.
20. Laing, W.A., Ogren, W.L., and Hageman, R.H.   (1974) Plant Physiol., 54, 678-685.
21. Zelitch, I.   (1976) In *CO₂ Metabolism and Plant Productivity* (Burris, R.H. and Black, C.C.,Jr., eds). University Park Press, Baltimore.
22. Chollet, R. and Anderson, L.L.   (1976) Arch. Biochem. Biophys., 176, 344-351.
23. Lorimer, G.H., Badger, M.R., and Andrews, T.J.   (1976) Biochemistry, 15, 529-536.
24. Ryan, F.J. and Tolbert, N.E.   (1975) J. Biol. Chem., 250, 4234-4238.
25. Lorimer, G.H. and Andrews, T.J.   (1973) Nature, 243, 359.
26. Hatch, M.D.   (1971) Biochem. J., 125, 425-432.
27. Carlson, P.S.   (1973) Proc. Natl. Acad. Sci. U.S.A., 70, 598-602.
28. Chaleff, R.S. and Carlson, P.S.   (1974) Annu. Rev. Genetics, 8, 267-278.
29. Carlson, P.S. and Polacco, J.C.   (1975) Science, 188, 622-625.
30. Linsmaier, E.M. and Skoog, F.   (1965) Physiol. Plant., 28, 100-127.
31. Pritchard, G.G., Whittingham, C.P., and Griffin, W.J.   (1963) J. Exptl. Bot., 14, 281-289.
32. Woolley, D.W.   (1960) J. Biol. Chem., 235, 3238-3241.
33. Berlyn, M.B. and Zelitch, I.   (1975) Plant Physiol., 56, 752-756.
34. Heichel, G.H.   (1973) Plant Physiol., 51 S, 42.

REGULATION OF PRODUCTS OF PHOTOSYNTHESIS
BY PHOTORESPIRATION AND REDUCTION OF CARBON *

N. E. TOLBERT
Department of Biochemistry,
Michigan State University,
East Lansing, Michigan, 48824

The ever intensifying shortage of oil and gas means that our other reserve of ancient photosynthate, coal, will have to be used to a greater degree until it too is gone. Instead, this last great reserve of reduced carbon should be carefully protected as a source of chemicals in the future rather than used now for fuel. Alternate sources of energy other than fossil or present day photosynthate must ultimately be developed. In the past decades we have been using the photosynthate produced over eons of time. The present total photosynthetic capacity in the United States is probably about equal to our present fossil fuel consumption, but it is needed for food and fiber. In the not too distant future we will have to produce biomass (plants) for food, fiber, and chemicals at the rate at which they are used, but our photosynthetic capacity as reduced carbon products is not likely to be sufficient for our fuel needs also.

PHOTOSYNTHETIC ASSIMILATORY POWER

In many parts of the world plant growth during the growing season is limited by the low CO_2 concentration of air. Light intensity is in excess. Other limiting factors such as water and nutrients can be partially corrected, but so far no practical way to increase the effective CO_2 concentration has been conceived. For most major crops and forests the situation is one of an excess photosynthetic assimilatory power over that needed to reduce the available CO_2 to carbohydrate. As a consequence a considerable part of the available biochemical energy of photosynthetic electron transport is used for a variety of other metabolic processes, some necessary for growth and some not, or the excess energy is simply wasted by respiratory processes. In this presentation some of these processes are to be cited.

The light absorption and electron transport during photosynthesis are integral parts of the chloroplast grana. In full sunlight most of these light traps are closed because there are insufficient electron acceptors. Otherwise photosynthetic electron transport is not highly regulated or limited. Regula-

*Support by research funds from the Michigan Agricultural Experiment Station, from NSF grant GB 32040X, and from Union Carbide Corporation.
Abbreviations: RuDP for ribulose diphosphate; PS I and PS II for photosystem I and II.

tion of the use of the photosynthetic assimilatory power that is produced in excess determines plant growth and composition. "Photosynthetic assimilatory power" is used to mean the reducing capacity of PS I and the ATP formed in that reduction. The reducing capacity may be reduced P_{430}, reduced Fd, or reduced NADP. The term photosynthetic assimilatory power was used by Arnon (1) for ATP and NADPH production. It has since been used also for the total gross reductive and synthetic capacity of the electron transport part of photosynthesis. Part of the assimilatory power may be used for processes other than NADP reduction, such as H_2 production or N_2 fixation. Likewise the use of the total assimilatory power may come in part after NADP reduction, as in glycolate biosynthesis by RuDP oxygenase, or prior to NADP reduction, as in glycolate biosynthesis by H_2O_2 oxidation of the thiamine pyrophosphate C_2-complex of transketolase.

LIMITATIONS ON PHOTOSYNTHESIS IMPOSED BY CO_2, O_2 CONCENTRATIONS

Photosynthesis by higher plants in air is a self-limiting process, from removal of one of its substrates, CO_2, and genera- tion of its own inhibitor, high O_2 concentrations. Net photosyn- thesis is limited by the available CO_2 in the air. The old photosynthetic literature illustrates this as a principle of a limiting factor. For C_3 plants net photosynthesis in air (0.033% CO_2) saturates around 1/4 to 1/3 full sunlight intensity, if other factors are not limiting. If CO_2 availability to the plant could be increased, net photosynthetic rates with avail- able sunlight can be increased 1 to 2 fold. With C_3 plants man has only accomplished this experimentally or by CO_2 gassing of commercial greenhouses. In the C_4 plants nature has achieved this by the addition of the C_4 dicarboxylic acid cycle to more efficiently trap the limiting CO_2.

In recent decades the amount of CO_2 in the air has been increasing from about 290 ppm to our 1975 level of about 331 ppm, presumably from greatly increased consumption of fossil fuels. This is being viewed as a serious environmental threat with further increases anticipated. However, all is not gloom and doom; plants will save us. Significant increased rates of photosynthesis per unit of plant has occurred and will continue to rise with increase in this indispensible limiting component. If it were not for the fact that the CO_2 concentration is the limiting factor in plant growth, the CO_2 content of the air would have shot up to unmanageable levels. Conferences on this subject should be held in which the environmentalist and the research investigators in photosynthesis can interact, rather than each group meeting separately as is now occurring. The paradox is that increasing the CO_2 in air is beneficial to photosynthesis and biomass production, yet too much of an increase in CO_2 will upset our temperature balance.

Photosynthesis in higher plants is also self-limiting by the high atmospheric content of O_2 which was created by the process itself. During electron transport PS II forms O_2 and PS I forms reduced components; these two products may self destruct by auto-oxidation. Partial prevention of this paradox is a very complex business. Over the years several terms have been used for phenomena associated with this situation. Warburg (32) demonstrated in 1920 that O_2 inhibited net photosynthesis and the phenomenon has been called the "Warburg Effect" (7, 30). In a series of papers in 1951, Mehler (16) studied O_2 uptake and H_2O_2 production by isolated chloroplasts, which has since been interpreted to be caused by auto-oxidation of the reduced components of PS I. This has been called the "Mehler reaction" or sometimes "pseudo-cyclic" electron transport. In many investigations during recent years O_2 uptake and CO_2 release during photosynthesis has been termed "photorespiration." The three terms are related if not similar, but the terms have confused investigators and evaluators alike. The term photorespiration is now most widely used and may serve as a general one for several reactions associated with photosynthesis and the production of excess reducing capacity in the presence of a high O_2 concentration. It is generally used for any respiration as O_2 uptake or CO_2 release that is associated with photosynthesis other than dark mitochondrial respiration of the tricarboxylic acid cycle. The term "Warburg Effect" is not descriptive nor specific. The Mehler reaction is better described as superoxide or H_2O_2 production as by Asada (2), or as O_2 uptake during photosynthesis as by Kok's group (20). Carbon metabolic processes which remove the photosynthetic reducing capacity by utilizing it for non-essential purposes may also be described by the specific reactions, such as the glycolate pathway. In this presentation I am using the term photorespiration for the general concept of reactions, which are associated with utilization of the photosynthetic assimilatory power but which otherwise do not seem essential for plant development.

PHOTORESPIRATION

After the discovery and development in the sixties of the concept of photorespiration as glycolate biosynthesis and metabolism, our dogma has not changed significantly since about 1971, at the time of several reviews and conferences on the subject (7, 8, 9, 27, 33). However, there remains a complete inability to regulate photorespiration in a practical way, which may be attributed to wrong hypotheses and gaps in our knowledge about it.

Photorespiration is the reduction of O_2 instead of CO_2 during photosynthesis. As examples there is O_2^- and H_2O_2 production and the synthesis of glycolate. Photorespiration disposes of excess reducing power from the chloroplast electron transport system. Ideally the photosynthetic capacity should all be used for CO_2 reduction and for other essential processes of plant growth which are directly or indirectly linked syntheses of essential components such as sugars, amino acids, fatty acids, etc. The use of excess photosynthetic capacity by reductive metabolic pathways which may or may not seem essential for plant growth will be classified as photosynthetic reduction of carbon, such as the production of glycerol and isoprene polymers. This concept also includes the biosynthesis of excess storage components such as hexose polymers and proteins which are eventually used for growth or by consumers of the plant. Consumers of plants might also use glycerol or isoprene polymers for fuel or chemicals if they were available in quantity. Therefore, photorespiration is restricted to processes which waste the photosynthetic capacity without the production of storage components. It is obvious that division of the total photosynthetic capacity among the various synthetic and respiratory processes will regulate growth and composition. How to regulate this division is the future research task for Plant Biochemistry. First we have to define all the various uses in the plant for the photosynthetic energy. The following discussion lists only those processes which may be classified as non-essential to plant growth, and not the well known essential syntheses.

Aerobic photorespiration during photosynthesis may be comparable in part to NADH oxidation during anaerobic glycolysis. For sustained ATP synthesis by substrate phosphorylation during glycolysis, it is necessary to remove the NADH. This may be accomplished by reduction of pyruvate to lactate or acetaldehyde to ethanol. Leaves do not do this for some reason, perhaps because they are always aerobic or do not have a mechanism of disposing or reclaiming the reduced carbon product. However, leaves are confronted by a somewhat similar situation during photosynthesis, namely an overproduction of NADPH and a need for more ATP synthesis. Cyclic photophosphorylation might account for excess ATP synthesis except that its occurrence in vivo is not well established. Photorespiratory wasting of the excess reducing capacity might be accomplishing the same purpose by allowing electron flow between PS II to PS I for more ATP synthesis. This concept has been used for pseudo-cyclic electron transport linked to the Mehler reaction.

PHOTORESPIRATION AS GLYCOLATE BIOSYNTHESIS AND METABOLISM

During photosynthesis in the chloroplasts CO_2 is reduced to sugars and during respiration the sugars are oxidized back to CO_2 while utilization of part of the energy for biological growth processes. This scheme has involved mitochondrial respiration or dark respiration for ATP synthesis in plants, animals, and bacteria. We are now expanding this scheme in plants to include reoxidation of the sugars by photorespiration. This other type of respiration only occurs in the light because it

is always associated in part with the assimilatory power of
the chloroplasts, although no direct photochemistry is involved.
Part of the oxidation associated with photorespiration also
occurs in the peroxisomes where the energy is lost as heat
rather than being converted to ATP and NADH. A similar type
of respiration also probably occurs in the ubiquitous animal
peroxisomes, but the details have not yet been elucidated.
There seem to be two respiratory processes associated with two
different subcellular organelles. One is energy conserving
during dark mitochondrial respiration and the other an apparently
energy wasting process during photorespiration in plants, which
is in part associated with peroxisomes. The mitochondria contain
the cytochrome electron transport system linked to ATP synthesis;
the peroxisomes contain the flavin linked oxidases which transfer
the electrons directly to O_2 to form H_2O_2, which is in turn
destroyed by the peroxisomal catalase. During the day this
photorespiration may be as much as half the rate of gross photosynthesis so that net photosynthesis is reduced by as much as
50%.
      In photorespiration as glycolate biosynthesis and metabolism,
the photosynthetic assimilatory power is wasted indirectly.
First the NADPH and ATP are used to reductively fix the CO_2 to
sugar phosphates. These newly formed sugar phosphates in turn
are reoxidized first to glycolic acid and then to CO_2. The
energy is lost as heat and CO_2 is regenerated for more photosynthetic CO_2 reduction. This futile process is like a clutch on
net photosynthetic sugar production. It has not yet been shown
to accomplish any other useful purpose, for the glycine and
serine produced in the cycle can be generated instead from
glycerate.

THE GLYCOLATE AND GLYCERATE PATHWAY

      The oxidative and irreversible biosynthesis and metabolism
of glycolate is a photorespiratory process. Glycolate is formed
in the chloroplasts only during photosynthesis and glycolate
is oxidized to glyoxylate and transaminated to glycine in the
peroxisomes (Fig. 1). Glycine–serine interconversion occurs in
the mitochondria. Details and many references are found in
other reviews (26, 27, 33). In the glycolate pathway there
are 5 physiologically irreversible enzymatic reactions: RuDP
oxygenase, P-glycolate phosphatase, glycolate oxidase, glutamate-
glyoxylate aminotransferase and serine glyoxylate aminotransferase. Energy is lost as heat during glycolate biosynthesis
and oxidation. O_2 uptake occurs during RuDP oxidation to
P-glycolate and glycolate oxidation to glyoxylate. CO_2 is
released during the conversion of 2 glycines to one serine in
the mitochondria. No energy is conserved on this glycolate
pathway of photorespiration except for one ATP in the conversion
of 2 glycine to 1 serine.

The glycerate pathway serves as an alternate parallel
pathway for serine and glycine formation from the photosynthetic
carbon cycle. This pathway is anaerobic, reversible and energy
is conserved as reduced NAD which can be shuttled out of the
peroxisomes for other reactions. However, P-glycerate is also
formed in the chloroplasts during photosynthesis followed by
P-glycerate phosphatase hydrolysis to glycerate. The conversion
of glycerate to serine occurs in the peroxisomes. These similarities suggest that whatever other function the glycolate
pathway performs besides wasting energy, the glycerate pathway
seems also capable of doing. The glycerate pathway is an
efficient pathway and whenever the glycolate pathway is inhibited,
as by low O_2 or light, the glycerate pathway is operational.
A balance between the two pathways provides a possibility for
metabolic regulation between photorespiration with energy loss
and serine synthesis from glycerate with energy conservation.

CO_2 RELEASE DURING PHOTORESPIRATION

The source of CO_2 released during photorespiration remains
controversial and there may well be several reactions involved.
Workers in the field agree that CO_2 release comes primarily
from the carboxyl group of glycolate and only much slower from
its alpha carbon. Other photorespiratory phenomenon may involve
O_2 uptake but not CO_2 release. The amount of CO_2 generated
can be greater than 25% of the carbon flow through the glycolate
pathway and may vary up to all of the glycolate carbon. In
the conversion of 2 glycine to one serine only one CO_2 is formed
representing 25% of the glycolate carbon. However, reconversion
of the serine to glycine and a C_1-tetrahydro folate complex
establishes a repeating system for converting all this carbon
to CO_2. The direct oxidation of glyoxylate to CO_2 either by
the strong oxidants in the chloroplasts or by H_2O_2 attack in
the peroxisomes have also been considered. More research on
these uncertainties is needed.

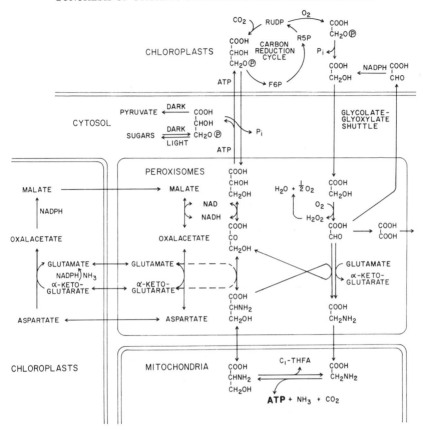

Fig. 1. Glycolate and glycerate pathways for glycine and serine formation.

GLYCOLATE BIOSYNTHESIS DURING PHOTOSYNTHESIS

Calvin's laboratory in 1950 observed glycolate biosynthesis during photosynthesis, but it was not an apparent component of the photosynthetic carbon cycle. Ever since investigators have tested hypotheses concerning its biosynthesis which may reach high rates comparable to photosynthesis. It is known that glycolate formation occurs in the chloroplasts, is light dependent and proportional to the O_2 concentration. A greater percentage of the total carbon is incorporated into glycolate at highest pH (above 8). All carbon atoms of glycolate and subsequent glycine and serine are equally labeled from labeled CO_2, but only one oxygen atom of the carboxyl group is labeled to the extent of 60% or more in 5 minutes by $^{18}O_2$ uptake during glycolate formation.

Data for proposals for glycolate formation have been reviewed (7, 28, 29, 33) and at this time two hypotheses are most prevalent. One is the oxidation by H_2O_2 of the thiamine pyrophosphate-C_2 complex formed in the transketolase reaction of the photosynthetic carbon cycle to free glycolic acid. This has been demonstrated with isolated chloroplasts (7). Otherwise this C_2 complex is transferred onto glyceraldehyde phosphate to form the precursor for RuDP, which is also easily oxidized and the products would be P-glycolate and P-glycerate. H_2O_2 generation as part of photorespiration is discussed in a subsequent section, and its oxidation of these sugar phosphates of the reductive photosynthetic CO_2 cycle to glycolate is a two-part hypothesis to explain photorespiration. H_2O_2 oxidation is an electron withdrawing reaction and ought not incorporate $^{18}O_2$ into the carboxyl group of glycolate. Therefore, at this time it is uncertain how much of the glycolate formed during photosynthesis in vivo could arise by this mechanism.

The other hypothesis is based upon the fact that RuDP carboxylase which catalyzes reaction 1 as the first step in CO_2 fixation also catalyzes a RuDP oxygenase cleavage of RuDP to P-glycolate and P-glycerate (reaction 2). In this enzymic reaction a partial carbanion at carbon #2 of the RuDP adds CO_2 or O_2, similarly to the reaction of these gases with a Grignard

1. RuDP $+ {}^{14}CO_2 \xrightarrow[\text{carboxylase}]{\text{RuDP}}$ 3-P-[1-^{14}C] glycerate + 3-P-glycerate.

2. RuDP $+ {}^{18}O_2 \xrightarrow[\text{oxygenase}]{\text{RuDP}}$ P-[1-^{18}O] glycolate + 3-P-glycerate + H_2^{18}O.

reagent, RMgX, to form either an acid with one more carbon or an alcohol. In the RuDP carboxylase/oxygenase reaction after addition of CO_2 or O_2 at the carbanion, an internal oxidation at carbon 3 and a reduction at carbon 2 occurs, which accounts for Calvin's dismutase term for the enzyme. In the case of the oxygenase reaction, during the dismutase cleavage, carbon #3 is oxidized to the carboxyl group of 3-P-glycerate from the lower 3 carbons of RuDP and the hypothetical peroxide intermediate on carbon #2 is reduced to a carboxyl group for P-glycolate. CO_2 and O_2 are competitive inhibitors in these two enzymatic reactions. The combination of carboxylase and oxygenase activities of this major protein fraction I of the chloroplasts seems to account for most of requirements known about glycolate biosynthesis.

COMBINATION OF THE PHOTOSYNTHETIC CARBON AND THE PHOTORESPIRATION CYCLE

A competition between the RuDP carboxylase and oxygenase reactions is dependent upon availability of CO_2 and O_2 and the enzymatic properties of this catalyst. A continuum of changing CO_2 and O_2 concentrations will produce changing rates of gross photosynthesis and photorespiration, the sum of which is net photosynthesis. As depicted in Fig. 2 the two reactions are balanced at the compensation point. No net hexose synthesis is occurring and all the fructose-6-P is converted to RuDP where part is oxidized to P-glycolate and then to CO_2. If more CO_2 or less O_2 were present excess fructose-6-P would be converted into storage carbohydrates. If even less CO_2 or more O_2 were present the continuous operation of oxygenase reaction generating the substrate for photorespiration pulls sugars from the storage reserves. Indeed continuous exposure of C_3 plants to high oxygen (over 40%) or CO_2 below the compensation point starves the plant to death by this photorespiration.

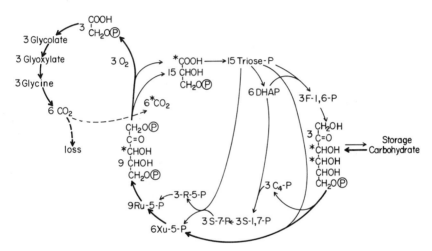

Fig. 2. Modification of the photosynthetic carbon cycle to include photorespiration. The heavy lines are for the postulated route of carbohydrate oxidation to P-glycolate and CO_2, when insufficient CO_2 is present. The asterisk indicates the predominate labeling with $^{14}CO_2$ from the air. The number of molecules are based upon the fixation of 6 CO_2 which produce one hexose by the reductive photosynthetic carbon cycle, if no photorespiration occurs. In this example all the CO_2 fixed is reoxidized to CO_2 via the glycolate pathway. From Ref. 29.

PROPERTIES OF RuDP CARBOXYLASE/OXYGENASE

The properties of this enzyme are critical, if this one enzyme controls the diversion of carbon flow between CO_2 fixation and glycolate synthesis. There are detailed recent reviews about this very large and complex enzyme (12, 15, 28, 29, 31), and there are laboratories working on its enzymatic properties, its biogenesis and its regulation. The K_m(RuDP), K_m(Mg^{++}), K_m(CO$_2$), K_m(O$_2$), and pH optima are similar for both the carboxylase and oxygenase reactions. Certainly research on this initiating enzyme must be continued to aid in ultimately regulating net photosynthesis.

Among the many present and future research programs on RuDP carboxylase/oxygenase are (a) characterization of the active sites for RuDP, CO_2 and O_2, (b) characterization of the activation sites which also involve CO_2 and Mg^{++}, (c) characterization of the several postulated effector sites for ATP, NADP, 6-P-gluconate, fructose bisphosphate, and ribose-5-P, (d) understand the changing K_m and V_m values, (e) determine the function of the small subunit, (f) study whether and how the carboxylase to oxygenase ratio may be varied, and (h) elucidate the molecular biogenesis of each subunit and the assembly of the holoprotein.

Recent advancements in 1976 by the people in our laboratory are studies which may lead to the elucidation of the nature of the substrate sites, which have not been described in the 23 years since the discovery of the enzyme. The CO_2 sites on the enzyme will form Schiff bases with pyridoxal 5'-phosphate (18). This compound and CO_2 compete for the same site on RuDP carboxylase, presumably an ε-amino group of a lysyl residue. There is also an apparent noncompetitive inhibition between pyridoxal 5'-phosphate and RuDP. The enzyme/pyridoxal 5'-phosphate complex has an extinction coefficient of 5800 ± 800 M^{-1} cm^{-1} at 432 nm. The titration of the amino groups with pyridoxal 5'-phosphate gives a biphasic double reciprocal plot, indicating 8-9 reactive sites at low pyridoxal 5'-phosphate concentration and another 8-9 reactive sites at high concentrations. Half of these amino groups were protected by RuDP. It is thought that 8 primary groups react with pyridoxal 5-phosphate, one at each of the 8 catalytic sites for formation of an enzyme/CO_2 complex. The second set of 8 primary amino groups which react with pyridoxal 5'-phosphate are postulated to be the CO_2 regulatory or activation sites. To account for the apparent noncompetitive inhibition with respect to RuDP, we postulate that the amine at the active site reacts with the carbonyl function of RuDP. This provides an explanation for the inactivation of RuDP carboxylase upon preincubation with RuDP in the absence of CO_2 and the inactivation of the enzyme at RuDP concentrations over 0.5 mM during the assay. Currently Dr. Paech in our group has succeeded in reducing the pyridoxal 5'-phosphate/enzyme complex with borohydride and is proceding with a study of the amino groups at this active center of the enzyme.

The RuDP substrate site is highly specific. However, xylitol diphosphate is a specific and potent inhibitor of the enzyme (21) because it binds at the RuDP site (unpublished). Thus both the two CO_2 binding sites and the RuDP site of the enzyme may now be more thoroughly investigated than thought possible only a year ago because specific inhibitors for each have been discovered.

REGULATION OF CARBOXYLASE/OXYGENASE ACTIVITIES

A major research target in this field is to determine whether the ratio of these two activities in vivo can be varied, and if so what causes this variation and how can this change be manipulated for controlling net photosynthesis. One school of thought is that the carboxylase/oxygenase ratio can not be varied, but that it is an inherent limitation of the primary carboxylation reaction in photosynthesis. In this view plant growth is ultimately limited by the composition of air which dictates a competition between CO_2 and O_2 for the two reactions catalyzed by this one enzyme (Fig. 2). Supportive of this argument is the fact that the enzyme from all photosynthetic tissues is both a carboxylase and an oxygenase with about the same 4 to 1 ratio of activities. Also it has been argued that the oxygen with its polar δ^+ end is simply attaching the δ^- on C_2 of the RuDP/enzyme complex without there being an O_2 binding site. As described in previous paragraphs there is now a defined CO_2 site, but no one has yet described an O_2 site.

There is, however, a contrary opinion and some hope that the RuDP carboxylase/oxygenase ratio of activities does vary in vivo. The ratio has been reported to vary from 1:1 in tobacco leaves to about 12:1 in marine sea weeds. A reevaluation of this ratio is needed with activated or low K_m form of the enzyme. All previous assays can be questioned, since they were run on homogenates or partial purified enzyme of varying degrees of activation (K_m values). There are abstract reports that the carboxylase/oxygenase ratio is high (10:1) in very young leaves and changes to a much lower value in mature or senescing tissue. These changes during development and senescence need to be carefully documented. The preliminary data so far suggests an increase in oxygenase activity in mature and senescing leaves.

A significant new development has been the demonstration that when the enzyme is activated by CO_2 and Mg^{++} at a site other than the substrate site, it has a low $K_m(CO_2)$ typical of the in vivo system (14). Different degrees of activation under previous assay conditions account in part for different K_m values and pH optima. We now routinely preactivate our enzyme for more consistent results. Under these conditions the broad pH optima for both the carboxylase and oxygenase are similar and around 8.1 to 8.3. Perhaps last year's terminology of low K_m form of the enzyme is being replaced by the term "CO_2 activated form."

Among the effectors of RuDP carboxylase/oxygenase are ATP, NADPH, fructose diphosphate, ribose-5-P and 6-P-gluconate, and recently Chollet et al. (4) have found that ribulose-5-phosphate is a powerful stimulator of the carboxylase at suboptimal bicarbonate concentrations. We had proposed that these effectors might regulate the carboxylase and oxygenase activity differently (29). A compound to preferentially inhibit the

oxygenase activity should stimulate plant growth because of
this inhibition of photorespiration. A reevaluation during
1976 indicates that differential regulation by these effectors
does not exist (4, Ryan et al., in manuscript). The degree of
regulation by an effector is pH dependent, and one must compare
the effector's activity on both the carboxylase and the oxygenase
at the same pH, and not at pH 7.8 for the carboxylase and 9.1
for the oxygenase, the pH optima of the high K_m or inactive
form of the enzyme. Plots of carboxylase activity as a function
of pH for three effectors (at 0.5 mM), fructose diphosphate,
ribose-5-P and 6-P-gluconate, indicate that each acted similarly
on both the carboxylase and oxygenase activities and that the
effect from each was very pH dependent.

FUTURE RESEARCH ON GLYCOLATE BIOSYNTHESIS AND METABOLISM

At this symposium are many speakers about photorespiration
and its relationship to net photosynthesis. This might
encourage one to think that maybe we understand this new
phenomenon well enough to apply the knowledge for regulating
net photosynthesis. However, those of us doing the research
know that is not true, although the principles may be correct.
There are many unsettled problems even for understanding the
glycolate pathway during photorespiration and some of these
are as follows.
1. The proposed different mechanisms for glycolate biosynthesis must be further studied to establish what portion
arises from each in vivo. There is no concensus among us on
this point.
2. Research on the regulation and control of glycolate
biosynthesis versus CO_2 fixation and reduction to sugars is
essential. These studies must range from enzymatic investigations on RuDP carboxylase/oxygenase to genetic and plant
breeding efforts to modify the process of glycolate formation.
3. The source of CO_2 released during photorespiration
remains controversial, and there well may be several reactions
involved.
4. The relationship of the glycolate pathway to glycerate
and sucrose synthesis and translocation needs to be elucidated.
Sucrose is synthesized from the intermediates of the reductive
photosynthetic carbon cycle. However, the amount of aerobic
sucrose synthesis is proportional to glycolate biosynthesis
and metabolism in some unknown manner.
5. Other functions for the glycolate pathway besides
disposing of excess energy may exist. As a general biological
principle most metabolic systems as large and complicated as
the glycolate pathway are interlinked with other processes.
As an example, glycolate excretion by algae or chloroplasts as
a membrane transport system needs to be investigated.

6. The glycolate and glycerate pathways have only been investigated in isolated peroxisomes from a C_3 plant, spinach. These subcellular organelles from leaves of C_4 and CAM plants should also be characterized.

7. Steps of the glycolate pathway in algae need further investigation. Photorespiration in algae differs in two major ways from that in the higher plant. Part of the glycolate is oxidized by a dehydrogenase rather than the peroxisomal flavin glycolate oxidase and part of the glycolate is excreted to the medium. The dehydrogenase system is inherently less wasteful of the photosynthetic assimilatory power.

REGULATION OF PHOTORESPIRATION IN C_4 PLANTS

When considering the control of net photosynthesis by regulating photorespiration, it is constructive to note how nature has done this in C_4 plants. A C_3 plant has a CO_2 compensation point of 45 to 75 ppm CO_2 which represents equilibrium between the rate of CO_2 fixation and CO_2 production by photorespiration. A C_4 plant can remove nearly all of the CO_2 from the air and its CO_2 compensation is near zero. Statements to the effect that C_4 plants have no photorespiration are incorrect. C_4 plants refix the CO_2 from photorespiration by the C_4 dicarboxylic acid cycle in the mesophyll cells before the CO_2 escapes from the leaf. As pointed out in the introduction, CO_2 is a limitation on photosynthesis and by more efficiently trapping the CO_2, the C_4 plant can utilize more of its total photosynthate for net photosynthesis. A C_4 plant has also reduced the magnitude of glycolate biosynthesis and photorespiration relative to a C_3 plant. This is achieved by locating the RuDP carboxylase/oxygenase and the reductive photosynthetic carbon cycle in the bundle sheath cells along with the peroxisomes. It appears that O_2 evolution occurs in the mesophyll cells and it may be somewhat less in the bundle sheath cells, but CO_2 is concentrated into the bundle sheath cells by the C_4 dicarboxylic acid cycle. This possible change in the relative concentration ratio of CO_2 to O_2 in the bundle sheath cells should favor the RuDP carboxylase reaction over the oxygenase reaction so that less glycolate is formed.

It is informative to note that nature over the years has not eliminated photorespiration in a C_4 plant by genetically modifying the RuDP carboxylase/oxygenase. Nature has only favored the carboxylase reaction over the oxygenase reaction by removal of the enzyme from a high O_2 environment. This suggests that reduction of photorespiration in C_3 plants will require genetic modification toward that of a dicarboxylic acid in C_4 plants, as discussed at these meetings by Dr. Zelitch. Efforts to modify photorespiration in C_3 plants by other means may not be successful short of changing the CO_2/O_2 ratio in the air, which we can not and ought not do.

PHOTORESPIRATION AS O_2 UPTAKE DURING PHOTOSYNTHESIS
AND O_2^- AND H_2O_2 FORMATION

As described in the previous sections O_2 uptake in the chloroplasts during glycolate biosynthesis directly competes with CO_2 reduction. In addition O_2 uptake by isolated chloroplasts occurs during auto-oxidation of reduced P_{430} and maybe reduced ferrodoxin components of PS I (Mehler reaction). From O_2 uptake experiments O_2^- and H_2O_2 production has been estimated to be in order of magnitude of photosynthesis. Superoxide dismutase of the chloroplasts detoxifies the extremely reactive O_2^- by converting it to H_2O_2. These reactions are a direct competition between O_2 and CO_2 for the photosynthetically generated reducing power. Many unresolved problems about this type of O_2 uptake remain. Should O_2 uptake by algae during photosynthesis, which was recently well demonstrated by Radmer and Kok (20), be considered as photorespiration? H_2O_2 must also be removed for it too inhibits photosynthesis (11). Should O_2^- and H_2O_2 generation and destruction be included in the term photorespiration?

Several possibilities for H_2O_2 removal probably occur simultaneously. (a) Since there is no catalase in the chloroplasts, rapid H_2O_2 diffusion from the chloroplasts down a steep concentration gradient to the peroxisomal catalase probably occurs where it would be dismutated to H_2O, $1/2$ O_2 and heat. Net O_2 gas exchange would be zero from O_2 uptake during O_2^- formation (equation 2) balanced by O_2 release at PS II (equation 1), and during H_2O_2 formation (equation 3) and destruction (equation 4). This combination of O_2 formation and reutilization during photosynthesis and photorespiration would not be detectable manometrically or by O_2 electrodes, but it can be studied by $^{18}O_2$ trace experiments. It represents a very futile cycle in which O_2 uptake wastes the photosynthetically generated reducing power. This by definition is photorespiration.

PS II	$2 H_2O \longrightarrow O_2 + 4 H^+ + 4 e^-$	1
PS I	$4 O_2 + 4 e^- \longrightarrow 4 O_2^-$	2
Superoxide dismutase	$4 O_2^- + 4 H^+ \longrightarrow 2 H_2O_2 + 2 O_2$	3
Peroxisomal catalase	$2 H_2O_2 \longrightarrow 2 H_2O + O_2$	4
Peroxidation	$2 H_2O_2 + 2 AH_2 \longrightarrow 4 H_2O + 2 A$	5
	$2 A + 4 e^- + 2 H_2O \longrightarrow 2 AH_2 + O_2$	6

(b) Reactive substrates AH_2 for H_2O_2 detoxification are present in the chloroplasts in large amounts, such as reduced

ascorbate and glutathione. After oxidation of the protectant
its reduction again could be linked to the photosynthetic assimilatory power. Surprisingly there is little data to support
this proposal in spite of the unknown function for a high ascorbate content in chloroplasts. An O_2 exchange balance can only
be proposed in equations 5 and 6 until more is known about
chloroplasts.

(c) Asada's group has also shown that H_2O_2 in isolated
chloroplasts will oxidize reduced plastocyanin, cytochrome f,
quinones, carotenoids and that Mn^{++} in chloroplasts destroys
O_2^- by a sequential oxidation to Mn^{+3} and $H2O_2$ which reacts to
regenerate Mn^{+2} and $O2$ (13). The oxidation of these components
of electron transport would be followed by further electron
transport to establish futile cycles, except for possible ATP
synthesis. Tight control over these possibilities would seem
to be necessary and, therefore, they may not even be allowed
in vivo.

(d) H_2O_2 may also oxidize one or more of the sugar phosphate components of the photosynthetic carbon cycle. Indeed,
H_2O_2 oxidation of the thiamine pyrophosphate C_2-complex of
transketolase produces glycolate (7). Since ribulose diphosphate is also a relatively unstable easily-oxidized molecule,
H_2O_2 might also oxidize it to P-glycolate which would be further
hydrolyzed by the chloroplastic P-glycolate phosphatase. Further
oxidation of the glycolate to CO_2 would occur outside of the
chloroplast.

RELATIONSHIP OF PHOTORESPIRATION TO PHOTOSYNTHETIC
PRODUCTION OF REDUCED STORAGE COMPONENTS

Excess photosynthate which is stored as starch or protein
or cell wall polysaccharides are dependent upon the availability
of adequate CO_2. Additional photosynthetic assimilatory power
is needed for the further reduction of the primary products of
photosynthetic CO_2 fixation to highly reduced compounds, such
as fatty acids and carotenoids. The production of these essential storage and structural components of the plant are all
part of the ultimate beneficial use of the photosynthate. The
production of very reduced components occurs with excess of
photosynthetic assimilatory capacity over that used in reducing
CO_2 to sugars. Photorespiratory processes compete with these
reductions for the excess assimilatory power. This means that
in order to increase these valuable storage components photorespiration ought to be inhibited. The reduction level of the
overall plant components ought to be increased by inhibiting
photorespiration, but there is not data available on this idea.
Two specific examples for increasing plant biomass production
of possible fuel or chemical components will be described next
to illustrate this point.

ISOPRENE POLYMER PRODUCTION

Production of C_5 polymers such as rubber does not seem beneficial to most plants, but it could be to the user of the plant. M. Calvin (3) has proposed modifying the isoprene polymeric product of Hevea from its normal molecular weight of 10^5 to 10^6 to about 10^4 so that it can be harvested as a fluid sap. Such an isoprene polymer would be energy rich and could be used directly as fuel or by the chemical industry. Before this will be possible more research by plant geneticists and on plant growth regulators will be required for the modification to produce the shorter chain polymer. In addition, more research will be needed on photosynthetic efficiency and on photorespiration, which should be competing against the isoprene biosynthesis.

In a series of papers Sanadze (22) of the USSR and Rassmussen (10) of the USA have been studying the phenomenon of evolution of isoprene gas by leaves of some plants such as the poplar tree. The biosynthesis of this single C_5 unit may become a model for further research on the production of C_5 polymers for industry. Isoprene production and glycolate synthesis during photorespiration have common characteristics. Both require light and O_2 and are maximal at high light intensity, high temperature and low CO_2 concentration. Sanadze (23) has found that poplar leaves evolve isoprene gas at or below the CO_2 compensation point, but that in air or higher levels of CO_2 free isoprene evolution is nil. Isoprene is labeled within seconds in the light upon exposure to $^{13}CO_2$ and in shortest times only one of the carbon atoms (number 2, 4 or 5) of isoprene has 70% of the label (22).

Most aspects of this isoprene biosynthesis can not be explained by current knowledge of isoprene biosynthesis and photorespiration. It seems logical to consider this phenomenon as another way to utilize excess photosynthetic assimilatory power which is in greatest excess in low CO_2. Then the available CO_2 is channeled into this very reduced product, isoprene. How this is done is unknown. The labeling pattern from CO_2 defies explanation from extensive knowledge about the labeling patterns of products from the photosynthetic carbon cycle and the role of acetyl CoA in isopentanyl pyrophosphate synthesis. There are no known enzymes for free isoprene formation. Although glycolate and glycine label isoprene the labeling pattern from CO_2 is inconsistent with a proposal that isoprene is formed from the glycolate pathway of photorespiration (10, 24).

GLYCEROL BIOSYNTHESIS

During photosynthesis glycerol containing lipids are very rapidly formed in the chloroplasts in large quantities (5). Some algae and the symbiotic algae of coral, the Zooxanthellae, excrete glycerol and some glycolate as the carbon source for the host (12). Glycerol production in the cold is a protective mechanism, and in the context of this evaluation, glycerol production occurs because of excess photosynthetic assimilatory power which is no longer needed for growth. Glycerol would be an ideal product for industrial chemicals, if it could be produced photosynthetically in large quantities. Handling of free glycerol by industry would not involve all the contaminants and other products encountered in utilization of wood products. Since glycerol is formed photosynthetically in large quantities by some systems, the challenge of the future is whether glycerol production can be bioengineered into a practical agriculture product. Glycerol is an unusual nontoxic compound in biological systems and ought to accumulate to a high percentage in solution before becoming toxic.

During photosynthetic CO_2 reduction the first sugar product is triose phosphate which is in equilibrium with dihydroxyacetone phosphate. The latter ought to be reduced to glycerol phosphate by NADH:dihydroxyacetone phosphate reductase or glycerol phosphate dehydrogenase. A phosphatase would be needed for free glycerol formation. This use of extra NADH would be another example of a metabolic reaction using excess photosynthetic assimilatory power for synthesis of a more reduced product. The process should also be increased upon inhibition of photorespiration or growth.

Glycerol formation during photosynthesis seems to be the simplest and most direct route for production of one of the best possible industrial chemicals. Nevertheless, so much research remains to be done before practical glycerol production by plants could be accomplished that we can only conjecture at this time what might be involved. Foremost is the puzzle why a NAD:glycerol phosphate dehydrogenase for glycerol phosphate synthesis is not known in leaves. The usual mitochondrial FAD-glycerol phosphate dehydrogenase for glycerol phosphate oxidation has been reported in plants (25). Also an α-D-glycerol phosphatase has not been reported in algae or plants. Certainly these two enzymes ought to be in the algae which excrete large amounts of glycerol and should be induceable or turned on in leaves during cold adaptation. It is not surprising to learn that glycerol synthesis is probably very tightly regulated for this pathway is potentially a direct carbon drain from the photosynthetic carbon cycle. It is vexing that we do not know enough biochemistry about it to use that drain for biomass production.

PHOTOREDUCTION OF INORGANIC COMPOUNDS

Excess photosynthetic capacity may be used for other processes, some beneficial, some of no apparent value to the plant, and some detrimental to both the plant and the environment. Direct or indirect photosynthetic reduction of N_2, nitrate, and nitrite to ammonia are beneficial examples. The reduction of H^+ to H_2 and sulfate and sulfite to H_2S are examples of a spin-off of excess photosynthetic reducing capacity into toxic components. It seems more appropriate to put these processes into a category of other reductions of photosynthesis rather than into photorespiration. There is probably no O_2 uptake or CO_2 release during these reductions. On the other hand polyphosphate accumulation or excess ATP formation may be associated with dissipation of the reducing capacity by a photorespiratory process involving O_2 uptake.

PHOTORESPIRATION AND N_2 METABOLISM

At a recent conference on Research Imperatives for Crop Productivity (34) the first two sections dealt with "nitrogen input" as nitrogen fixation and nitrate reductase and with "carbon input" as photosynthesis and respiration. Both topics are related in the whole metabolism of the plant. N_2 fixation and nitrate reduction utilize the products of net photosynthesis as the energy source. Thus, if photorespiratory processes reduce net photosynthesis, it consequently reduces N_2 fixation and nitrate reduction. Hardy's group (19) has shown that if one stimulates net photosynthesis by higher CO_2 or lower O_2 levels, the nitrogen fixation rate can be increased in soybeans by as much as 5 fold and prolonged over a longer period of growth. This is attributed to more photosynthetic sucrose synthesis which upon being transported to the root provides more energy for N_2 fixation. Likewise, photorespiration ought to compete directly with NADH nitrate reduction by leaves. Consequently, research to reduce photorespiration and increase photosynthetic efficiency are of greatest importance to the area of nitrogen metabolism in plants.

REFERENCES

1. Arnon, D.I., F.R. Whatley, and M.B. Allen. 1958. Assimi-
    latory power in photosynthesis. Science, 127: 1026.

2. Asada, K., Kiso, K., and Yoshikawa, K. 1974. Univalent
    reduction of molecular oxygen by spinach chloroplasts on
    illumination. J. Biol. Chem. 249: 2175.

3. Calvin, M. 1976. Photosynthesis as a resource for energy
    and materials. Photochem. and Photobiol., 23: 425.

4. Chollet, R. and L. L. Anderson. 1976. Regulation of
    ribulose 1,5-bisphosphate carboxylase-oxygenase activities
    by temperature pretreatment and chloroplast metabolites.
    Arch. Biochem. Biophysics, 176: 344.

5. Ferrari, R.A. and A.A. Benson. 1961. The path of carbon
    in photosynthesis of the lipids. Arch. Biochem. Biophysics,
    93: 185.

6. Frederick, S.E. and E.H. Newcomb. 1969. Cytochemical
    localization of catalase in leaf microbodies (peroxisomes).
    J. Cell. Biol., 43: 343.

7. Gibbs, M. 1969. Photorespiration, Warburg effect and
    glycolate. Ann. N.Y. Acad. Sci., 168: 356.

8. Hatch, M.D., C.B. Osmond, and R.O. Slatyer. 1971. Photo-
    synthesis and photorespiration. Wiley-Interscience. 565 p.

9. Jackson, W.A. and R. J. Volk. 1970. Photorespiration.
    Ann. Rev. Plant Physiol., 21: 385.

10. Jones, C.A., and R.A. Rasmussen. 1975. Production of
    isoprene by leaf tissue. Plant Physiol., 55: 982.

11. Kaiser, W. 1976. The effect of hydrogen peroxide on CO_2
    fixation of isolated intact chloroplasts. Biochem. Bio-
    physica Acta, 440: 476.

12. Kawashima, N. and S.G. Wildman. 1970. Fraction I protein.
    Annual Rev. Plant Physiol., 21: 325.

13. Kono, Y., M-A. Takahashi and K. Asada. 1976. Oxidation of
    manganous pyrophosphate by superoxide radicals and illumina-
    ted spinach chloroplasts. Arch. Biochem. Biophys., 174:
    454.

14. Lorimer, G.H., M.R. Badger, and T. J. Andrews. 1976. The activation of ribulose-1,5-phosphate carboxylase by carbon dioxide and magnesium ions. Equilibrium, kinetics, a suggested mechanism and physiological implication. Biochem., 15: 529.

15. McFadden, B.A. 1973. Autotrophic CO_2 assimilation and the evolution of ribulose diphosphate carboxylase. Bact. Rev., 37: 289.

16. Mehler, A.H. 1951. Studies on reactions of illuminated chloroplasts. II. Stimulation and inhibition of the reaction with molecular oxygen. Arch. Biochem. Biophys., 34: 339.

17. Muscatine, L., R.R. Pool and E. Cernichiari. 1972. Some factors influencing selective release of organic material by Zooxanthellae from reef corals. Marine Biol., 13: 298.

18. Paech, C., F.J. Ryan and N.E. Tolbert. 1977. Essential primary amino groups of ribulose bisphosphate carboxylase/oxygenase indicated by reaction with pyridoxal 5'-phosphate. Arch. Biochem. Biophysics, in press.

19. Quebedeaux, B. and R.W.F. Hardy. 1975. Reproductive growth and dry matter production of Glycine max (L.) Merr. in response to oxygen concentration. Plant Physiol., 55: 102.

20. Radmer, R.J. and B. Kok. 1976. Photoreduction of O_2 primes and replaces CO_2 assimilation. Plant Physiol., 58: 336.

21. Ryan, F.J., R. Barker, and N.E. Tolbert. 1975. Inhibition of ribulose diphosphate oxygenase by xylitol 1,5-diphosphate. Biochem. Biophys. Res. Commun., 65: 39.

22. Sanadze, G. 1971. On the incorporation of carbon into an isoprene molecule from $^{13}CO_2$ assimilated during photosynthesis. In Second International Congress on Photochemistry, ed. C. Forti. p. 1958.

23. Sanadze, G.A. and A.L. Kursanov. 1966. On certain conditions of evolution of the diene C_5H_8 from poplar leaves. Fiziol. Rast., 13: 201.

24. Shah, S.P.J. and L.J. Rogers. 1969. Compartmentation of terpene biosynthesis in green plants. Biochem. J., 114: 395.

25. Stumpf, P.K. 1955. Fat metabolism in higher plants. III. Enzymic oxidation of glycerol. Plant Physiol., 30: 55.

26. Tolbert, N.E. 1963. Glycolate pathway. In Photosynthesis mechanisms in green plants. Natl. Acad. Sci.-N.R.C. publication 1145, p. 648.

27. Tolbert, N.E. 1971. Microbodies - Peroxisomes and glyoxysomes. Ann. Rev. Plant Physiol., 22: 45.

28. Tolbert, N.E. 1973. Glycolate biosynthesis and metabolism. In B.L. Horecker and E.R. Stadtman (eds.), Current topics in cellular regulation. p. 21. Academic Press, New York.

29. Tolbert, N.E. and F.J. Ryan. 1976. Glycolate biosynthesis and metabolism during photorespiration. In R.H. Burris and C.C. Black (eds.), CO_2 metabolism and plant productivity. p. 141. University Park Press.

30. Turner, J.S. and E.G. Brittain. 1962. Oxygen as a factor in photosynthesis. Biol. Rev., 37: 130.

31. Walker. D.A. 1973. Photosynthetic induction phenomena and the light activation of ribulose diphosphate carboxylase. New Phytol., 72: 209.

32. Warburg, O. 1920. Über die Geschwindiglseit der photochemischen Kohlensaurezersetzung in lebenden Zellen. Biochem. Z., 103: 188.

33. Zelitch, I. 1971. Photosynthesis, photorespiration, and plant productivity. Academic Press, New York. 347 p.

34. 1975. Crop productivity-research imperatives. eds. A.W.A. Brown, T.C. Byerly, M. Gibbs and A. San Pietro. Michigan Agricultural Exp. Station, E. Lansing, MI, 399 p.

EFFECTS OF LIGHT INTENSITY ON THE RATES OF PHOTOSYNTHESIS AND PHOTORESPIRATION IN C3 AND C4 PLANTS

RYUICHI ISHII, TOSHIROH TAKEHARA AND YOSHIO MURATA

Faculty of Agriculture, University of Tokyo, Bunkyo-ku, Tokyo, 113, Japan

AND

SHIGETOH MIYACHI

Institute of Applied Microbiology, University of Tokyo, Bunkyo-ku Tokyo, 113, Japan

Summary

In rice, a C_3 plant, the apparent photosynthesis-light intensity curve made in air bent sharply at the light compensation point (about 400 lux) showing a clear Kok effect. The depressed rate continued up to 10 klux. At above 10 klux the rate increased steeply and then gradually to attain the saturating rate. Thus the curve could be divided into two parts at about 10 klux (two step-increase). In contrast, C_4 plants such as maize, barnyard millet and eulalia showed neither Kok effect nor two step-increase. Likewise, in the atmosphere containing 2% O_2, rice did not show Kok effect or the two step-increase. At light intensities lower than 10 klux, CO_2 compensation points determined in air with rice increased from 55 to 103 ppm according to the decrease in light intensities, while those in C_4 plants kept the constant value (8 ppm) at all the intensities applied. Based on these results we assumed that the photorespiratory activity in rice increased in parallel with the light intensities from the light compensation point to about 10 klux and kept the relatively constant level at higher intensities, and that both Kok effect and the two step-increase in apparent photosynthesis are incurred by the above-mentioned changes in photorespiratory activity. We also concluded that the percent loss of photosynthetically fixed CO_2 due to photorespiration is significantly higher under low light intensities than under the intensities higher than 10 klux.

Introduction

We have been studying the effects of environmental factors on

265

photosynthesis in higher plants. Special emphasis has been laid on the elucidation of those factors which limit the rate of photosynthesis under various environmental conditions. In this paper the effects of light intensity on the rate of apparent photosynthesis in C_3 and C_4 plants will be reported.

It has been assumed that the apparent photosynthesis is determined by the rate of photosynthetic CO_2 fixation and that of respiratory CO_2 release. It has also been confirmed that photorespiratory activity is enhanced as the intensity of incident light is increased (1,5,7,8,11,13,15,16). However, the relation between photorespiration and photosynthesis as influenced by the intensity of incident light has not been studied. The present experimental results which scrutinized the rate of apparent photosynthesis under the atmospheres containing 21% and 2% O_2 indicated that, in rice, a C_3 plant, the ratio of photosynthetic CO_2 fixation to photorespiratory evolution of fixed CO_2 was greatly influenced by the intensities of incident light.

Materials and Methods

Materials used were maize (*Zea mays* L. cv. Okuzuruwase), barnyard millet (*Echinochloa crus-gali* L. cv. Hida-akabie), eulalia (*Miscanthus sinensis* Anderss), C_4 plants, and rice (*Oryza sativa* L. cv. Nihonbare), a C_3 plant. All plants except eulalia were grown in the experimental field at the University of Tokyo during summer period. Eulalia was obtained from the natural habitat. The youngest and fully expanded leaves from adult plants were used for the experiment.

The apparatus for measuring gas exchange consisted of an air pump, a regulator for gas flow rate, a humidifier, an acrylic plastic leaf chamber (7 x 10 x 0.5 cm) and an infrared gas analyzer (Beckman, Model 864) which were connected with tygon tubes. The system could be instantaneously changed from open to closed circuit, and *vice versa* by means of a magnetic valve. The infrared gas analyzer was connected to a recorder and the amount of carbon dioxide in the gas leaving from the leaf chamber was continuously recorded. Five excised leaves were placed in the leaf chamber and the chamber was illuminated from above with a metal halide lamp (Toshiba, Yōkō lamp). The light intensity at the chamber was monitored with a photoelectric photometer (Toshiba, PSI-5).

To determine CO_2 compensation point (CCP), the leaf chamber was illuminated at 52,000 lux under a continuous flow of ordinary air containing about 400 ppm CO_2 according to the open circuit. After about 30 minutes the system was changed to a closed circuit. The CO_2 concentration in the circuit started to decrease and attained CCP after about 20 minutes. After the CO_2 concentration attained CCP at 52,000 lux, the light intensity was gradually lowered to determine the CCP at each intensity.

To determine the rate of apparent photosynthesis the leaf chamber was first illuminated at 52,000 lux under a continuous flow
of gas containing about 400 ppm CO_2 according to the open circuit. The system was then changed to a closed circuit. After
a couple of minutes the system was again changed to a open circuit. This was repeated untill the decrease in CO_2 concentration
in a closed circuit showed a constant rate. Finally the system
was kept closed untill the CO_2 concentration decreased lower than
300 ppm and the rate of apparent photosynthesis was determined
from the slope at the point crossing the level of 300 ppm. After
the rate of apparent photosynthesis at 52,000 lux was determined,
the system was changed to an open circuit and then the light intensity was lowered. After 15 minutes the rate of apparent
photosynthesis under lower light intensity was determined by the
same principle mentioned above. These procedures were repeated
to attain the complete darkness. Different materials were used
for the determination of CCP and the apparent rate of photosynthesis. Gas flow rate was 3.5 l/min and the temperature in the
leaf chamber was 25°C.

Results

Fig. 1 shows the apparent photosynthesis-light intensity curves
determined in air for various plants. The curve for rice showed

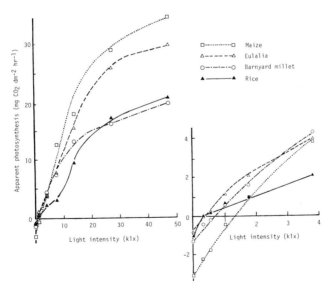

*Fig. 1. Effects of light intensity on the rates of apparent
photosynthesis for rice, maize, barnyard millet and eulalia
leaves. (The rates under low light intensities are shown in the
right side of the figure.)*

Kok effect. It bent sharply at the light compensation point
(right side of the figure) and kept the depressed rate up to 8
klux. Above 8 klux the rate increased steeply and this was again
followed by a gradual increase (left side of the figure). Thus
the apparent photosynthesis-light intensity curve for rice could
be divided into two parts at about 8 klux. This trend of change
will be referred to two step-increase. In contrast, C_4 plants
showed no Kok effect, consequently no two step-increase.

When the rate of apparent photosynthesis were determined in the
gas containing 2% O_2, neither Kok effect nor two step-increase
was observed in rice (Fig. 2). Assuming that the difference between the rate of apparent photosynthesis in the atmosphere containing 2% O_2 and that in the atmosphere containing 21% O_2 (air)
can be accounted for by photorespiration, the rate of photorespiration was calculated. We found that, up to 10 klux, the rate of
photorespiration increased proportionally with the increase in
light intensity. The increase became sluggish at intensities
higher than 10 klux and attained the constant level at 30 klux
(Fig. 2). It will also be seen from Fig. 2 that at intensities
lower than 10 klux, about 70 to 90% of photosynthetically fixed
carbon is released as CO_2 by photorespiration. The percent loss
decreases as the light intensities are elevated higher than 10
klux to attain 40%-loss at 40 klux.

Fig. 3 shows that CCP of rice, a C_3 plant, kept the constant
value of 55 ppm at light intensities higher than 10 klux. However, at the intensities lower than 10 klux the values increased
with the decrease in light intensity and attained 103 ppm at the

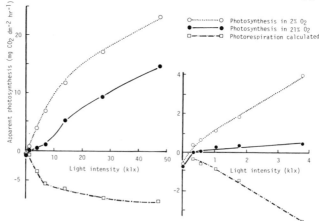

*Fig. 2. Effects of light intensity on the rates of apparent
photosynthesis of rice leaves in the atmosphere containing 2%
and 20% O_2. (The rates under low light intensity are shown in the
right side of the figure. The rate of photorespiration was calculated from the difference between the rate of apparent photosynthesis in the atmosphere containing 2% O_2 and that in the
atmosphere containing 21% O_2.*

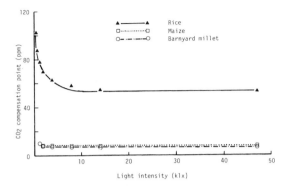

Fig. 3. Effects of light intensity on the CO_2 compensation points of rice, maize and barnyard millet leaves.

intensity just above the light compensation point (2,15).   In contrast, CCP of C_4 plants such as maize and barnyard millet kept the constant value of about 8 ppm throughout all the intensities above light compensation points (12).   It is assumed that CCP reflects the photorespiratory activity of plant (17).   The results with C_4 plants therefore agree with the generally accepted concept that C_4 plants do not exhibit photorespiration.   The changes in CCP in rice seems to indicate that its photorespiratory activity is enhanced when the light intensity was lowered below 10 klux.   However, from the results shown in Fig. 2, we concluded that the photorespiratory activity in rice decreases under such conditions. We assume that, under the light intensities lower than 10 klux the decrease in the capacity for photosynthetic CO_2 fixation is more significant than that in photorespiration and this difference brought about the observed increase in CCP.

Discussion

   Kok effect has been studied mainly with algae (4,6,9,10). There are conflicting data in higher plants: Van der Veen (14) observed Kok effect with tobacco leaf, while Gabrielsen and Vejlby (3) failed to observe it with potato and some other plant leaves.   The present result showed Kok effect in rice leaves.   To see whether the plant at certain growth stage exhibits the Kok effect, we also examined the rice leaves at various growth stages. Although data are not shown here, all showed Kok effect.   Thus it is clear that rice exhibits Kok effect.   Whether Kok effect is observable in C_3 plants in general is the problem to be studied in future.   In this connection it should be mentioned that the measurements by Van der Veen, and Gabrielsen and Vejlby were carried out under high CO_2 concentrations (above 3%).   Photorespiration is known to be inhibited by high CO_2 concentration and the present data indicated that Kok effect is closely correlated with photorespiration. It is therefore possible that the past experiments were carried out under the conditions which were difficult to induce Kok effect.

Literature cited

1. Brix.H. 1968. Influence of light intensity at different temperatures on rate of respiration of Douglas-fir seedlings. Plant Physiol. 43: 389-393.
2. Bulley,N.R., C.D.Nelson and E.B.Tregunna. 1969. Photosynthesis: Action spectra for leaves in normal and low oxygen. Plant Physiol. 44: 678-684.
3. Gabrielsen,E.K. and K.Vejlby. 1959. On the Kok-phenomenon in photosynthesis of leaves interaction of excess carbon dioxide and temperature on photosynthesis in weak light. Physiol. Plant. 12: 425-440.
4. Healey,F.P. and J.Myers. 1971. The Kok effect in *Chlamydomonas reinhardi*. Plant Physiol. 47: 373-379.
5. Hew,C.H., G.Krotkov and D.T.Canvin. 1969. Determination of the rate of CO_2 evolution by green leaves in light. Plant Physiol. 44: 662-670.
6. Hoch,G., O.v.H.Owens and Kok,B. 1963. Photosynthesis and respiration. Arch. Biochem. Biophys. 101: 171-180.
7. Holmgren,P. and P.G.Jarvis. 1967. Carbon dioxide efflux from leaves in light and darkness. Physiol. Plant. 20: 1045-1051.
8. Jolliffe,P.A. and E.B.Tregunna. 1968. Effect of temperature, CO_2 concentration and light intensity on oxygen inhibition of photosynthesis in wheat leaves. Plant Physiol. 43: 902-906.
9. Jones,L.W. and J.Myers. 1963. A common link between photosynthesis and respiration in a blue-green alga. Nature 199: 670-672.
10. Kok,B. 1949. On the interrelation of respiration and photosynthesis in green plants. Biochim. Biophys. Acta 3: 625-631.
11. Poskuta,J. 1968. Photosynthesis, photorespiration and respiration of detached spruce twigs as influenced by oxygen concentration and light intensity. Physiol. Plant. 21: 1192-1136.
12. Poskuta,J. 1969. Photosynthesis, respiration and post-illumination fixation of CO_2 by corn leaves as influenced by light and oxygen concentration. Physiol. Plant. 22: 76-85.
13. Tregunna,E.B., G.Krotkov and C.D.Nelson. 1966. Effect of oxygen on the rate of photorespiration in detached tobacco leaves. Physiol. Plant. 19: 723-733.
14. Veen,R.van der. 1949. Induction phenomena in photosynthesis, 1. Physiol. Plant. 2: 217-234.
15. Zelawski,W. 1967. A contribution to the question of the CO_2-evolution during photosynthesis in dependence on light intensity. Bull. Acad. Polonaise Sci. 15: 565-570.

16. Zelitch,I. 1959. The relationship of glycolic acid to respiration and photosynthesis in tobacco leaves. J.Biol. Chem. 234: 3077-3081.
17. Zelitch,I. 1971. Photosynthesis, photorespiration and plant productivity. Academic Press, New York. pp. 162-169.

SECTION III. Nitrogen Fixation and Production of Single Cell Protein

ENERGETICS OF BIOLOGICAL N_2 FIXATION

R. H. Burris

Department of Biochemistry

College of Agricultural and Life Sciences

University of Wisconsin-Madison

Madison, Wisconsin 53706

The recent surge of interest in biological N_2 fixation can be attributed to several factors. First, there is the realization that if the population of the world continues to increase at its present rate, many will starve unless food production is increased substantially. Second, there is a recognition that fixed nitrogen commonly limits food production and that leguminous plants can satisfy much of their need for nitrogen by fixing N_2 from the air. Third, there has been enough information disseminated about the energy crisis to convey to a broad audience the fact that the Haber process of chemical N_2 fixation demands a high energy input and that our rapidly dwindling supplies of natural gas are being used as feedstock for this process. Biological N_2 fixation also has a high energy demand, but energy is supplied by the sun via photosynthesis. Recognition that biological N_2 fixation is inherently interesting brought few practitioners to the field in the past, but the combined inpact of the food and energy crises has made many new converts.

To understand where energy demands originate in biological N_2 fixation, it will be necessary to discuss the mechanism of the process. Nitrogenase, the enzymic system that reduces N_2, consists of 2 proteins, a Fe protein and a Mo-Fe protein. The Fe protein has 4 Fe and 4 acid labile sulfurs and is a dimer of 55,000-65,000 molecular weight. Its properties are rather similar to those of a ferredoxin. The Fe protein from some organisms is cold labile under certain circumstances. It is irreversibly inactivated by O_2 and always must be handled anaerobically. The Mo-Fe protein is somewhat less sensitive to O_2 than is the Fe protein. It has a molecular weight of about 220,000 and is generally believed to contain 2 Mo and 24 Fe per molecule. It has 4 subunits of 2 different types. The Mo-Fe protein is thought to serve as the substrate-binding unit of nitrogenase. Both the Fe and Mo-Fe proteins are acidic.

Nitrogenase has an absolute requirement for MgATP and a strong reductant to drive the reaction. The most common physiological reductant is ferredoxin, but $Na_2S_2O_4$ usually is substituted experimentally. As shown in Figure 1, the MgATP binds specifically to the Fe protein of nitrogenase, and the Mo-Fe protein neither binds MgATP nor influences its binding to the Fe protein. When MgATP is bound to the Fe protein, the binding is accompanied by a marked lowering of the oxidation-reduction potential of the Fe protein to about -400 mv. This confers on the Fe protein·MgATP complex the unique ability to reduce the Mo-Fe protein which in turn can reduce substrate. Although $Na_2S_2O_4$ has a very low potential, it is incapable of reducing the Mo-Fe protein directly.

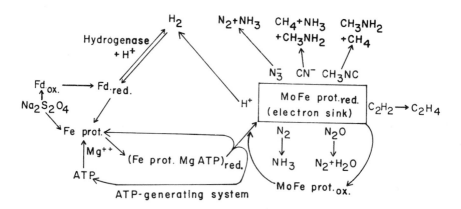

Fig. 1. Working hypothesis for electron flow and substrate reduction in biological N_2 fixation (from Winter and Burris, 1976).

The electrons used in the nitrogenase system originate in reduced ferredoxin or in $Na_2S_2O_4$ in reconstructed systems. The MgATP·Fe protein complex is reduced (the oxidized Fe protein apparently can bind MgATP) and acquires a potential low enough to reduce the Mo-Fe protein. The Mo-Fe protein binds substrate and reduces it and the reoxidized Mo-Fe and Fe proteins can recycle.

Note that nitrogenase can reduce a variety of substrates other than N_2. It can reduce N_2, N_2O, C_2H_2, CN^-, CH_3NC, N_3^- and analogues of some of these compounds. In addition, it reduces $2H^+$ to H_2 and it hydrolyzes ATP to ADP + P_i. A step in the electron transfer, rather than ATP binding or hydrolysis, appears to be rate limiting. Although the specific point of ATP hydrolysis has not been established, it may well be at the point where electrons are passed from the Fe protein to reduce the Mo-Fe protein.

The reduction of $2H^+$ to H_2 will be of importance to our subsequent discussion. Note that N_2 fixing organisms often are endowed with hydrogenase as well as with nitrogenase. Such organisms can produce H_2 with nitrogenase by a mechanism that is ATP dependent but insensitive to CO, or with hydrogenase by a mechanism that is ATP independent and CO sensitive.

Evidence for the pathway of electron flow has been provided primarily by measurement of the EPR spectra of the components under various conditions (Orme-Johnson et al., 1972). The Mo-Fe protein and the Fe protein each has a characteristic EPR spectrum at liquid helium temperatures. The EPR signal of the Fe protein changes when the Fe protein binds MgATP, whereas the addition of MgATP does not affect the EPR signal of the Mo-Fe protein. The Mo-Fe protein is isolated anaerobically from N_2-fixing organisms in a partially reduced state, and this state exhibits the characteristic EPR signal. The signal disappears upon complete oxidation of the Mo-Fe protein with ferricyanide or with complete reduction of the protein by the reduced Fe protein·MgATP complex. When the reductant, e.g. $Na_2S_2O_4$, is exhausted, the signal for the partially reduced Mo-Fe protein returns. The shuttle between the partially and fully reduced states appears to be the physiologically active transfer, and there is no evidence that the completely oxidized state has any physiological significance. The fully reduced Mo-Fe protein transfers electrons to reduce the various substrates.

For a time there was disagreement on the stoichiometry of MgATP hydrolysis in N_2 fixation, but most workers now will agree that about 4 MgATP are required per pair of electrons transferred in isolated nitrogenase systems. As the reduction of N_2 to 2 NH_3 requires 6 electrons, the minimal requirement for reduction of N_2 to 2 NH_3 becomes 12 MgATP. The use of MgATP is not always this efficient, so the requirement may be higher than 12 MgATP.

Hill (1976) reported that the MgATP/N_2 molar ratio is 29 in
<u>Klebsiella</u> <u>pneumoniae</u> grown anaerobically in a chemostat under
glucose-limited conditions. The requirement was no lower in an
O_2-limited chemostat. Earlier Daesch and Mortenson (1968) and
Dalton and Postgate (1969), respectively, had reported MgATP/N_2
molar ratios of 20 for growing <u>C</u>. <u>pasteurianum</u> and 4 to 5 for
growing <u>Azotobacter</u> <u>chroococcum</u> (no other investigators have
approached the low values reported by Dalton and Postgate).

Ljones and Burris (1972a) and Davis <u>et</u> <u>al</u>. (1975) reported
that the ratio of Mo-Fe protein to Fe protein in reconstructed
nitrogenase systems can influence the MgATP requirement. When
increasing Fe protein is titrated against a constant amount of
Mo-Fe protein the rate of acetylene reduction increases until it
reaches a plateau value. In contrast, when Fe protein is held
constant and Mo-Fe protein is increased to reconstruct nitrogenase, the nitrogenase activity increases to a maximal rate and then
decreases with excess Mo-Fe protein. Examination of the hydrolysis of MgATP by the mixture with excess Mo-Fe protein shows that
there is a great increase in hydrolysis of MgATP relative to N_2
fixed. The usual ratio of 4 MgATP/2 electrons may increase to 20
MgATP/2 electrons with excess Mo-Fe protein. Apparently the
excess Mo-Fe protein catalyzes an uncoupled hydrolysis of MgATP,
although Mo-Fe protein by itself does not hydrolyze MgATP.

The stoichiometry of the oxidation of $Na_2S_2O_4$ in reduction
of N_2 appears to be straightforward. Ljones and Burris (1972b)
described a method for measuring nitrogenase activity by following
the oxidation of $Na_2S_2O_4$ spectrophotometrically. Oxidation of
$Na_2S_2O_4$ appeared to correlate directly with reduction of C_2H_2 or
N_2. Ferredoxins functional in N_2 fixation may be of the one
electron-carrying 4 Fe, 4 acid labile S type characteristic of
<u>Bacillus</u> <u>polymyxa</u> or of the 2 electron-carrying 8 Fe, 8 acid
labile S type characteristic of <u>Clostridium</u> <u>pasteurianum</u>.

We have mentioned the dissipation of energy accompanying the
hydrolysis of excessive MgATP when the Mo-Fe protein is in excess
of the Fe protein. Another serious loss for N_2 fixation can be
incurred when nitrogenase transfers electrons to reduce H^+ to H_2
rather than to reduce N_2 to NH_3. I recognize that many of you are
working to improve production of H_2 from microorganisms, but for
those concerned with efficient reduction of N_2 a loss of electrons
to H_2 is undesirable.

As indicated in Figure 1, H_2 often can arise by two pathways
in N_2-fixing organisms, as many of them have a hydrogenase.
Reduced ferredoxin or $Na_2S_2O_4$ can serve as a source of electrons
either for hydrogenase or for nitrogenase. Hydrogenase evolves

H_2 through a pathway that requires no MgATP but is inhibited by CO. In contrast, nitrogenase evolves H_2 through a pathway that requires MgATP but is insensitive to CO. The ATP requirement and CO sensitivity furnish means for distinguishing the two modes of H_2 evolution.

The balance between reduction of H^+ and N_2 will vary with the nitrogenase system. It is apparent that for organisms with an active hydrogenase the balance can vary widely. For organisms that evolve H_2 through nitrogenase action only, the observed net production of H_2 also varies among organisms. Some contend that there is an absolute stoichiometry between H_2 and NH_3 production and that for each 6 electrons used for reduction of N_2 two are used for production of H_2 via nitrogenase. The observed difference in net production of H_2 is attributed to recycling of the H_2 to recapture energy from it (Dixon, 1967).

Schubert and Evans (1976) have presented interesting data on H_2 production by root nodules. Nodules from angiospermous nonlegumes produced very little H_2 while reducing N_2. Among the legumes, cowpea nodules produced little H_2, whereas soybean nodules evolved substantial amounts of H_2. As Schubert and Evans (1976) pointed out, there is a real possibility that one can obtain bacterial-plant associations that are relatively efficient by selecting those that produce little H_2.

Jones and Bishop (1976) have studied the H_2 and O_2 metabolism of Anabaena. Peterson (1976) in my laboratory also has been studying the hydrogen metabolism of isolated heterocysts from an axenic culture of Anabaena 7120. The organism was grown on an N-free medium in a 70 liter illuminated fermentor at 34°C. The vegetative cells were disrupted by sonication or with lysozyme under anaerobic conditions, and all subsequent operations were anaerobic. The heterocysts could be recovered almost free from debris by passing them through an interface into 5% Ludox[R], a colloidal silica. Heterocysts prepared in this way retain their metabolic activities. H_2 output or uptake was measured manometrically or with the hydrogen electrode described by Wang, Healey and Myers (1971).

The isolated heterocysts catalyzed oxidative phosphorylation, as they produced $AT^{32}P$ from ADP and $^{32}PO_4^{-3}$. The ATP was trapped with hexokinase plus 2-deoxy-D-glucose as phosphate acceptor. The heterocysts also catalyzed the uptake and evolution of H_2.

Figure 2 shows H_2 production by heterocysts as measured amperometrically. The evolution of H_2 was initiated when ATP was added to the reaction mixture, and the rate then remained linear for some time.

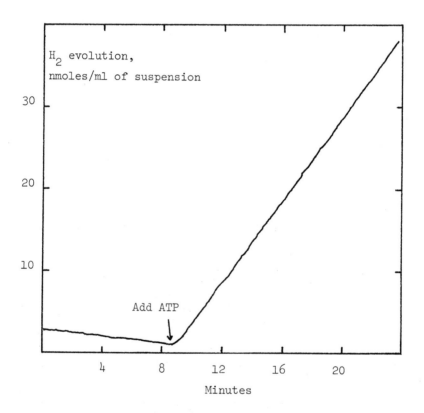

Fig. 2. Evolution of H_2 by isolated heterocysts from Anabaena 7120; a strip chart recording from a H_2 electrode is shown. The data are from Peterson (1976).

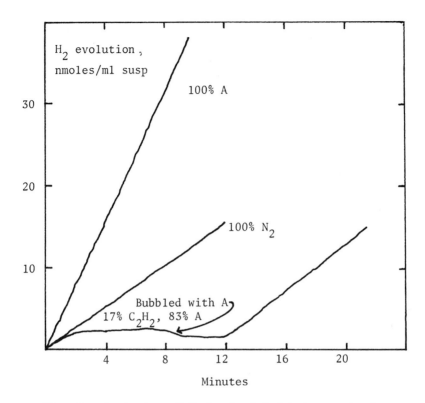

Fig. 3. Evolution of H_2 by isolated heterocysts from Anabaena 7120 and its inhibition by N_2 and by C_2H_2. The inhibition by C_2H_2 is partially reversed by bubbling with argon. The strip chart recording from a H_2 electrode is taken from Peterson (1976).

Figure 3 shows that production of H_2 is vigorous in argon, is inhibited 68% by N_2 and is abolished by 17% C_2H_2. The inhibition by C_2H_2 was reversed partially by sparging with argon. The data of Figures 2 and 3 indicate that the H_2 evolution being measured originates from the activity of nitrogenase not hydrogenase. The evolution requires ATP and it is suppressed by N_2 or C_2H_2. The data emphasize another point of some importance; Rivera-Ortiz and Burris (1975) indicated that inhibition of H_2 evolution by N_2 and C_2H_2 are qualitatively different, as C_2H_2 abolishes evolution of H_2, whereas N_2 even at very high pressures never completely abolishes evolution of H_2.

The isolated heterocysts also take up H_2 if a suitable acceptor is furnished. Dichlorophenolindophenol (DCPIP), methylene blue (MB), potassium ferricyanide and O_2 function well as electron acceptors, whereas NAD^+, $NADP^+$ and methyl viologen function poorly.

Ferredoxin plus NADP$^+$ has intermediate activity. Figure 4 shows an amperometric trace of H_2 uptake with and without O_2; it is apparent that O_2 is a highly effective oxidant in this oxyhydrogen reaction. The reaction was inhibited by high levels of O_2. In the oxyhydrogen reaction of heterocysts, electrons appear to be transferred through the cytochrome system, as uptake of $H_2 + O_2$ is inhibited by CO and the inhibition is partially light reversible.

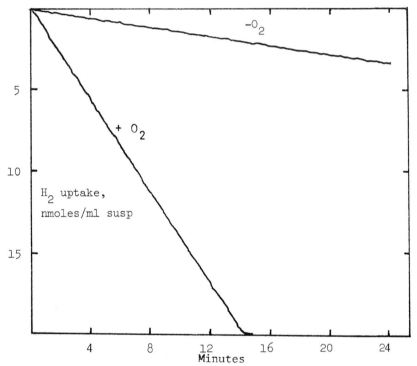

Fig. 4. Uptake of H_2 by isolated heterocysts from <u>Anabaena</u> 7120 and the influence of O_2 as the oxidant in the <u>reaction</u>. The strip chart recording from a H_2 electrode is taken from Peterson (1976).

One of Peterson's (1976) most interesting observations was that electron flow in the oxyhydrogen reaction could be coupled to ATP formation. This means that although the organisms dissipate energy by evolving H_2, they also have the potential to recover a part of the energy by recycling H_2 to form ATP (Dixon, 1967).

When we examine the overall efficiency of biological N_2 fixation and find that it requires at least 12 ATP and perhaps as many as 30 ATP per N_2 reduced, that it requires reduced ferredoxin to serve as reductant, and that it dissipates H_2, it is evident that the biological system is not highly efficient. The

deficiences of the Haber process for producing ammonia and its
high energy demand are often decried. However, the energy balance
sheets indicate that biological N_2 fixation does little better than
the Haber process. The obvious advantage of biological N_2 fixation
is that it can tap the energy of sunlight via photosynthesis and
thus put no drain on our limited supplies of fossil fuels.

How can we improve the efficiency of biological N_2 fixation?
It can be improved by selecting systems that produce minimal
amounts of H_2. The work of Schubert and Evans (1976) indicates
the wide variation in H_2 production among nitrogenase systems. In
addition we can attempt to identify nitrogenase systems that use
minimal amounts of ATP. Again, data in the literature suggest
that nitrogenases in growing microorganisms vary in their ATP
utilization per unit of N_2 fixed.

The energy crisis and the accompanying awareness that
chemical fixation of N_2 constitutes a drain on our energy
resources has prompted a renewed search for N_2 fixing organisms
capable of making a practical contribution to agricultural practice.
Spirillum lipoferum particularly has attracted attention.
Beijerinck (1925) first described this organism as a nitrogen
fixer, but the careful work of Schröder (1932) raised questions
regarding its ability to fix N_2. Becking (1963) used $^{15}N_2$ to
establish the validity of its N_2 fixation but work on the
organism languished until Döbereiner and Day (1975) reported that
it fixed N_2 in association with the roots of Digitaria
decumbens. They found that it invaded the roots, and that roots
of D. decumbens not only yielded cultures of S. lipoferum but
also reduced acetylene vigorously after a period of preincubation
at low pO_2. Interest was increased further with the report of
von Bülow and Döbereiner (1975) that S. lipoferum reduced acetylene
in association with maize roots. Yaacov Okon and Steve Albrecht
have studied S. lipoferum in my lab and in the field and have
observed some of its properties.

S. lipoferum is a microaerophilic organism when fixing N_2
but functions as a normal aerobe when it is supplied ammonia as a
source of nitrogen. On ammonia it will grow with a doubling time
of about an hour, as shown in Figure 5, if air is sparged
vigorously through the culture. When a heavy inoculum of the
organism growing on ammonia is transferred to a culture maintained
at a low pO_2, its doubling time will increase to about 2 hours
until the ammonia introduced with the inoculum is exhausted, and
then it will adapt quickly to N_2 fixation and establish a doubling
time of 5.5 to 7 hours. The organism grows well on succinate,
malate, lactate or pyruvate, and its culture on N_2 poses no
problems other than maintaining a low pO_2.

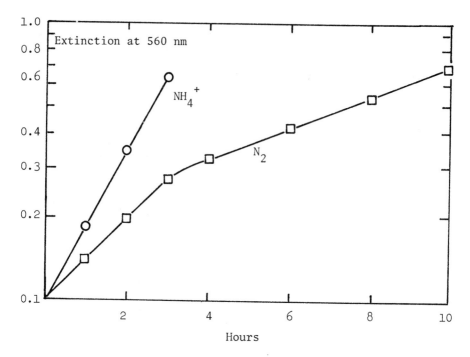

Fig. 5. Growth curves for <u>Spirillum lipoferum</u> on N_2 and on NH_4^+. The culture supplied NH_4^+ was grown on air and the N_2 culture received 0.005 to 0.007 atm O_2 through an oxygenstat. During the initial rapid growth phase, the N_2 culture was utilizing residual NH_4^+ from a heavy inoculum. The data are from Okon et al. (1977).

Okon <u>et al</u>. (1977) used an oxygenstat to maintain a constant dissolved oxygen concentration in cultures of <u>S</u>. <u>lipoferum</u> on a malate medium. A sterilizable O_2 electrode immersed in the culture signaled when the dissolved O_2 was too low or too high and the oxygenstat control unit opened or closed a solenoid admitting O_2 or blocking its entry. Table 1 shows that during the period of exponential growth the optimal pO_2 for <u>S</u>. <u>lipoferum</u> fixing N_2 was 0.005 to 0.007 atm. The pO_2 for greatest efficiency, based on cell nitrogen produced per gram of malate consumed during exponential growth, was 0.002 atm.

As one might expect from its growth response, <u>S</u>. <u>lipoferum</u> oxidizes malate, succinate, lactate and pyruvate rapidly (Okon, Albrecht and Burris, 1976). It oxidizes sugars poorly and grows very slowly on them.

Table 1. Effects of the pO_2 on the generation time,
rate of C_2H_2 reduction and efficiency of utilization
of malate by Spirillum lipoferum. The data are
from Okon et al. (1977).

pO_2 in atm	doubling time in h	nmoles C_2H_2 red h x ml culture	mg N assimil g malate used
0.002-0.003	9.0-10.0	250-280	11.9-12.5
0.005-0.007	5.5- 7.0	280-300	8.0-10.0
0.009-0.011	7.5- 8.5	80-100	2.5- 4.0
0.014-0.018	8.5- 9.5	50- 75	1.2- 2.1

S. lipoferum reduces C_2H_2 and it reduces $^{15}N_2$ effectively.
N_2-fixing cell-free extracts can be prepared and purified by
techniques much like those used with Rhodospirillum rubrum. The
Mo-Fe protein and Fe protein have not been recovered in a
homogeneous state. It is interesting that the Fe protein recovered
from S. lipoferum is in an inactive state but can be activated
with its own activating factor or the activating factor from
R. rubrum (Ludden and Burris, 1976). In our experience, only the
Fe proteins from R. rubrum and S. lipoferum require activation
from the form in which they are isolated.

S. lipoferum has been isolated by J. Döbereiner from European,
African, North American and South American soils. We have recovered
the organism after it overwintered in a sandy Wisconsin soil, and
we have isolated it from several areas in Wisconsin. Tests with
aseptic plants grown in cotton stoppered bottles established that
the organism invaded the roots of maize (surface sterilized roots
yielded S. lipoferum when the roots were crushed) and reduced small
amounts of acetylene in association with roots.

As our tests for an effect of S. lipoferum on yield of maize
and other plants grown in growth chambers or in a greenhouse were
equivocal, we grew inoculated and uninoculated maize in the crossed
gradient room of the University of Wisconsin Biotron. The room
provided 4 temperatures (day temperatures of 40, 36, 32 and 28° and
night temperatures 10° below these) and 4 light intensities at a
right angle to the temperature gradients (500, 1250, 2400 and 3000
foot candles); this provided a grid with 16 different light-
temperature conditions. The suggestion had been made, that the
association between plants and S. lipoferum was more effective in
tropical than in temperate climates. The response in total N
assimilation and in dry weight production was rather disappointing
as shown in Table 2, in the sense that there was little evidence
of benefit to maize by inoculation with S. lipoferum and little
variation in response among the light and temperature conditions.

Table 2. Yield of total N and dry weight of maize plants
grown for 94 days in a crossed-gradient room under various
light and temperature conditions.

Day Temp.*	Light, foot candles		500	1250	2400	3000
40°	Inoc.	D.W.[†]	27.8	34.5	32.4	32.3
		% N	1.46	1.04	1.00	1.01
	Uninoc.	D.W.	25.0	30.8	28.3	33.8
		% N	1.66	1.20	1.07	1.03
36°	Inoc.	D.W.	31.0	34.6	33.8	29.5
		% N	1.07	0.90	0.94	0.89
	Uninoc.	D.W.	26.7	35.7	31.1	36.0
		% N	1.38	0.90	0.93	0.84
32°	Inoc.	D.W.	31.2	33.5	36.8	41.0
		% N	1.08	1.01	0.62	0.75
	Uninoc.	D.W.	32.6	38.6	38.8	36.2
		% N	0.96	0.74	0.65	0.67
28°	Inoc.	D.W.	34.2	42.2	44.2	43.0
		% N	1.12	0.69	0.67	0.59
	Uninoc.	D.W.	30.7	43.1	44.1	43.6
		% N	1.09	0.66	0.62	0.57

*Night temperatures were 10° lower than day temperatures.

[†]D.W. = dry weight as g/total plant.

The only yield data in the U.S. recording benefit to
plants from inoculation with S. lipoferum was reported by Smith
et al. (1976). They found that a small supplement of fixed N in
the field enhanced N_2 fixation by S. lipoferum in association
with pearl millet and guinea grass.

Most field trials with S. lipoferum to date have depended upon
measurements of C_2H_2 reduction by excised roots that have been
incubated overnight at a reduced pO_2. Although this method may
measure the potential of the system for N_2 fixation, serious
questions can be raised about its applicability as a quantitative
measure of N_2 fixation in the field. Okon, Albrecht and Burris
(1977) have observed proliferation of S. lipoferum during the
period of preincubation. Our mass spectrometric data show that the

excised roots rapidly deplete the limited O_2 in their reaction
vessel; it seems probable that the roots and bacteria then form
lactate or other organic acids anaerobically and that the acid
supports proliferation of S. lipoferum.

When soil cores, including substantial amounts of the root
system, are incubated with C_2H_2 they give a far less rapid reduction
of C_2H_2 than preincubated root samples. The production of C_2H_4 by
the cores usually is linear with time, and this is reassuring for
the validity of the method as a measure of N_2 reduction.

Although our field work in 1975 showed very little yield
benefit to maize or other plants from inoculation with S. lipoferum, the tests to date in our laboratory and elsewhere have
been very limited. It is quite possible than an intensive search
for the proper plant cultivar and the most effective culture
of S. lipoferum will uncover a plant-bacteria combination highly
effective for N_2 fixation. The system is a relatively simple one
and hence is amenable to selection and genetic alteration. It
does not involve the organizational complexies of the leguminous
nodule and this makes it appealing.

The search for new and improved biological agents capable of
N_2 fixation has intensified in the past few years, and the
potential of such systems for food production and energy conservation warrant further intensification of the search.

REFERENCES

1. Becking, J. H., Antonie van Leeuwenhoek 29, 326 (1963).

2. Beijerinck, M. W., Zentbl. Bakt. Parasitkde. II, 63, 353-359 (1925).

3. Daesch, G. and L. E. Mortenson, J. Bacteriol. 96, 346-351 (1968).

4. Dalton, H. and J. R. Postgate, J. Gen. Microbiol. 56, 307-319 (1969).

5. Davis, L. C., V. K. Shah and W. J. Brill, Biochim. Biophys. Acta 403, 67-78 (1975).

6. Dixon, R. O. D., Annals of Botany N.S. 31, 179-188 (1967).

7. Döbereiner, J. and J. M. Day, In "Nitrogen fixation by free-living micro-organisms", W. D. P. Stewart, ed., pp. 39-56 (1975). Cambridge University Press, Cambridge.

8. Hill, S., J. Gen. Microbiol. 95, 297-312 (1976).

9. Jones, L. W. and N. I. Bishop, Plant Physiol. 57, 659-665 (1976).

10. Ljones, T. and R. H. Burris, Biochim. Biophys. Acta 275, 93-101 (1972a).

11. Ljones, T. and R. H. Burris, Anal. Biochem. 45, 448-452 (1972b).

12. Ludden, P. W. and R. H. Burris, Science 194, 424-426 (1976).

13. Okon, Y., S. L. Albrecht and R. H. Burris, J. Bacteriol. 128, 592-597 (1976).

14. Okon, Y., S. L. Albrecht and R. H. Burris, Applied and Environ. Microbiol. (in press, 1977).

15. Okon, Y., J. P. Houchins, S. L. Albrecht and R. H. Burris, J. Gen. Microbiol. (in press, 1977).

16. Orme-Johnson, W. H., W. D. Hamilton, T. Ljones, M.-Y. W. Tso, R. H. Burris, V. K. Shah and W. J. Brill, Proc. Natl. Acad. Sci. U.S.A. 69, 3142-3145 (1972).

17. Peterson, R. B., Ph.D. Thesis, Department of Biochemistry, University of Wisconsin-Madison, 1976.

18. Rivera-Ortiz, J. M. and R. H. Burris, J. Bacteriol. 123, 537-545 (1975).

19. Schröder, M., Zentbl. Bakt. Paraṣitkde. II, 85, 177-212 (1932).

20. Schubert, K. R. and H. J. Evans, Proc. Natl. Acad. Sci. U.S.A. 73, 1207-1211 (1976).

21. Smith, R. L., J. H. Bouton, S. C. Schank, K. H. Quesenberry, M. E. Tyler, J. R. Milam, M. H. Gaskins and R. C. Littell, Science 193, 1003-1005 (1976).

22. Von Büllow, J. F. W. and J. Döbereiner, Proc. Natl. Acad. Sci. U.S.A. 72, 2389-2393 (1975).

23. Wang, R., F. P. Healey and J. Myers, Plant Physiol. 48, 108-110 (1971).

N_2 FIXING HYDROGEN BACTERIA

JIRO OOYAMA

Fermentation Research Institute, Inage-Higashi, Chiba-city, Japan

Summary

Culture conditions of hydrogen bacteria having N_2 fixing abilities were described. Emphasis was laid on effects of O_2 upon the cell growth in N_2 fixing and non N_2 fixing autotrophic cultures. Possible roles of hydrogen bacteria as well as N_2 fixing bacteria were also discussed.

Introduction

Several methods for production of SCP have been proposed up to now. The most leading ones seem to be that from petroleum hydrocarbons or from their derivatives such as methanol. An industrialization of these SCP is almost ready now in terms of the technique, although whether the industrialization is reasonable or not is controversial from a practical point of view.

But, supposing that production of SCP from hydrocarbons is to meet an anticipated food shortage in future, it can hardly be believed that the hydrocarbons will be a dependable material for SCP in the far future, as the hydrocarbons will run out in several decades, although they may play an important role in near future. When the hydrocarbons on the earth would have been consumed, CO_2 would remain as the only available carbon resources to be converted to the protein foods. CO_2 is even now being fixed photochemically by plants and algea. The photosynthesis is, however, thought to have some restrictions, such as geographical limitation of location or spatial restriction from a viewpoint of its industrial applications, so far as it is solar energy depending. Besides photosynthetic organisms, there are chemolithoautotrophic microorganisms that could be future SCP producers. They are able to fix CO_2 by using energy derived from inorganic chemical reactions. Among chemolithoautotrophic bacteria, hydrogen bacteria seem to be the most probable SCP producer in future, as their growth rates are exceptionally high. They can grow vigorously by a supply of CO_2, H_2 and O_2 in addition to water and inorganic minerals including proper nitrogen compounds. It is, therefore, not necessary for their culture, to reserve vast spaces for sunlight, however it will do to build compact fermentors, which is much more convenient to control, especially against biological contaminations. Besides, hydrogen bacteria have many advantages as SCP producers. That is, first, the resulted product of chemical reaction on which hydrogen bacteria depends to get energy for their growth is just water being clean, having no fear for environmental pollutions. Secondary, the growth rates of hydrogen bacteria are high, the highest one is about 2 hrs in terms of doubling time. This is comparable to that of SCP from n-paraffin. Further, in some strains, the protein content of the cells reaches 80% in dried base.

On the other hand, hydrogen bacteria has also disadvantages from a practical point of view. The most outstanding one of them is the very thing that they need H_2 for their growth. So long as H_2 is produced industrially by pyrolysis of hydrocarbons such as methane, it is no wonder that the SCP by hydrogen bacteria would be disadvantageous in cost compared with that from hydrocarbon itself. However, the situation might be widely changed half a century later. By that time, hydrocarbons would become invaluable, and most of the fuel hydrocarbons will have been replaced by inexpensive H_2 produced by biological or pyrochemical decomposition of water. On that occasion, the hydrogen bacteria might find the place to be SCP producer.

Such being the case, hydrogen bacteria is regarded as one of the possible SCP producers in the far future.

As regards N_2 fixations, several questions had been kept unsolved in my mind. Those were: is it an inevitable fact that N_2 fixing potencies should be restricted to a few particular genus of microorganisms such as *Rhizobium*, *Azotobacter* or *Clostridium*, etc.?, and, is it unreasonable to assume that there might be some inadequate factors in the experimental condition which prevent the researchers from detecting the possible microbial N_2 fixations? Suppose these two questions are denied, a widespread N_2 fixations by microorganisms in great variety must be revealed, just by excluding these interfering factors from the experiment. Then an intensive promotion of agricultural production might be possible by stimulating the widely distributed N_2 fixing activities.

Such being the case, an isolation of N_2 fixing hydrogen bacteria and the development of their culture methods were planned out as a preliminary experiment so as to find out a SCP producer of new type, and to acertain a part of the above views of the N_2 fixation as well.

The planned experiment was, however, unlikely to be easy. Because, as it was known, since P. W. Wilson, some forty years ago, failed in his N_2 fixing experiments with a plant in an atmosphere containing H_2, H_2 had been shown to be toxic for N_2 fixation. The reasons for this inhibitory action of H_2 have not been elucidated so far. Therefore, the proposed experiment was carried out anyway, and we succeeded in demonstrating the presence of such bacteria.

Isolation of microorganisms

The organisms were isolated from oily soils in northern district of Japan. Mineral agar plates containing no nitrogen compounds were sprayed with diluted soil suspensions, and put in a glass container, which was filled with a gas mixture of H_2, O_2, CO_2 and N_2 as the sole energy, carbon, and nitrogen sources. Grown colonies on the plates were purified, and some 40 strains of bacteria as possible N_2 fixers were obtained, unexpectedly at one time by the first screening test (6). Strain N34 was one of the fastest growing group of the isolates.

Culture vessel

1 liter suction flask was employed for the culture. The top of the flask was fitted with a silicon rubber stopper, which was equipped with a photometric tube upright. The side arm of the flask was connected, with a short silicon rubber tube, to a glass tube containing sterile cotton wool for the gas exchange. 10 ml of culture medium was added. The flask was sterilized and inoculated with organisms. Then, flask was evaccuated, refilled with proper gas mixture, the side arm was closed, and the flask was shaken at 30°C. The degree of the cell growth in the culture fluid was measured photometrically by turning the flask upside down during the culture.

Proof of N_2 fixation

N_2 fixation by the isolated hydrogen bacterium was proved as follows. Each 10 ml of mineral solution which did not contain either carbon or nitrogen compounds was put in two flasks, and inoculated with small amount of cells of st. N34. Each flask was filled with different gas mixture. That was, the first flask contained all four kinds of gases, while, in the second flask, N_2 being replaced with Ar. A remarkable cell growth occured only in the first flask after 138 hrs incuvation. 7.0 g of the culture fluid of the first flask was taken up and centrifuged, precipitated cells were dried and yielded 16.7 mg, of which the nitrogen content was 6.82%. Thus, N_2 fixation by the hydrogen bacterium st. N34 was proved (6).

Effects of minerals

A wide range of the pH values, from 5 to 9, was applicable to the starting culture medium for st. N34. pH of the cultures was actually kept around 6 to 7 during the culture period without positive pH adjusting, since CO_2 and N_2 were the sole sources of carbon and nitrogen. Fe^{2+} was recognized to be better for the

cell growth than Fe^{3+} as the culture medium component, same to
the results that had been shown by R. Repaske with *Alcaligenes
eutrophus* (9). Including other mineral components in addition
to Fe^{2+} and pH, a basal mineral solution for batch culture medium
was tentatively proposed as follows: KH_2PO_4 300 mg, K_2HPO_4 400 mg,
$MgSO_4 \cdot 7H_2O$ 200 mg, $FeSO_4 \cdot 7H_2O$ 50 mg, $ZnSO_4 \cdot 7H_2O$ 50 µg, $Na_2MoO_4 \cdot$
$2H_2O$ 100 µg, distilled water up to 1 liter, pH 7.0. This
solution was used in the following experiments.

Among the elements in this basal medium, Mo and Fe had at
least close relations to N_2 fixation. A subtraction of Mo from
the culture medium caused a remarkable suppression of the cell
growth. When NH_4Cl had been, however, added to the culture
solution in advance so that the ammonium nitrogen might be taken
into the cells instead of N_2, the above growth suppression did
not occur. The effect of Fe subtraction from the culture
solution was also similar to that of Mo subtraction. These
positive effects of Mo and Fe on N_2 fixation may imply that N_2
fixing process in st. N34 is basically the same to that in N_2
fixing bacteria already known (4).

Effects of O_2

O_2 was the most critical factor for the N_2 fixing autotrophic
cultures of st. N34. The growth rate, in general, decreased
remarkably with increasing initial Po_2. In batch culture the
highest specific growth rate was observed at 12 mmHg of starting
Po_2, equivalent to 1/12 of atmospheric O_2, and actually no
growth occurred with an initial Po_2 over 100 mmHg (4). To elucidate the details of this inhibitory effect of O_2, an experiment
was performed in connection with nitrogenase activity in the
cell (5). Nitrogenase activities were assayed by acetylene
reductivity. The culture was started with a proper gas mixture
composed of H_2, O_2, CO_2 and N_2, whereas Po_2 was 30 mmHg. After
40 hrs cultivation, the gas mixture in the culture flask was
replaced by that containing 90 mmHg of Po_2. Right after the gas
substitution, nitrogenase activity in the cell was decreased
suddenly and remarkably, and the cell growth also almost ceased.
However, in case NH_4Cl had been added in the medium beforehand,
a ceasing of the growth did not occur. These facts would mean
that the nitrogenase was inactivated by the high Po_2 if a rapid
turnover of nitrogenase was occuring in the cell. In any case
the discontinuation of the cell growth under 90 mmHg of Po_2
seemed to be ascribed mainly to the inhibitory effect of O_2 on
the N_2 fixing process itself. This may also imply that N_2 fixations in general are intrinsically sensitive to O_2 as H. Dalton
pointed out (1), and that of N34 is not exceptional.

The abovementioned nitrogenase activity once decreased by high
Po_2 was, however, rapidly reactivated again about 10 hrs after
replacing the gas phase, in spite of Po_2 being still around 90
mmHg, and an active cell growth also started. Those facts may
be attributed to the aquisition of some protecting or tolerating

mechanisms against high Po2. This may further suggest that st.
N34 would grow under high Po2 in a continuous culture.

On the other hand, the effects of O2 was quite different in
case that nitrogen compounds such as urea or NH4Cl had been
added in the culture medium. In that case nitrogenase in terms
of acetylene reductivity could not be detected at all in the
grown cell. Apparently the nitrogen compounds in the medium were
exclusively used as the nitrogen source. The growth suppression
by O2 did not appear in this case up to 80 mmHg of Po2 as was
expected. When the initial Po2 was higher than 80 mmHg, however,
a starting of the exponential cell growth was extremely delayed.
The degree of the delay was proportional to that of Po2, while
the growth rate of the cells in the exponential phase was again
in high level even after the lag phase of growth had been prolonged enough. In these cases also, a protective mechanisms or
something toward O2 must have been organized in the cells during
the long period of lag phase of the culture. This phenomenon
has, of course, no relations to the N2 fixations. Studies are
being carried on to elucidate the details of this mechanisms.

Culture example

An analytical result of st. N34 culture is shown in Table I,
where the factors described above are taken into consideration.
In case the starting gas composition was varied largely from the
values written in this table, the maximum specific growth rate
became lower. In this example, the cell growth ceased at 0.9 g/
liter of cell concentration due to the shortage of O2. The cell
growth could be still continued by successive introducing of O2
and other gases into the flask.

Table I   *A culture result by st. N34*

Gas changes (mmHg)		starting	final
	CO_2	16	11
	N_2	540	–
	H_2	99	69
	O_2	12	0
Absorbed gas ratio	$CO_2 : H_2 : O_2 = 1 : 6 : 2.4$		
Culture time	89.5 hr		
μmax	$0.097\ hr^{-1}$		
Max. cell growth	0.9 g dried cell / liter		

Heterotrophic growth

Organic compounds which were recognized to support N_2 fixing
heterotrophic growth of st. N34 were: *n*-butane, *n*-pentane, *n*-

hexane, n-heptane, n-decane, methanol, ethanol, n-propanol, n-butanol, iso-butanol, sec-butanol, n-hexanol, formic acid, acetic acid, propionic acid, butyric acid, succinic acid, gluconic acid, citric acid (7). Autotrophic and heterotrophic cultures of st. N34 seemed to be completely reversible.

N_2 fixing, CO oxidizing growth

Some strains, among the N_2 fixing hydrogen bacteria isolated, could utilize CO as the sole carbon source. There are two types of N_2 fixing CO utilization. Strain numbered S17 was typical of the first type. The culture medium was substantially the same as described before. St. S17 could obtain the growth energy from the oxidation of H_2, just as the cultures of usual hydrogen bacteria. When CO was omitted from the gas mixture, or, N_2 was replaced by Ar, the cell growth did not take place in either case. In this culture, a low partial pressure of CO was essential. The cell growth rate was fairly slow (8).

The second type of N_2 fixing CO utilization was carried by an isolate designated A305. In this case the growth energy was supplied by the oxidation of CO itself, so the presence of H_2 was not neccessary. CO is not directly fixed by the cells, but is primarily oxidized to CO_2. When CO_2 was eliminated from the starting gas mixture, growth of st. A305 was extremely delayed. Further, when CO_2 was removed during the culture by placing a potassium hydroxide solution in the culture vessels, bacterial growth ceased. The low partial pressures of CO and O_2 were also essential for CO oxidizing, CO_2 and N_2 fixing growth of A305 (10).

In both types of CO utilization by hydrogen bacteria, with and without H_2, N_2 fixations were suppressed by supply of NH_4Cl in the culture media, just as in the case of st. N34 culture with CO_2.

Comparison of isolates

St. N34, S17 and A305 were isolated separately. Although they are not identified yet, their growth substrate specificity differs from one another, as shown in Table II.

Table II *Comparison of substrate specificities of the isolated N_2 fixers*

Gas system	N34	S17	A305
CO_2-N_2-H_2-O_2	+++	+++	+
CO -N_2-H_2-O_2	±	+++	+++
CO -N_2-O_2-(CO_2)	−	++	+++

Discussions

Researches on the N_2 fixing hydrogen bacteria have also begun
recently in West Germany (3) and in the Netherlands (2). Some of
the bacteria they are dealing with used to be regarded as usual
hydrogen bacteria, of which N_2 fixities were found anew and some
are newly isolated as N_2 fixers through our works. Natures of
these bacteria, however, make no great difference from that of
our isolates described above, that is, their growth rates, their
culturing characteristics, especially their sensitiveness to O_2,
heterotrophic growing capacities, etc. are fairly similar to that
of our isolates.

As for N_2 fixations, my original intention to find out N_2
fixing microorganisms in wide variety of microbial genus, therefore, seeds to have been realized partly within a group of
hydrogen utilizing bacteria. It would also be possible to prove
the N_2 fixing abilities in other microorganisms than a group of
hydrogen bacteria, if studies are forwarded with due consideration that there might also be some inadequate experimental conditions other than low Po_2 condition which might prevent discover-
ies of the N_2 fixing abilities.

With regard to the possibility for the hydrogen bacteria to be
SCP producers in the far future, we do not have special opinions,
but, there must be several problems to be solved for them to be
future SCP producer, for instances, an assessment of their
utility and safety as cattle feed or food material, a development
of industrial production of inexpensive H_2 and of appropriate
fermentation systems and others.

As for the N_2 fixing hydrogen bacteria, its industrial application for SCP production, however, might be restricted to
particular situations where the synthesis of ammonia is unfavorable for certain reasons, even if the bacterial growth rates,
their sensitiveness to O_2, etc. are improved to some extent.

REFERENCES

(1) Dalton, H. and J. R. Postgate: Effect of oxygen on growth of
    Azotobacter chroococcum in batch and continuous cultures.
    J. gen. Microbiol. 54: 463-473 (1969).

(2) de Bont, J. A. M. and M. W. M. Leijten: Nitrogen fixation by
    hydrogen-utilizing bacteria. *Arch. Microbiol.* 107: 235-240
    (1976).

(3) Gogotov, J. N. and H. G. Schlegel: N_2-fixation by chemoauto-
    trophic hydrogen bacteria. *Arch. Microbiol.* 97: 359-362
    (1974).

(4) Nakamura, Y. and J. Ooyama: Culture condition for N_2 fixing

autotrophic growth of *Hydrogenomonas*-like bacterium N34, and metals relating to N_2 fixation. *Report Ferm. Res. Inst. 45*: 37-50 (1974).

(5) Nakamura, Y. and J. Ooyama: Effect of oxygen on nitrogenase of N_2-fixing hydrogen bacteria. *J. Agr. Chem. Soc. Japan 50*: 439-441 (1976).

(6) Ooyama, J.: Simultaneous fixation of CO_2 and N_2 in the presence of H_2 and O_2 by a bacterium. *Report Ferm. Res. Inst. 39*: 41-44 (1971).

(7) Ooyama, J. and T. Shibuya: N_2 fixing growth on the organic compounds by a *Hydrogenomonas*-like bacterium N34. *Report Ferm. Res. Inst. 42*: 87-105 (1972).

(8) Ooyama, J. and T. Shinohara: Simultaneous fixation of CO and N_2 in the presence of H_2 and O_2 by a bacterium. *Report Ferm. Res. Inst. 40*: 1-5 (1971).

(9) Repaske, R.: Nutritional requirements for *Hydrogenomonas eutropha*. *J. Bact. 83*: 418-422 (1962).

(10) Shinohara, T. and J. Ooyama: N_2 fixation by a CO oxidizing bacterium. *Report Ferm. Res. Inst. 42*: 81-85 (1972).

POSSIBLE ROUTES TO INCREASE THE CONVERSION
OF SOLAR ENERGY TO FOOD AND FEED BY GRAIN
LEGUMES AND CEREAL GRAINS (CROP PRODUCTION):
CO_2 AND N_2 FIXATION, FOLIAR FERTILIZATION,
AND ASSIMILATE PARTITIONING

R. W. F. Hardy and U. D. Havelka
E. I. du Pont de Nemours and Company
Central Research and Development Department
Experimental Station
Wilmington, Delaware 19898

I. INTRODUCTION

Increasing the conversion of solar energy to food and
feed or in more conventional agricultural terms increasing
crop production at an annual rate of 3% is the major key to feeding the expanding and more affluent billions in the future as it
has been in the past. The opportunity for increase is shown by
a current efficiency of only about 1% in the conversion of solar
energy to grain (Cooper, 1975). Combinations of improved
varieties, increased fertilizer and plant protectant chemical
use, expanded irrigation, and possibly mechanization enabled
world cereal grain production to increase from 600-700 million
metric tons to 1300-1400 million metric tons during the past
25 years. Increasing inputs of the above as well as supplementation with additional technologies will be required to meet
the doubling of crop production during the next 25 years
(Hardy, 1976).

In addition, the declining ratio of high protein grain
legume:low protein cereal grain production must be reversed.
Decreased fossil energy inputs are a secondary goal relative to
increased production as a primary goal. Agronomic crop production consumes less than 3% of U. S. total fossil energy and
the current return of two to five calories of grain for each
calorie of fossil energy used is quite favorable.

In this report the current status of four areas of crop
physiology research that may provide the basis for technologies
to increase the conversion of solar energy to food and feed will
be summarized. These areas are:
1. Carbon dioxide fixation by cereal grains and grain
legumes

• CO_2 enrichment and yield of field-grown photosynthetically inefficient crops
2. Nitrogen fixation by grain legumes
    • photosynthate as a major limitation
    • nodular energy efficiency including H_2 metabolism
    • summary of limitations
3. Foliar fertilization of grain legumes
    • multiple applications of N, P, K, and S mixtures
4. Assimilate partitioning in cereal grains and grain
legumes
    • O_2 process regulating reproductive growth
II. CARBON DIOXIDE FIXATION BY CEREAL GRAINS AND
GRAIN LEGUMES

Carbon dioxide enrichment of field-grown crops is a
useful technique to assess the quantitative significance of
aspects of photosynthesis such as photorespiration on yield
limitation. Desirable attributes of the technique include (1)
minimal extraneous perturbation of plant, (2) utility for
multi-plot experiments and scale-up, (3) facile measurement
of CO_2 concentration in plots, (4) option to alter CO_2 concentration for any desired time period, and (5) relatively
simple equipment and inexpensive as an experimental technique enabling use in the major crop production areas throughout the world. In view of these characteristics it is surprising that the technique was not used in the field until recent
years. Wittwer (1967) stated that "this is one of the most
surprising deficiencies of modern research effort in the plant
sciences."

The first report of increased growth of plants exposed
to atmospheres enriched in CO_2 (de Saussure, 1804) appeared
over 170 years ago. In this case pea plants exposed to 8%
CO_2 grew better than the controls exposed to ambient air.
Other reports on the effect of CO_2 supplementation on plant
growth in laboratories and greenhouses appeared regularly,
e.g., Thaer, 1880; Kruesler, 1885; Demoussy, 1904;
Cummings and Jones, 1918; Reinau, 1927; Arthur, et al.,
1930; Richter, 1938) and these results led to the practical use
of CO_2 enrichment of greenhouse crops early in the 20th century and the practice continues to be used.

During the past five years data have begun to be
accumulated on responses of yield and related parameters of
field-grown crops to CO_2 enrichment. Both legume and non-

legume crops have been examined. The information available from other laboratories as well as our own will be summarized. Only data from field-grown crops are considered since increasing the conversion of solar energy to food and feed will be achieved by identifying and overcoming limitations in the field where the crop is naturally grown rather than in the artificial conditions of the greenhouse.

Some results on CO_2 enrichment of four grain legumes-soybeans, peanuts, peas, and beans-have been obtained (Table 1). Yield increases of about 50% were produced in all of these legumes by enrichment to 1000-1500 ppm CO_2 around

Table 1. Carbon Dioxide Enrichment from Initial Anthesis to Senescence and Yield of Field-Grown Grain Legumes*

	Soybeans	Peanuts	Peas	Beans
	(CO_2 Enriched as % of Air Controls)			
Yield (kg/ha)				
Dry Matter	160	160	148	151
Vegetative	147	144	137	142
Reproductive	198	130	152	165
Seed		**	153	159
Harvest Index			103	101
Plant Density	111	109		
Senescence	Delayed	**	Delayed	Delayed

*   Data available for (1) soybeans from Hardman and Brun, 1971; Hardy and Havelka, 1973, 1975; Havelka and Hardy, 1974, 1976a; Shivashankar, et al., 1976; (2) peanuts from Havelka and Hardy, 1976b; (3) peas from Havelka and Hardy, 1976c; and (4) beans from Havelka and Hardy, 1976c.
** Frost killed prior to complete maturity.

the canopy from initial anthesis to senescence. The increased yield is attributed to decreased photorespiration, delayed senescence, increased plant density, and retention of more reproductive structures. Harvest index was not altered.

Some results on CO_2 enrichment of five cereal grains-wheat, rice, barley, oats, and cotton-have been obtained (Table 2). Average yield increases of about 10-50% were produced by pre-anthesis CO_2 enrichment. Photosynthesis prior to anthesis may be more limited than after anthesis in cereal grains while photosynthesis after anthesis is strongly limiting in grain legumes. The increased yield in cereal grains is attributed to decreased photorespiration and in some cases increased plant density and harvest index. Senescence was not delayed and was enhanced in some cases; the enhancement may have been produced by a chamber rather than a CO_2 enrichment effect.

Table 2. Pre-Anthesis and Post-Anthesis Carbon Dioxide Enrichment and Yield of Field-Grown Cereal Grains*

	Wheat	Rice	Barley	Oats	Cotton
	(CO_2 Enriched as % of Air Controls)**				
Yield (kg/ha)					
Dry Matter				110/	126/
Vegetative				113/	
Reproductive				104/	
Seed	116/114	131/118	150/128		
Harvest Index	105/	108/106		96/	
Plant Density	123/123			100/	
Senescence	No effect to earlier			earlier	

*   Data available for (1) wheat from Fischer and Aguilar, 1976; Havelka and Hardy, 1976c; Krenzer and Moss, 1975; (2) rice from Cock and Yoshida, 1973; Havelka and Hardy, 1976c; Yoshida, 1973; (3) barley from Gifford, et al., 1973; (4) oats from Criswell and Shibles, 1972; and (5) cotton from Harper, et al., 1973a,b.
**   ___/___ is for CO_2 enriched pre-anthesis and post-anthesis, respectively.

The above results are the only data available to support the hypothesis that decreased photorespiration during specific periods of the development of field-grown photosynthetically inefficient grain legumes and cereal grains will produce substantial increases in yield. They were obtained with only a few crops. Of greater significance is the fact that the data for only two of the seven crops were obtained at multiple sites. We recommend the collection of data in the different climatic areas of substantial production of the major cereal grains and grain legumes. All of the reported data are from photosynthetically inefficient crops and comparable data are needed for photosynthetically efficient crops.

Standardization of the CO_2 enrichment technique is essential to enable comparison of results between laboratories and sites. In addition, a useful technique must enable the establishment of a similar environment in experimental plots and the open field except for pCO_2 and this goal has not been achieved in many of the experiments. Closed chambers used in some experiments produced elevated temperatures and earlier senescence. CO_2-Enriched chambers have been compared in some experiments with the open field rather than with air chambers. Some chambers that have been used have covered only a small area, e.g., $0.3 \ m^2$. We recommend an open-top, side-enclosed chamber of adequate size as shown in Figure 1 and used in all of the experiments on grain legumes summarized in Table 1. This chamber circumvents the above limitations (Hardy and Havelka, 1975; Havelka and Hardy, 1976a).

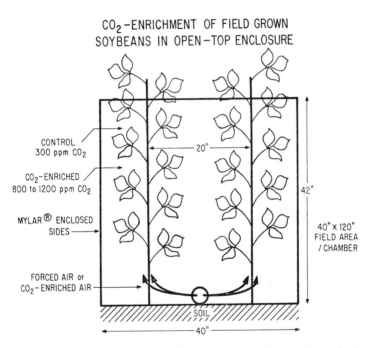

Figure 1. Schematic of open-top side-enclosed chamber for CO_2 enrichment of field-grown crops (Hardy and Havelka, 1975).

It is recommended that a variety of parameters be measured for CO_2-enriched crops. These include dry matter, vegetative, reproductive and seed yield, number of flowers and other reproductive characteristics, harvest index, plant density, and senescence. Such information will provide guidance

in seeking practical solutions to increase the conversion of
solar energy to feed and food by crops.

CO_2 Enrichment should only be viewed as a useful
experimental tool rather than a practical possibility for
agronomic crops. Addition of sufficient CO_2 at the soil level
in normal dense corn to increase the CO_2 content at soil level
by 45-fold produced little change in the CO_2 content of the upper
canopy (Lemon, et al., 1971). A computer simulation for CO_2
enrichment of cotton indicated that it would require a total of
70 metric tons of CO_2 per ha delivered over a 55-day period
to increase cotton yield by 30% (Harper, et al., 1973a, b).

III. NITROGEN FIXATION BY GRAIN LEGUMES (2/1)

Grain legumes, because of their generally higher protein content than cereal grains, require more fixed nitrogen
per unit of yield. For example, soybeans require 0.1 kg of
fixed nitrogen for each kg of seed produced. Only 25-50% of
this nitrogen is provided by biological nitrogen fixation in
annual grain legumes, e.g., soybeans, peanuts, and peas,
grown under normal field-production conditions (Hardy, et al.,
1968, 1971, 1973, 1977; Hardy and Havelka, 1975; Havelka
and Hardy, 1976a, b, c; Johnson, et al., 1975). The average
of all reported data for soybeans is 31% for a total of 86 kg N_2
fixed/ha·season (Criswell, et al., 1976a). The remainder is
provided from the fixed nitrogen reserves in the soil.

A recent simulation model led to the hypothesis that
soybeans are self-destructive since they need to translocate
large amounts of nitrogen from vegetative tissues during seed
fill. It was suggested that increased nitrogen is required to
extend seed filling and increase yields of soybeans (Sinclair
and de Wit, 1975, 1976). The following section provides evidence that increasing the photosynthate available to the nodule
increases nitrogen input and yield while a subsequent section
provides evidence that foliar fertilization with a mixture of N,
P, K, and S also increases nitrogen input and yield. The
former suggests that photosynthate is the primary limitation,
while the latter suggests that nutrients including N are the
primary limitation. Clearly there is a strong interdependence
between photosynthesis and nitrogen assimilation.

1. Evidence for Photosynthate as a Major Factor Limiting N_2
     Fixation (1/1)

Several lines of evidence have been accumulated to
identify photosynthate as a major factor limiting N_2 fixation by

several grain legumes. Time courses of biological N_2 fixation have been determined for field-grown soybeans, peanuts, and peas with those for soybeans obtained at eight different locations and with several cultivars (Criswell, et al., 1976b; Hardy, et al., 1968, 1971, 1973; Hardy and Havelka, 1975; Harper, 1974, 1976; Havelka and Hardy, 1976a, b, c; Lawn and Brun, 1974a; Mague and Burris, 1972; Shivashankar, et al., 1976; Sloger, et al., 1975; Thibodeau and Jaworski, 1975; Weber, et al., 1971). In general they are all similar with most N_2 fixation occurring post-anthesis but the exponential phase of increase in N_2 fixation usually terminates about mid-pod filling when the demand of the reproductive structure for photosynthate becomes large.

The carbon flow in N_2-fixing vegetative peas grown in the greenhouse has been determined (Minchin and Pate, 1973). Of 100 net units of carbon fixed by photosynthesis, 32 units move to the nodule where 12 units are respired as CO_2, 6 are used for nodule growth, and 15 are transported back to the shoot. These results show that 4 kg of carbohydrate are consumed to fix 1 kg N_2 and indicate that the nodule may only be as inefficient as is in vitro nitrogenase.

Biological N_2 fixation is the most energy-consuming process recognized in nature with a direct ATP requirement of in excess of 12 molecules of ATP per molecule N_2 fixed and ADP is an inhibitor so that the ratio of ATP/ADP may regulate N_2-fixing activity (Burris, 1977). In the soybean subjected to variable light treatments, the ATP/ADP ratio in the nodule is directly related to N_2 fixation (Ching, et al., 1975) as is the ATP/ADP ratio and bacteroid N_2-fixing activity in the presence of increasing amounts of the inhibitor N phenylimidazole (Appleby, et al., 1975).

Eight factors that affect photosynthate available to the nodule produce parallel effects on N_2 fixation (Table 3). Discussions of light quantity, source size, photosynthetic type, competitive sinks, and translocation are available (Brun, 1976; Hardy and Havelka, 1975; Weil and Ohlrogge, 1975). A recent report demonstrates that part of the decrease in N_2 fixation at low water potential arises from decreased photosynthesis with CO_2 enrichment of the canopy capable of overcoming some of the decrease (Huang, et al., 1975a, b). The inhibition of legume N_2 fixation by the addition of fixed nitrogen may also be mediated in part by photosynthesis since translocation of fresh

Table 3. Photosynthesis and Symbiotic N_2 Fixation by Legumes*

	Effect on N_2 Fixation	
Factor	Increase	Decrease
Light Quantity	Day	Night
	Long Days	Short Days
	Supplemental Light	Shading
Source Size	Additional Foliage	Defoliation
	Low Planting	High Planting
	Density	Density
CO_2/O_2 Ratio in Canopy	Increased pCO_2	Decreased pCO_2?
	Decreased pO_2	Increased pO_2
Photosynthetic Type	C_4 (?)	C_3
Competitive Sinks	Removal of	Development of
	Reproductive	Reproductive
	Structures	Structures
Translocation		Girdling
Water		Low Water Potential
Fixed Nitrogen		Nitrogen Fertilization

* Hardy and Havelka, 1975, and Silver and Hardy, 1976.

photosynthate to nodules of various legumes decreases following fertilization with nitrate (Gibson, 1975; Small and Leonard,
1969). In the latter case it is not clear whether decreased
photosynthate translocation to the nodule is the cause of or the
response to decreased N_2 fixation.

In the experiments in our laboratory on field-grown soybeans, peanuts, and peas used to assess the response of yields
to canopy CO_2 enrichment, nitrogen input parameters including nitrogen fixation were measured also (Table 4). Total

Table 4. Carbon Dioxide Enrichment from Initial Anthesis to Senescence
and Nitrogen Input of Field-Grown Legumes*

	Soybeans	Peanuts**	Peas
	(Air/CO_2 Enriched)		
Nitrogen Input			
Total Nitrogen (kg/ha)	295/511		
N_2 Fixed (kg/ha)	76/427	58/102	87.5/128
% N from N_2	26/83		45/45
Nodule Characteristics			
Mass	Increased	Increased	Increased
Specific N_2-Fixing Activity	Increased		Increased
Senescence	Delayed	Delayed	Delayed

* Data for (1) soybeans from Hardy and Havelka, 1973, 1975; Havelka
and Hardy, 1974, 1976a; Shivashankar, et al., 1976; (2) peanuts
from Havelka and Hardy, 1976b; and (3) peas from Havelka and
Hardy, 1976c.
** Senescence produced by low temperature.

nitrogen input in soybeans was increased 73% and N_2 fixation 460% by CO_2 enrichment so that 83% of the total nitrogen input was provided by N_2 fixation in the CO_2-enriched vs. only 26% in the air plants. Significant but less dramatic increases of 76 and 46% in N_2 fixation were produced by CO_2 enrichment of peanuts and peas, respectively. The increased N_2 fixation by all three crops arose from increased nodule mass and delayed nodule senescence. In addition, the specific N_2-fixing activity of soybean and pea nodules was increased. Greenhouse experiments on peas (Phillips, et al., 1976) and alfalfa and red clover (Wilson, et al., 1933) also show substantial increases in N_2 fixation from CO_2 enrichment.

It took forty years before it was discovered that the favorable results of CO_2 enrichment of forage legumes grown in bottles in the laboratory were relevant to grain legumes in the field. Unfortunately, general failure of exploratory plant biologists to attempt to translate their work to the field continues to exist today as it did forty years ago.

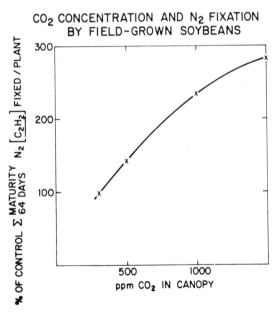

Figure 2. Relationship of N_2 fixed by field-grown soybeans and pCO_2 around canopy from anthesis to senescence (Hardy, et al., 1977).

We suggest that the major effect of CO_2 enrichment is

to decrease photorespiration. Two experiments support this conclusion. The first determined the relationship between canopy pCO_2 and N_2 fixed by field-grown soybeans. Enrichment was from initial anthesis to senescence. The total amount of N_2 fixed per season increased in a linear manner as canopy CO_2 increased from 300 to 1000 ppm and was curvilinear as it increased from 1000 to 1500 ppm (Figure 2) (Hardy, et al., 1977; Havelka and Hardy, 1974). In the second experiment the CO_2/O_2 ratio around the canopy of soybeans grown in a high light intensity growth room was altered by either CO_2 enrichment at a constant pO_2 of 21% or by O_2 depletion or enrichment at a constant pCO_2 of 300 ppm (Quebedeaux, et al., 1975). During vegetative growth including the early flowering stage there was a close correlation between CO_2/O_2 ratios and nitrogen fixation providng strong support that the effect of CO_2 enrichment was mediated by decreased photorespiration (Figure 3) (Hardy, et al., 1977). The relationship was less well correlated at maturity and this is attributed to the arresting of reproductive development by subatmospheric O_2 during the reproductive growth phase.

It is concluded that chemicals that decrease photorespiration or naturally occurring or scientist-made mutants with decreased photorespiration should lead to large increases in N_2 fixation in legumes. Other approaches that would increase the amount of solar energy (photosynthate) available to the nodule or the efficiency of the nodule's use of energy should also lead to increases in N_2 fixation. These same conclusions will apply to other N_2 fixation systems such as associative symbioses with cereal grains. Increased photosynthate production by CO_2 enrichment also increases nitrogen input in non-nodulating soybeans (Havelka and Hardy, 1974; Harper, 1976), demonstrating the importance of energy for assimilation of nitrate as well as N_2. In the next section one possible aspect of inefficiency in nodular use of energy will be described.

2. Nodular Energy Efficiency Including H_2 Metabolism (1/1)

The reaction catalyzed by nitrogenase wastes energy in at least three known ways: (1) The direct use of 12 molecules of ATP to reduce 1 molecule of N_2 utilizing reduced ferredoxin or flavodoxin is not required by thermodynamic considerations. (2) Both H^+'s and N_2 are reduced by nitrogenase and 4 ATP molecules are used to reduce 2 H^+'s to H_2

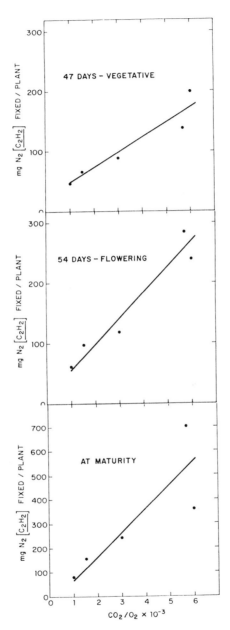

Figure 3. Relationship of N_2 fixed by soybean and CO_2/O_2 around canopy up to indicated stages of development. CO_2/O_2 Ratio altered by O_2 depletion or enrichment at

constant pCO_2 and CO_2 enrichment at constant pO_2 (Hardy, et al., 1977).

with electrons that are already at the redox potential of a hydrogen electrode. (3) The H_2 produced by the nitrogenase reaction may be lost to the atmosphere unless a hydrogenase is present. Major improvements in nodule energy efficiency would be possible by overcoming the waste described in (1) and (2) and avoidance of (2) would eliminate any concern about (3). Unfortunately no data are yet available from basic studies to suggest that (1) and (2) can be manipulated in a beneficial way. Some recent data suggest that (3) can be manipulated and that information and what it might mean are given below.

Nitrogenase reduces H^+'s to H_2 as well as N_2 to $2NH_3$ and at saturating concentrations of N_2 about 25% of the electrons transferred by nitrogenase in vitro are wasted in the reduction of H^+'s (Bulen and LeComte, 1966; Burris, 1977; Burns and Hardy, 1972; Mortensen, 1966). Hydrogen evolution by soybean nodules has been known for some time (Hoch, et al., 1960; Bergersen, 1963) and in some cases hydrogenase may be present in the bacteroid to enable utilization of this evolved H_2 as a reductant or possibly in generation of ATP by oxidative phosphorylation (Dixon, 1968, 1972). Although it has been "realized that factors such as... reduction of H_3O^+ to H_2 which may compete with N_2 for electrons, etc., may... limit nitrogenase activity" in legumes (Hardy and Havelka, 1975) it was measurements by Schubert and Evans (1976) of hydrogen evolution by various legumes that led to their proposal of hydrogen evolution as a major factor affecting the efficiency of nitrogen fixation by nodulated symbionts.

Nodular efficiency in H_2 metabolism is defined as the following ratio:

$$1 - \frac{\text{electrons evolved as } H_2}{\substack{\text{total electrons transferred by nitrogenase to} \\ N_2 + H^+}}$$

determined at ambient pN_2. The reported nodular efficiency of most legumes ranges from only 0.40-0.61 with one notable exception of 0.99 for cowpea (Schubert and Evans, 1976).

We have calculated the theoretical effect of a 1.0 vs. 0.5 nodular efficiency in H_2 metabolism on N_2 fixation, nitrogen input and yield of a field-grown legume crop. For the 0.5 efficiency case 42 ATP equivalents are consumed to evolve

$3H_2$ and fix $1N_2$, while in the 100% efficiency case 33 ATP equivalents are consumed to fix 1 N_2. In the latter case the same amount of H_2 is evolved but hydrogenases may enable the nodule to recover the ATP equivalents of the H_2. The theoretical increase in N_2 fixation for 1.0 vs. 0.5 nodular efficiency in H_2 metabolism is 27%. For comparison, it would be 100% if one could stop H^+ reduction by nitrogenase (wasteful process (2) as described above) and it would be 390% if one could eliminate both H^+ reduction and the direct ATP requirement for N_2 reduction (wasteful processes (1) and (2) as described above). Since N_2 fixation provides only 25-50% of total nitrogen in field-grown soybeans, peas, or peanuts, the theoretical increase in nitrogen input for 1.0 vs. 0.5 nodular efficiency is 6 to 14%. Since extra photosynthate is required to increase yield, the theoretical increase in yield may be less than 6 to 14% for 1.0 vs. 0.5 nodular efficiency.

A recent newspaper report of work at Oregon State University (New York Times, September 26, 1976) indicates a possible 10% yield increase for greenhouse-grown soybeans with 0.9 vs. 0.5 nodular efficiency in hydrogen metabolism. The increased nodular efficiency was presumably obtained by selecting rhizobia possessing hydrogenase(s) to enable utilization of the evolved H_2 and indicates an approach for selection of improved rhizobia.

However, experiments are needed under field light intensity to assess the significance of nodular efficiency in hydrogen metabolism over the complete growth cycle during both the night and day. The partitioning of electrons between H^+ and N_2 reduction by in vitro nitrogenase changes with ATP concentration. Proton reduction is favored at low ATP concentration and N_2 reduction at high ATP concentration (Silverstein and Bulen, 1970). Since most of the experiments on nodular efficiency have been conducted under the low light intensity of the greenhouse, the reported nodular efficiencies in H_2 metabolism may be lower than those occurring in the field. We are unable to judge the potential significance of nodular efficiency in H_2 metabolism under normal production conditions until the indicated field information is obtained.

3. Summary of Limitations for N_2 Fixation by Field-Grown Legumes (1/1)

Our judgment of the significance of the various limitations described above as well as equilibrium and kinetic

rhizosphere temperature effects (Hardy and Criswell, 1976),
subatmospheric rhizosphere pO_2 (Criswell, et al., 1975, 1976c,
1977; Hardy and Criswell, 1976), and intensity of reproductive
sinks for fixed nitrogen (Quebedeaux, et al., 1975) are shown
in Table 5.

Table 5. Summary of Significance of Limitations for N_2 Fixation by
Field-Grown Legumes

Significance	Limitation
Severe	Photosynthate Available to Nodule
	Low Rhizosphere pO_2 Combined with Low Temperature
Intermediate	Continuous Low or High Rhizosphere Temperature
Minor	Reproductive Sinkness
	Low Rhizosphere pO_2
Unknown*	Direct ATP Requirement for N_2 Reduction
	Reduction of Protons by Nitrogenase
	Metabolism of H_2

* Significance not assessable because of lack of data from plants grown
  under field or similar conditions.

IV. FOLIAR FERTILIZATION (2/1)

In contrast to cereal grains, grain legumes such as
soybeans have failed to respond consistently and substantially
to nitrogenous fertilizers (Hardy, et al., 1971, 1973;
Harper, 1974, 1976; Lawn and Brun, 1974b; Wilson and
Wagner, 1935) since the added fertilizer substitutes for rather
than supplements symbiotic nitrogen fixation (Table 6). Even
multiple foliar applications of urea at various times during
pod filling inhibited N_2 fixation and failed to increase yields.

However, multiple foliar fertilization of soybean during
pod filling with a mixture of N:P:K:S as occurs in the mature
seed was reported this year to increase seed yield by 600-1200
kg/ha (Table 7) (Garcia and Hanway, 1976).

Many field tests have been made in 1976 utilizing up to
a total of 100 tons of fertilizer to assess further this potentially significant discovery and within a year or two the practical
impact of this new technology should be clear. A major question to be answered concerns economics of the treatment. To
avoid foliar injury, low-biuret urea as a nitrogen source and
polyphosphate as a phosphate source are necessary with indicated treatment costs of $50+/ha.

V. ASSIMILATE PARTITIONING (2/1)

Recognition that "The mechanisms by which photosynthates are distributed in plants...are as important determinants of productivity as is photosynthetic capacity" (Evans,

Table 6. N Fertilizers: N_2 Fixation and Yield*

Form**	Application Locus***	Age (days)	N_2 Fixation Control (%)	Yield
NO_3^-	S	0	43	98
NH_4^+	S	0	51	104
Urea	S	0	48	109
Protein	S	0	112	123
NO_3^-	S	40-50	50	106
NH_4	S	40-50	37	106
Urea	S	40-50	44	97
Urea	F	50-71	29	100
Urea	F	78-99	53	100

*   Hardy, et al., 1973.
**   135 kg N/ha of indicated form.
*** S, soil; F, foliar.

Table 7. Foliar Fertilization of Soybeans*

N	P	K	S	Applications (No.)	Yield (kg/ha)
-	-	-	-	-	2980
80	8	24	4	4	3550
-	8	24	4	2	2810
80	-	24	4	4	3090
80	8	-	4	5	2840
80	8	24	-	4	3310
40	4	12	2	4	2940
120	12	36	6	5	3220

(Total Nutrient Application, N P K S in kg/ha)

* Garcia and Hanway, 1976.

1975) is increasing. Knowledge of the mechanisms which control assimilate partitioning and ultimately harvest index are poorly understood (Evans, 1975; Wareing and Patrick, 1975). Consequently plant breeding and chemical approaches to increase harvest indices have been forced to use empirical approaches.

In 1971 it was discovered that subatmospheric pO_2 around reproductive structures arrests reproductive growth of all crops tested - soybean, peanut, wheat, sorghum, rice, and cotton (Akita and Tanaka, 1973; Hardy, et al., 1976; Quebedeaux and Hardy, 1973, 1975a,b, 1976). Just as altered CO_2 provides a non-surgical method to manipulate source intensity, altered pO_2 provides the first non-surgical method to manipulate sink intensity. The characteristics of this O_2

process that regulates sink activity have been determined
extensively through growth of whole plants under altered pO_2's
Table 8).

Table 8. Characteristics of O_2 Process Regulating Sink Intensity
of Reproductive Structures

- Essential for reproductive growth
- Similar for both C_3 and C_4 plants and therefore unrelated to
    photorespiration
- Occurs at all stages of reproductive growth
- Seed development more sensitive than pods
- Early exposure arrests pod while later arrests seed development
- Localized to the reproductive structure
- Irreversible except for short exposure of $\leq$ 3 days
- Independent of pCO_2
- Independent of light on reproductive structure
- Coincident with light on vegetative structure
- O_2 Concentration for maximal reproductive growth $>$ 21% vs.
    1 to 3% to saturate dark respiration in leaves
- Fertilization is normal
- Dark respiration not decreased in intact plants by 5% O_2
- ATP Concentration not decreased in seeds by 5% O_2
- Rapid alteration in translocation of fresh assimilate, e.g.,
    six-hour exposure to 5% O_2 decreases by 50+% and 30% O_2
    increases by 50+% ^{14}C-assimilate to reproductive sinks
- Physical process or chemical reaction

In brief, the partitioning of assimilate between reproductive and vegetative parts of higher plants is regulated by
the pO_2 surrounding the reproductive structures throughout all
phases of reproductive growth. Short exposures of the reproductive structures to supra-atmospheric pO_2 increases the
transport of freshly synthesized photosynthate to them while
subambient pO_2 decreases the transport. The mechanism of
this O_2 process is undefined but its discovery suggests the
opportunity to manipulate harvest indices.

VI. CONCLUSION (2/1)

It is indicated that there are several promising research routes which could lead to technologies for much improved (20-50%) conversion of solar energy to food and feed by
grain legumes and cereal grains.

VII. REFERENCES (2/1)

Akita, S., and I. Tanaka. 1973. Studies on the mechanism of differences in photosynthesis among species. III.
Influence of low oxygen concentration on dry matter production
and grain fertility of rice plant. Nippon Sakumotsu Gakkai
Kiji 42:18-23.

Appleby, C. A., G. L. Turner, and P. K. Manicol.

1975. Involvement of oxyleghaemoglobin and cytochrome P-450 in an efficient oxidative phosphorylation pathway which supports nitrogen fixation in Rhizobium. Biochim. Biophys. Acta 387:461-474.

Arthur, J. M., J. D. Guthrie, and J. M. Newell. 1930. Some effects of artificial climates on the growth and chemical composition of plants. Am. J. Bot. 17:416-482.

Bergersen, F. J. 1963. The relationship between hydrogen evolution, hydrogen exchange, nitrogen fixation, and applied oxygen tensions in soybean root nodules. Aust. J. Biol. Sci. 16:669-680.

Brun, W. A. 1976. The relation of N_2 fixation to photosynthesis. In: L. D. Hill, ed., World Soybean Research Conference, The Interstate Printers and Publishers, Inc., Danville, Illinois, pp. 135-150.

Bulen, W. A., and J. R. LeComte. 1966. The nitrogenase system from Azotobacter: two enzyme requirement for N_2 reduction, ATP-dependent H_2 evolution and ATP hydrolysis. Proc. Natl. Acad. Sci. U.S. 56:979-986.

Burns, R. C., and R. W. F. Hardy. 1972. Purification of nitrogenase and crystallization of its Mo-Fe protein. Methods in Enzymology 24B:480-496.

Burris, R. H. 1977. The energetics of N_2 fixation: In: These Proceedings.

Ching, T. M., S. Hedtke, S. A. Russell, and H. J. Evans. 1975. Energy state and dinitrogen fixation in soybean nodules of dark grown plants. Plant Physiol. 55:796-798.

Cock, J. H., and S. Yoshida. 1973. Changing sink and source relations in rice (Oryza sativa L.) using carbon dioxide enrichment in the field. Soil Sci. Plant Nutr. 19:229-234.

Cooper, J. P. 1975. Control of photosynthetic production in terrestial systems. In: J. P. Cooper, ed., Photosynthesis and Productivity in Different Environments, Cambridge University Press, London, pp. 593-621.

Criswell, J. G., and R. M. Shibles. 1972. Influence of sink-source on flag-leaf net photosynthesis in oats. Iowa State J. Sci. 46:405-415.

Criswell, J. G., U. D. Havelka, B. Quebedeaux, and R. W. F. Hardy. 1975. Nitrogen fixation under altered rhizosphere pO_2 by excised and intact plants of soybeans. Agron. Abstr., p. 131.

Criswell, J. G., R. W. F. Hardy, and U. D. Havelka. 1976a. Nitrogen fixation in soybeans: measurement techniques and examples of applications. In: L. D. Hill, ed., World Soybean Research Conference, The Interstate Printers and Publishers, Inc., Danville, Illinois, pp. 108-134.

Criswell, J. G., D. J. Hume, and J. W. Tanner. 1976b. Effect of anhydrous ammonia and organic matter on components of nitrogen fixation and yield of soybeans. Crop. Sci. 16:400-404.

Criswell, J. G., U. D. Havelka, B. Quebedeaux, and R. W. F. Hardy. 1976c. Adaptation of nitrogen fixation by nodules of intact soybean plants to altered rhizosphere pO_2. Plant Physiol. 58: in press.

Criswell, J. G., U. D. Havelka, B. Quebedeaux, and R. W. F. Hardy. 1977. Effect of rhizosphere pO_2 on nitrogen fixation by excised and intact nodulated soybean roots. Crop Sci.: in press.

Cummings, M. B., and C. H. Jones. 1918. The aerial fertilization of plants with carbon dioxide. Vt. Agr. Exp. Sta. Bull. 211.

Demoussy, E. 1904. Sur la vegetation dans des atmospheres riches en acide carbonique. Compt. Rend. Acad. Sci. 139:883-885.

de Saussure, Th. 1804. Recherches Chemiques sur la Vegetation, Paris.

Dixon, R. O. D. 1968. Hydrogenase in pea root nodule bacteroids. Arch. Mikrobiol. 62:272-283.

Dixon, R. O. D. 1972. Hydrogenase in legume root nodule bacteroids: occurrence and properties. Arch. Mikrobiol. 85:193-201.

Evans, L. T. 1975. Beyond photosynthesis - the role of respiration, translocation, and growth potential in determining productivity. In: J. P. Cooper, ed., Photosynthesis and Productivity in Different Environments, Cambridge University Press, London, pp. 501-507.

Fischer, R. A., and I. Aguilar M. 1976. Yield potential in a dwarf spring wheat and the effect of carbon dioxide fertilization. Agron. J. 68:749-752.

Garcia, R., and J. J. Hanway. 1976. Foliar fertilization of soybeans during the seed-filling period. Agron. J. 68:653-657.

Gibson, A. H. 1975. Recovery and compensation by

nodulated legumes to environmental stress. In: P. S. Nutman, ed., Symbiotic Nitrogen Fixation in Plants, Cambridge University Press, London, pp. 385-403.

Gifford, R. M., P. M. Bremner, and D. B. Jones. 1973. Assessing photosynthetic limitation to grain yield in a field crop. Aust. J. Agric. Res. 24:297-307.

Ham, G. E., I. E. Liener, S. D. Evans, R. D. Frazier, and W. W. Nelson. 1975. Yield and composition of soybean seed as affected by N and S fertilization. Crop Sci. 67:293-297.

Hardman, L. L., and W. A. Brun. 1971. Effect of atmospheric carbon dioxide enrichment at different developmental stages on growth and yield components of soybeans. Crop Sci. 11:886-888.

Hardy, R. W. F. 1976. Chemical and biological research will provide the new technologies to increase world crop productivity to feed the expanding billions. Am. Chem. Soc. Centennial Mtg. Symposium on Energy, Food, Population, and World Interdependence, New York, Am. Chem. Soc., in press.

Hardy, R. W. F., and J. G. Criswell. 1976. Assessment of environmental limitations of symbiotic $N_2[\,C_2H_2]$ fixation: temperature and pO_2. Agron. Abstr. p. 72.

Hardy, R. W. F., and U. D. Havelka. 1973. Symbiotic N_2 fixation: multifold enhancement by CO_2 enrichment of field-grown soybeans. Plant Physiol. 48(S):35.

Hardy, R. W. F., and U. D. Havelka. 1975. Photosynthate as a major factor limiting nitrogen fixation by field-grown legumes with emphasis on soybeans. In: P. S. Nutman, ed., Symbiotic Nitrogen Fixation in Plants, Cambridge University Press, London, pp. 421-439.

Hardy, R. W. F., R. D. Holsten, E. K. Jackson, and R. C Burns. 1968. The acetylene-ethylene assay for N_2 fixation: laboratory and field evaluation. Plant Physiol. 43:1185-1205.

Hardy, R. W. F., R. C. Burns, R. R. Hebert, R. D. Holsten, and E. K. Jackson. 1971. Biological nitrogen fixation: a key to world protein. Plant Soil (Special Volume), pp. 561-590.

Hardy, R. W. F., R. C. Burns, and R. D. Holsten. 1973. Applications of the acetylene-ethylene assay for measurement of nitrogen fixation. Soil Biol. Biochem. 5:47-81.

Hardy, R. W. F., U. D. Havelka, and B. Quebedeaux. 1976. Opportunities for improved seed yield and protein production: N_2 fixation, CO_2 fixation, and oxygen control of reproductive growth. In: Genetic Improvement of Seed Protein, National Academy of Sciences, Washington, D. C., pp. 196-228.

Hardy, R. W. F., J. G. Criswell, and U. D. Havelka. 1977. Investigations of possible limitations of nitrogen fixation by legumes: (1) methodology, (2) identification, and (3) assessment of significance. In: W. E. Newton, J. R. Postgate, and C. Rodriguez-Barrueco, eds., Proceedings 2nd International Symposium on Nitrogen Fixation, Academic Press, New York, in press.

Harper, J. E. 1974. Soil and symbiotic nitrogen requirements for optimum soybean production. Crop Sci. 14:255-260.

Harper, J. E. 1976. Contribution of dinitrogen and soil or fertilizer nitrogen to soybean (Glycine max L. Merr.) production. In: L. D. Hill, ed., World Soybean Research Conference, The Interstate Printers and Publishers, Inc., Danville, Ill., pp. 101-107.

Harper, L. A., D. N. Baker, J. E. Box, Jr., and J. D. Hesketh. 1973a. Carbon dioxide and the photosynthesis of field crops: a metered carbon dioxide release in cotton under field conditions. Agron. J. 65:7-11.

Harper, L. A., D. N. Baker, and J. E. Box, Jr. 1973b. Fertilize the air over a field. Crops & Soils Magazine, pp. 8-9, November.

Havelka, U. D., and R. W. F. Hardy. 1974. Agron. Abstr. p. 133.

Havelka, U. D., and R. W. F. Hardy. 1976a. Legume N_2 fixation as a problem in carbon nutrition. In: W. E. Newton and C. J. Nyman, eds., Proceedings of the 1st International Symposium on Nitrogen Fixation, Washington State University Press, Pullman, Washington, pp. 456-475.

Havelka, U. D., and R. W. F. Hardy. 1976b. $N_2[C_2H_2]$ fixation, growth, and yield response of field-grown peanuts (Arachis hypogea L.) when grown under ambient and 1500 ppm CO_2 in the foliar canopy. Agron. Abstr. p. 72.

Havelka, U. D., and R. W. F. Hardy. 1976c. Unpublished results.

Hoch, G. E., K. C. Schneider, and R. H. Burris.

1960. Hydrogen evolution and exchange and conversion of N_2O to N_2 by soybean root nodules. Biochim. Biophys. Acta 37:273-279.

   Huang, C. Y., J. S. Boyer, and L. N. Vanderhoef. 1975a. Acetylene reduction (nitrogen fixation) and metabolic activities of soybean having various leaf and nodule water potentials. Plant Physiol. 56:222-227.

   Huang, C. Y., J. S. Boyer, and L. N. Vanderhoef. 1975b. Limitation of acetylene reduction (nitrogen fixation) by photosynthesis in soybean having various leaf and water potentials. Plant Physiol. 56:228-232.

   Johnson, J. W., L. F. Welch, and L. T. Kurtz. 1975. Environmental implications of N fixation by soybeans. J. Environ. Qual. 4:303-306.

   Krenzer, E. G., and D. N. Moss. 1975. Carbon dioxide enrichment effects upon yield and yield components in wheat. Crop Sci. 15:71-74.

   Kruesler, U. 1885. Über eine methode zur beobachtung der assimilation und athmung der pflanzen. Landw. Jahrb. 14:913-965.

   Lawn, R. J., and W. A. Brun. 1974a. Symbiotic nitrogen fixation in soybeans. I. Effect of photosynthetic source sink manipulation. Crop Sci. 14:22-25.

   Lawn, R. J., and W. A. Brun. 1974b. Symbiotic nitrogen fixation in soybeans. III. Effect of supplemental nitrogen and intervarietal grafting. Crop Sci. 14:22-25.

   Lemon, E., D. W. Stewart, and R. W. Shawcroft. 1971. The sun's work in a cornfield. Science 174:371-378.

   Mague, T. H., and R. H. Burris. 1972. Reduction of acetylene and nitrogen by field-grown soybeans. New Phytologist 71:275-286.

   Minchin, F. R., and J. S. Pate. 1973. The carbon balance of a legume and the functional economy of its root nodules. J. Exp. Bot. 24:259-271.

   Mortensen, L. E. 1966. Components of cell-free extract of Clostridium pasteurianum required for ATP-dependent H_2 evolution from dithionite and for N_2 fixation. Biochim. Biophys. Acta 127:18-25.

   Phillips, D. A., K. D. Newell, S. A. Hassell, and C. S. Felling. 1976. The effect of CO_2 enrichment on root nodule development and symbiotic N_2 reduction in Pisum sativum L. Am. J. Bot. 63:356-362.

Quebedeaux, B., and R. W. F. Hardy. 1973. Oxygen as a new factor controlling reproductive growth. Nature 243:477-479.

Quebedeaux, B., and R. W. F. Hardy. 1975a. Reproductive growth and dry matter production of Glycine max (L.) Merr. in response to oxygen concentration. Plant Physiol. 55:102-107.

Quebedeaux, B., and R. W. F. Hardy. 1975b. O_2 Effects on transport and accumulation of photosynthate from leaves to reproductive structures in the light. Plant Physiol. 55(S):17.

Quebedeaux, B., and R. W. F. Hardy. 1976. Oxygen concentration: regulation of crop growth and productivity. In: R. H. Burris and C. C. Black, eds., CO_2 Metabolism and Crop Productivity, University Park Press, Baltimore, Md., pp. 185-204.

Quebedeaux, B., U. D. Havelka, K. L. Livak, and R. W. F. Hardy. 1975. Effect of altered pO_2 in the aerial part of soybean on symbiotic N_2 fixation. Plant Physiol. 56:761-764.

Reihau, E. 1927. Praktische kohlensäuredüngung in gärtnerei und landwirtschaft. Berlin.

Richter, A. A. 1938. On the practice of aerial fertilization with carbon dioxide. Compt. Rend. (Doklady) Acad. Sci. SSSR 18:59-62.

Schubert, K. R., and H. J. Evans. 1976. Hydrogen evolution: a major factor affecting the efficiency of nitrogen fixation in nodulated symbionts. Proc. Natl. Acad. Sci. U.S. 73;1207-1211.

Shivashankar, K., K. Vlassak, and J. Livens. 1976. A comparison of the effect of straw incorporation and CO_2 enrichment on the growth, nitrogen fixation, and yield of soybeans. J. Agric. Sci. Camb. 87:181-185.

Silver, W. S., and R. W. F. Hardy. 1976. Newer developments in biological dinitrogen fixation of possible relevance to forage production. In: C. S. Hoveland, ed., Biological Nitrogen Fixation in Forage-Livestock Systems, American Society for Agronomy, Madison, Wisconsin, pp. 1-36.

Silverstein, R., and W. A. Bulen. 1970. Kinetic studies of the nitrogenase catalyzed hydrogen evolution and nitrogen reduction reactions. Biochemistry 9:3809-3815.

Sinclair, T. R., and C. T. de Wit. 1975. Comparative analysis of photosynthate and nitrogen requirements in the production of seeds by various crops. Science 189:565-567.

Sinclair, T. R., and C. T. de Wit. 1976. Analysis of the carbon and nitrogen limitations to soybean yield. Agron. J. 68:319-324.

Sloger, C., D. Bezdicek, R. Milberg, and N. Boonkerd. 1975. Seasonal and diurnal variations in N_2 [C_2H_2]-fixing activity in field soybeans. In: W. D. P. Stewart, ed., Nitrogen Fixation by Free-Living Micro-Organisms, Cambridge University Press, London, pp. 271-284.

Small, J. G., and O. A. Leonard. 1969. Translocation of C^{14}-labelled photosynthate in nodulated legumes as influenced by nitrate-nitrogen. Am. J. Bot. 56:187-194.

Thaer, A. 1880. Grundsätze der rationallen landwirtschaft. Berlin.

Thibodeau, P. S., and E. G. Jaworski. 1975. Patterns of nitrogen utilization in the soybean. Planta 127:133-147.

Wareing, P. F., and J. Patrick. 1975. Source-sink relations and the partition of assimilates in the plant. In: J. P. Cooper, ed., Photosynthesis and Productivity in Different Environments, Cambridge University Press, London, pp. 481-494.

Weber, D. F., B. E. Caldwell, C. Sloger, and G. H. Vest. 1971. Some USDA studies on the soybean-Rhizobium symbiosis. Plant Soil (Special Volume) pp. 293-304.

Weil, R. R., and A. J. Ohlrogge. 1975. Seasonal development of and the effect of inter-plant competition on soybean nodules. Agron. J. 67:487-490.

Wilson, P. W., E. B. Fred, and M. R. Salmon. 1933. Relation between carbon dioxide and elemental nitrogen assimilation in leguminous plants. Soil Sci. 35:145-165.

Wilson, P. W., and F. C. Wagner. 1935. Combined nitrogen and the nitrogen fixation process in leguminous plants. Transactions Wisconsin Academy of Sciences, Arts and Letters, Madison, Wisconsin, 30:43-50.

Wittwer, S. H. 1967. Carbon dioxide and its role in plant growth. Proc. XVII Int. Hort. Cong., 3:311-322.

Yoshida, S. 1973. Effects of CO_2 enrichment at different stages of panicle development on yield components and yield of rice (Oryza sativa L.). Soil Sci. Plant Nutr. 19:311-316.

DISTRIBUTION OF NITROGEN-FIXING BLUE-GREEN ALGAE IN THE FORESTS

ATSUSHI WATANABE

Biological Laboratory, Seijo University, Setagaya-ku, Tokyo, Japan

Summary

The present observations show that the nitrogen-fixing blue-green algae are present even in the forests of cold northern districts such as Hokkaido. Moreover, it was indicated that the algal fixers are present in higher frequency of occurrence in the forests than in fields in the tropics such as Malaysia, and that they must be participating in the improvement of the nitrogen fertility of not only the soils of fields but also those of forests.

Introduction

In the earlier report (1) of this line of research, report was made of nitrogen-fixing blue-green algae in the four main islands of Japan, in which it was stated that no algal fixers were detected from the samples collected from the paddy mud and field soil of the northernmost island, Hokkaido, while these algae were found in the southern islands, Honshu, Shikoku and Kyushu. Recently the author had an opportunity to collect specimens from the forests of Hokkaido and found that there exist some algal fixers in the natural forests.

Several years ago the author had a chance to investigate the natural forests of Australia, New Guinea and Malaysia and found that the algal fixers are present in higher frequency of occurrence in the forest than in paddy mud and field soil.

Jurgensen and Davey (2) have reported that the algal fixers are generally absent in acid forest soils but abundantly present in acid nursery soils. No other reports have been made about the algal fixers in the forests.

The present paper reports on the distribution of the algal fixers in the forests of Hokkaido, Australia, New Guinea and Malaysia, and the relationship between their occurrence and some conditions of those forests.

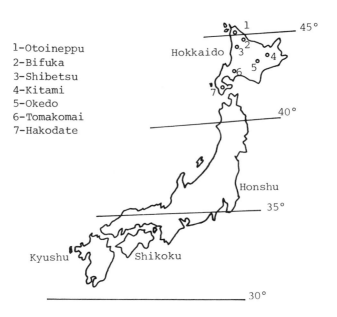

1-Otoineppu
2-Bifuka
3-Shibetsu
4-Kitami
5-Okedo
6-Tomakomai
7-Hakodate

Fig. 1. Map of Japan.

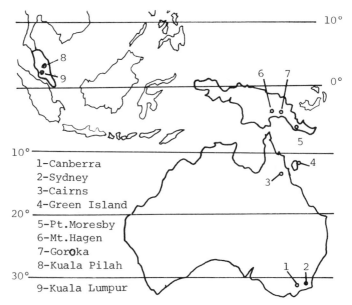

1-Canberra
2-Sydney
3-Cairns
4-Green Island
5-Pt.Moresby
6-Mt.Hagen
7-Goroka
8-Kuala Pilah
9-Kuala Lumpur

Fig. 2. Map of Australia, New Guinea and Malaysia.

Experimental

At each location three types of sample were collected: a soil
sample and a rotted, fallen leaf sample taken from just under the
tree, and a moss sample gathered from the surface of the bark near
the root.
One loopful or small piece of the sample was inoculated into an
Ehrlenmeyer flask of 300 ml capacity containing 100 ml of a nitrogen-free culture medium which consisted of per liter: 0.25 g
K_2HPO_4 ; 0.25 g $MgSO_4 \cdot 7H_2O$; 0.1 g NaCl ; 0.01 g $CaCl_2$; 0.01 g
$FeSO_4 \cdot 7H_2O$; Arnon's A_5-solution 1 ml. The inoculated flasks were
kept in a greenhouse at about 25°C. After repeated subculturing
in this nitrogen-free medium, the growth of nitrogen-fixing blue-
green algae, if any, was obtained.
A small portion of the algal mass from a crude culture was first
vigorously shaken for 20 minutes in sterilized culture medium with
the use of an electric vibrator. The disintegrated algal mass was
after exposure to ultraviolet ray for 1-5 minutes at a distance of
30 cm from the light source (10 W sterilizing lamp) transfered to
a silica-gel plate soaked with the same nitrogen-free medium. On
incubating them for one or two weeks at 25°C, under the weak light
of the greenhouse, the colonies of the blue-green alga were usually found growing on the surface of the plate. The algal samples
obtained from these colonies were subjected to taxonomic determination of their genera and species.
The conditions of the forests, such as the state of plantations
kind of trees and pH value of soils were recorded.

Results and Discussion

The results obtained are summarezed in Tables 1, 2, 3 and 4.
Respective localities are indicated in Figs. 1 and 2.

A) Hokkaido

From the 116 samples examined, 15 forms of algal fixers were isolated, which were found to consist of 3 species, *Nostoc punctiforme*
(Kütz.) Hariot, *N. paludosum* Kützing and *Anabaena cylindrica* Lemm.
The samples which contain algal fixers represent 12.9 percent of
the total samples collected.
In the previous paper (1), it was reported that no algal fixers
was found in 53 samples collected from the fields of Hokkaido.
This island is situated between the 41° and 46° north latitude and
has severe winters. The existence of these algal fixers in the
forests of Hokkaido may be, at least partly, due to the sheltering
effects of the canopy and leaf litter of the forests, which are
lacking in the fields.
It is evident further from Table 1 that all algal fixers were
those isolated from the samples collected from natural forests and
that all these forests were those containing broad leaved trees.

Concerning the planted forests investigated it has been recorded that *Larix leptolepis* Gord. was transplanted in 1957 at Bifuka, *Abies sachalinensis* Masters in 1927 at Shibetsu, *Picea Glehnii* Masters in 1932 at Kitami, *Picea Glehnii* Masters in 1962 at Tomakomai, and *Cryptomeria japonica* D. Don in 1914 and *Abies sachalinensis* Masters in 1923 at Hakōdate. From the samples of these 5 plantations no algal fixers were detected. These plantations had all acidic soils.

The pH-values of the soils of natural forests were more or less higher than those of planted forests. The soil samples which had the high frequency of occurrence of algal fixers showed pH-values near neutral, as seen in the case of the Okedo region.

B) Australia

From the 21 samples examined, 4 forms of algal fixers were isolated, which were found to consist of 4 species, *Nostoc paludosum* Kuetzing, *N. sphaericum* Vaucher and 2 *Nostoc* sp. The samples which contain algal fixers represent 19.0 per-cent of the total samples collected.

C) New Guinea

From the 19 samples examined, 7 forms of algal fixers were isolated, which were found to consist of 5 species, *Nostoc punctiforme* (Kuetz.) Hariot, *N. pruniforme* C.A. Agardh, *Calothrix atricha* Fremy, *Hapalosiphon confervaceus* Borzi. and *Plectonema purpureum* Gomont. The samples which contain algal fixers represent 37.3 per-cent of the total samples collected.

D) Malaysia

From the 22 samples examined, 6 forms of algal fixers were isolated, which were found to consist of 5 species, *Nostoc punctiforme* (Kuetz.) Hariot, *N. muscorm* C.A. Agardh, *Tolypothrix distorta* Kuetzing, *Calothrix parietana* (Naeg.) Thuret, *Plectonema nostocrum* Bornet. The samples which contain algal fixers represent 27.2 per-cent of the total samples collected.

In the earlier research (1), report was made of the distribution of algal fixers in Malaysia, in which it was stated that from 12 samples collected from the fields, 2 forms of algal fixers were isolated(16.6%). It is, therefore, observed that the percentage of the forests was higher than that of the fields.

Table 1. *Distribution of N₂-fixing blue-green algae in the forests of Hokkaido.*

I. Locality: Otoineppu

Consecutive number samples (No.)	State of forests	Kind of trees*	pH value of soils	Type of samples	Number of samples	Existence of N₂-fixing algae**	Number of N₂-fixing species
1	Natural	Broad leaved	4.50	Moss	6	–	1
2				Leaf		–	
3				"		+	
4				Soil		–	
5				"		–	
6				Moss		–	
7	Natural	Broad and needle leaved	5.25	Moss	10	+	2
8				Leaf		–	
9				Soil		–	
10				Moss		+	
11				Leaf		+	
12				Soil		–	
13				"		–	
14				Leaf		–	
15				"		–	
16				Soil		–	

* The species of trees, under which the samples were collected, were as follows:

    No. 1-2 *Magnolia obovata* Thunb.   3-4 *Fagus crenata* Blume
         5-6 *Ostrya japonica* Sargent   7-9 *Fagus crenata* Blume
       10-12 *Magnolia obovata* Thunb. 13-14 *Picea Glehnii* Masters
       15-16 *Quercus serrata* Thunb.

** The isolated nitrogen-fixing blue-green algae were identified to be as follows:
    No. 3,7 and 11 *Anabaena cylindrica* Lemm.

II. Locality: Bifuka

No.			pH	Substrate			
17	Natural	Broad leaved	5.50	Leaf	6	–	1
18				Moss		–	
19				Leaf		–	
20				Moss		+	
21				Soil		–	
22				Leaf		–	
23	Planted	Needle leaved	5.63	Leaf	10	–	0
24				"		–	
25				Soil		–	
26				Leaf		–	
27				"		–	
28				Soil		–	
29				Moss		–	
30				Leaf		–	
31				Soil		–	
32				"		–	

No. 17-18 *Tilia japonica* Engl.   19-20 *Fagus crenata* Blume
     21-22 *Acer mono* Maxim.   23-32 *Larix leptolepis* Gord.

No. 20 *Nostoc punctiforme* (Kütz.) Hariot

III. Locality: Shibetsu

No.							
33	Planted	Needle leaved	5.15	Moss		—	
34				Leaf		—	
35				Moss		—	
36				Soil		—	
37				"	10	—	0
38				"		—	
39				Leaf		—	
40				Moss		—	
41				Soil		—	
42				"		—	

No. 33–42 *Abies sachalinensis* Masters

IV. Locality: Kitami

No.							
43	Planted	Needle leaved	5.82	Moss		—	
44				Leaf		—	
45				"		—	
46				Soil	8	—	0
47				"		—	
48				Leaf		—	
49				"		—	
50				Soil		—	

No. 43–50 *Picea Glehnii* Masters

V. Locality: Okedo

No.			pH	Substrate			
51	Natural	Broad and needle leaved	6.92	Leaf		−	
52				Moss		−	
53				Soil	6	+	1
54				"		−	
55				"		−	
56				"		−	
57	Natural	Broad and needle leaved	6.87	Moss		+	
58				Leaf		−	
59				Soil	9	−	5
60				Moss		+	
61				Leaf		+	
62				Soil		−	
63				"		+	
64				"		−	
65				"		+	

No. 51-53 *Acer mono* Maxim.   54 *Abies sachalinensis* Masters
55 *Ulmus davidiana* Planch var. *japonica* Nakai
56 *Picea Glehnii* Msters   57-59 *Quercus serrata* Thunb.
60-62 *Betula Ermani* Cham. var. *nipponica* Maxim.
63 *Tilia japonica* Engl.   64 *Acer mono* Maxim.
65 *Ostrya japonica* Sargent

No. 53,57,60,61,63 and 65 *Nostoc punctiforme* (Kütz.) Hariot

VI. Locality: Tomakomai

No.	Type	Vegetation	pH	Substrate			
66	Natural	Broad and needle leaved		Moss		–	
67				Leaf		–	
68				Soil		–	
69				Moss		–	
70				Leaf		+	
71			5.80	Soil	15	–	3
72				Moss		–	
73				Leaf		–	
74				Soil		–	
75				Moss		–	
76				Leaf		+	
77				Soil		–	
78				Moss		–	
79				Leaf		+	
80				Soil		–	
81	Natural	Broad leaved		Moss		–	
82				Leaf		+	
83			5.80	Soil	6	–	1
84				Moss		–	
85				Leaf		–	
86				Soil		–	
87	Planted	Needle leaved		Moss		–	
88				Leaf		–	
89			5.50	Soil	6	–	0
90				Moss		–	
91				Leaf		–	
92				Soil		–	

331

No. 66-68 *Magnolia obovata* Thunb.   69-74 *Abies sachalinensis* Masters
75-77 *Ulmus davidiana* Planch var. *japonica* Nakai
78-80 *Quercus serrata* Thunb.   81-83 *Ostrya japonica* Sargent
84-86 *Quercus serrata* Thunb.   87-92 *Picea Glehnii* Masters
No. 70 *Nostoc punctiforme* (Kütz.) Hariot
76,79 and 82 *Nostoc paludosum* Kützing

VII. Locality: Hakodate

No.							
93	Planted	Needle leaved	5.15	Moss		—	
94				Leaf		—	
95				Soil		—	
96				Moss		—	
97				Leaf		—	
98				Soil	9	—	0
99				Moss		—	
100				Leaf		—	
101				Soil		—	
102	Natural	Broad leaved	6.45	Moss		—	
103				Leaf		—	
104				Soil		—	
105				Moss		—	
106				Leaf		—	
107				Soil	15	—	1
108				Moss		—	
109				Leaf		—	
110				Soil		—	
111				Moss		—	
112				Leaf		—	
113				Soil		—	
114				Moss		—	

	115	116
Leaf	+	
Soil	−	

No. 93–98 *Cryptomeria japonica* D. Don
99–101 *Abies sachlinensis* Masters
105–107 *Magnolia obovata* Thunb.
111–113 *Fagus crenata* Blume
102–104 *Fagus crenata* Blume
108–110 *Acer mono* Maxim.
114–116 *Quercus serrata* Thunb.

No.115   *Nostoc punctiforme* (Kütz) Hariot

Table 2. *Distribution of N_2-fixing blue-green algae in the forests of Australia.*

Consecutive number of samples (No.)	Place of sampling	State of forests	Kind of trees	Type of samples	Number of samples	Existence of N_2-fixing algae	Number of N_2-fixing species
1				Moss		−	
2				Soil		−	
3				"		−	
4				Moss		−	
5	Canberra	Natural	Broad leaved	Soil	12	−	2
6				"		−	
7				"		−	
8				Moss		+	
9				Soil		−	
10				Moss		+	
11				Moss		−	
12				Soil		−	
13				Moss		+	

No.	Location						
14	Sydney	Natural	Broad leaved	Soil	3	–	1
15				"		–	
16	Cairns	Natural	Broad leaved	Moss	3	–	0
17				Soil		–	
18				"		–	
19	Green island	Natural	Broad leaved	Soil	3	–	1
20				"		–	
21				"		+	

No. 1 *Eucalyptus gummifera* Hochr. 2 *E. trifora* Blackely
3 *E. floribunda* F. Muell. 4 *E. mannifera* A. Cunn.
5 *E. Rossei* R. T. Baker 6 *E. globoidea* Blakely
7 *E. macrorhyncha* F. Muell. 8 *E. umbra* R. T. Baker
9 *Acacia dealbata* Link 10 *A. decurrens* Willd.
11 *A. dealbata* Link 13 *A. cultiformis* Cunn.
14 *Eucalyptus macrorhyncha* F. Muell.
15 *E. trifora* Blakely

No. 8 *Nostoc* sp. 10 *N. paludosum* Kuetzing
13 *N.* sp. 21 *N. sphaericum* Vaucher

Table 3. *Distribution of N₂-fixing blue-green algae in the forests of Papua New Guinea.*

No.	Location			Substrate		Sign	
1	Pt. Moresby	Natural	Broad leaved	Soil		–	
2				"		–	
3				Moss		–	
4				Soil	7	–	1
5				"		–	
6				"		–	
7				Moss		+	
8	Mt. Hagen	Natural	Broad leaved	Moss		+	
9				"		–	
10				Soil		–	
11				"		+	
12				"	9	+	5
13				Moss		–	
14				Soil		+	
15				"		–	
16				"		+	
17	Goroka	Natural	Broad leaved	Soil		+	
18				"	3	–	1
19				Moss		–	

No. 7 *Nostoc punctiforme* (Kuetz.) Hariot 11 *Calothrix atricha* Frémy
     8 *N. pruniforme* C. A. Agardh 12 and 14 *Hapalosiphon confervaceus* Borzi 17 *Nostoc punctiforme* (Kuetz.) Hariot
     16 *Plectonema purpureum* Gomont

Table 4. *Distribution of N₂-fixing blue-green algae in the forests of Malaysia.*

No.	Location	Forest	Substrate		+/−	
1	Kuala Pilah	Natural	Broad leaved	Soil	−	
2				"	−	
3				Moss	−	
4				Soil	+	
5				"	+	
6				"	−	
7				Moss	14	+	5
8				Soil	+	
9				"	+	
10				"	−	
11				"	−	
12				"	−	
13				Moss	−	
14				Soil	−	
15	Kuala Lumpur	Natural	Broad leaved	Soil	−	
16				"	−	
17				"	−	
18				"	8	−	1
19				"	−	
20				"	+	
21				Moss	−	
22				"	−	

No. 4 *Plectonema nostocorum* Bornet   5 *Tolypothrix distorta* Kuetzing
     7 *Calothrix parietana* (Naeg.) Thuret
     8 *Plectonema nostocorum* Bornet   9 *Nostoc punctiforme* (Kuetz.) Hariot
    20 *N. muscorum* C. A. Agardh

336

Literature cited

1. Watanabe,A. 1959. Distribution of nitrogen-fixing blue-green algae in various areas of south and east Asia. J. Gen. Appl. Microbiol. 5: 21.
2. Jurgensen,M.F. and C.B.Davey. 1968. Nitrogen-fixing blue-green algae in acid forest and nursery soils. Can. J. Microbiol. 14: 1179.

GENETIC DEREPRESSION OF NITROGENASE -

MEDIATED H_2 EVOLUTION BY *KLEBSIELLA PNEUMONIAE*

Kjell Andersen, K. T. Shanmugam and R. C. Valentine
Plant Growth Laboratory
Department of Agronomy and Range Science
University of California
Davis, California 95616

Introduction

The efficient conversion of water to hydrogen using solar energy represents one of the greatest challenges facing scientists today:

$$\text{Solar Energy} + H_2O \rightarrow H_2 + \tfrac{1}{2}O_2$$

Biologists have long recognized that "light is equivalent to H_2" in photosynthetic organisms where precursors of H_2, such as reduced ferredoxin, are generated in great quantities. With few exceptions these powerful reductants are used for CO_2 fixation and other cellular reactions rather than H_2 formation. Genetic engineers eye this problem and wonder if it might be possible to genetically "derepress" the H_2-forming genes resulting in increased synthesis of H_2 during photosynthesis. Unfortunately, there is very little background information on the molecular biology of H_2 production. In connection with our NSF-RANN-supported program on the applications of molecular biology of N_2 fixation we have begun some work in this area. This is a progress report.

Derepression of Nif Genes Results in Derepression of Nitrogenase Mediated H_2 Evolution

For this study it was necessary to construct a triple-mutant of *Klebsiella* with the following properties: The first two mutations derepress nitrogenase synthesis in the presence of NH_4^+ and blocks the

339

assimilation of NH_4^+ into glutamate, resulting in ex-
port of fixed nitrogen (1,2,3). The third mutation
blocks the conventional hydrogen evolution and uptake
system.

Selection for mutants blocked in conventional hydrogen evolution and uptake activity involved the use of
the nitrate analogue, chlorate. Chlorate resistant
clones used here are pleiotropic, being nitrate assimilation negative as well as blocked for conventional
hydrogen evolution activity. The selection and properties of these strains will be described in detail in a
separate publication (4).

The apparatus used for studying nitrogenase-mediated H_2 evolution (Fig. 1A) as well as the time course
of H_2 evolution by the mutant strains in the presence
and absence of NH_4^+ (Fig. 1B) are illustrated in Fig.1.

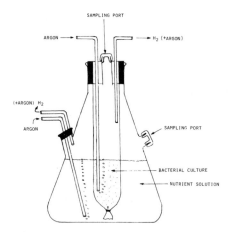

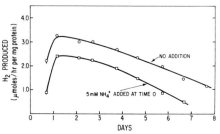

Fig.1A Apparatus (Artificial Nodule) for measuring
nitrogenase-catalyzed H_2
evolution.
(See text for detail)

Fig.1B Time course of H_2
evolution by the Nif derepressed mutant N-20,
blocked in conventional
H_2 evolution.

For this experiment, Nif derepressed mutants of *Kleb-siella* were grown in a dialysis bag (25ml), bathed in nutrient solution (250ml glucose-salts medium with 0.1 M phosphate; Ref.3). Both the bacterial culture and the nutrient solution were sparged constantly with argon. Samples were withdrawn aseptically with syringes through sampling ports closed with serum stoppers. Usually, 50μg/ml of glutamine or glutamate (inside the bag only) is supplied as N-source which gives a total cell yield of 2-3mg cell protein per flask. (The strains require glutamine or glutamate for growth.) The bacterial cultures normally stop dividing during the first 1 to 1.5 days of incubation, after which time they are in an essentially non-growing state. Incubation was at 25°. H_2, flushed from the reactor bag with the sparging gas (argon) was analyzed using a gas chromotograph equipped with a molecular sieve 5A column and a thermal conductivity detector. Fig.1B shows:

i) H_2 evolution is not severely repressed by addition of NH_4^+ (15mM) at the start of the experiment (to be discussed in more detail below);

ii) H_2 evolution continued for almost one week in the reactor.

Some of the properties of strains derepressed (in the presence of NH_4^+) for nitrogenase mediated H_2 evolution are summarized in Table 1.

TABLE I

GENETIC DEREPRESSION OF
NITROGENASE-MEDIATED H_2 EVOLUTION

	Parental	Nitrogenase Activity* (μmoles/hr.mg cell protein)		H_2 Production** (μmoles/hr.mg cell protein)	
Strain	Strain	$-NH_4^+$	$+NH_4^+$ (15mM)	$-NH_4^+$	$+NH_4^+$ (15mM)
N-25	SK-24	0.77	1.50	1.90	2.20
N-27	SK-24	1.95	1.60	2.70	2.00
N-15	M5A1	0.94	<0.001	7.50	0.07

* See Ref. 3 for nitrogenase assay conditions.
** Assayed as described for Fig. 1 after incubating
for 2 days in the reactor.

Note that H_2 evolution catalyzed by the wild-type
organism genetically blocked for conventional H_2 evolution (strain N-15) is almost completely repressed by
NH_4^+; as expected, nitrogenase activity (measured as
acetylene reducing activity) is also strongly repressed. In contrast, nitrogenase-mediated H_2 evolution is not repressed by NH_4^+ at concentrations as high
as 15mM in Nif derepressed strains (N-25 and N-27).
In further experiments (data not shown) it was found
that amino acids, which behave as repressors of nitrogenase in these strains, also completely repressed H_2
evolution.

We conclude that derepression of Nif genes results
in derepression of nitrogenase-mediated H_2 evolution.
In the future it may be possible to genetically manage
the H_2-evolution genes with the goal of "molecular
farming" for H_2.

References

1. Shanmugam, K. T., I. Chan and C. Morandi. 1975.
    Regulation of nitrogen fixation: Nitrogenase
    derepressed mutants of *Klebsiella pneumoniae*.
    Biochem. Biophys. Acta 408:101-111.

2. Nagatani, H., M. Shimizu, and R. C. Valentine.
    1971. The mechanism of ammonia assimilation
    in nitrogen fixing bacteria. Arch. Mikrobiol.
    79:164-175.

3. Shanmugam, K. T. and R. C. Valentine. 1975.
    Microbial production of ammonium ion from
    nitrogen. Proc. Natl. Acad. Sci. USA.
    72:136-139.

4. Andersen, K., and K. T. Shanmugam. 1976. Ener-
    getics of biological nitrogen fixation: De-
    termination of the H_2/NH_4^+ ratio catalyzed
    by nitrogenase of *Klebsiella in Vivo*. Bio-
    chem. Biophys. Acta (in preparation).

SECTION IV. Practical Applications

POSSIBILITIES OF BIOMASS FROM THE OCEAN
THE MARINE FARM PROJECT

WHEELER J. NORTH
W. M. Keck Engineering Laboratories
California Institute of Technology
Pasadena, California   91125

The Marine Farm Project is an undertaking designed to grow Giant Kelp, Macrocystis, in surface waters of the deep ocean. The plant organic matter created would be processed and converted to fuels such as methane. Food might also be produced on the oceanic farms. This paper is primarily concerned with problems involved in growing Macrocystis in a deep water setting.

We have studied adult Macrocystis transplants moored at three semi-oceanic locations (Figure 1). Depth of the attachment was about 12 meters. The transplants were fastened to rope grids which in turn were anchored at depths of 50 to 150 meters (Figure 2). The three grid sites were more than one kilometer from shore and their surface waters were only occasionally enriched with nutrients by coastal upwelling. Strong currents typically occurred at the grid sites. We reasoned that perhaps the rapid flux of nutrients sweeping past the kelp transplants might compensate for

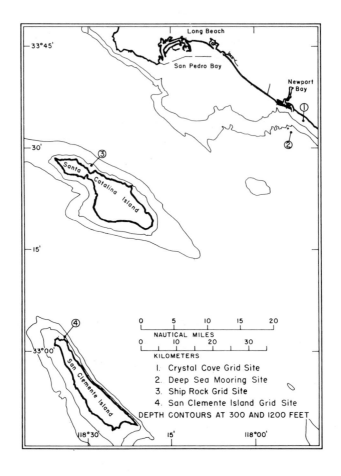

Figure 1. Chart of a portion of southern California showing locations of our grid structures for mooring kelp transplants in deep sea settings (1, 3, 4). The chart also shows the site where we pump up deep water for use in our laboratory experiments and where we installed the wave pump and 300 meter pipe shown in Figure 6 (2).

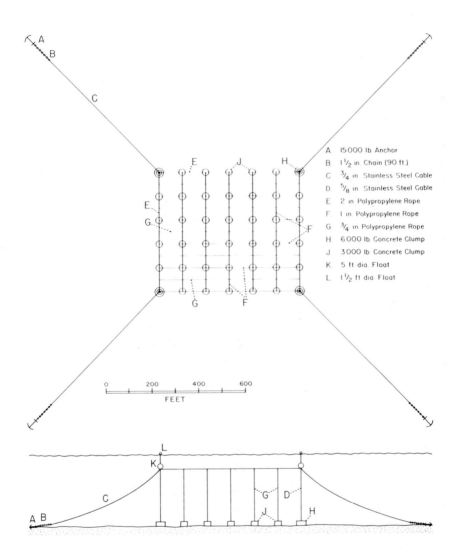

A   15000 lb. Anchor
B   1½ in. Chain (90 ft.)
C   ¾ in. Stainless Steel Cable
D   ⅝ in. Stainless Steel Cable
E   2 in. Polypropylene Rope
F   1 in. Polypropylene Rope
G   ¾ in. Polypropylene Rope
H   6000 lb. Concrete Clump
J   3000 lb. Concrete Clump
K   5 ft. dia. Float
L   1½ ft. dia. Float

Figure 2.   Horizontal and plan views of the grid system constructed by the U. S. Naval Undersea Center off San Clemente Island.   The grid served as a mooring for kelp transplants.

the rather low concentrations in the surface waters. The test plants, however, grew slowly and their nitrogen contents resembled those of plants from natural beds in low-nutrient situations. We measured significant increases in growth rates when artificially upwelled water from depths of 30 to 50 meters was dispersed among our test plants. It appeared that kelp plants on marine farms in oceanic settings would need fertilizing. Deep water might serve as a source of necessary nutrients. Deep water can be raised to the sea surface with relatively small energy expenditure.

Growth was variable when we tested effects of deep water on juvenile kelp plants under static (batch) culturing conditions in the laboratory. Growth rates were high among test plants grown in water collected from depths of 200 to 300 meters in December 1975 and January 1976 (Figure 3). Growth became mediocre or negligible in water from the same depths collected several months later (i.e. May and June, Figure 4; poor growth also occurred in deep water collected during summer and fall).

Failure of deep water to support growth by Macro-cystis might be due to inhibitory substances or to inadequate amounts of nutrients. Calculations indicated that supplies of nitrogen and phosphorus in our culturing media were adequate. If poor growth resulted from an insufficiency, elements such as trace metals were probably responsible. Culturing in a flowing system could probably determine whether failure of deep water to support kelp growth arose from inhibition or inadequacy. Trace substances would be more available to test organisms in flowing systems than in batch systems. Inhibitory compounds would presumably rapidly lead to very poor health in a test plant in a flowing

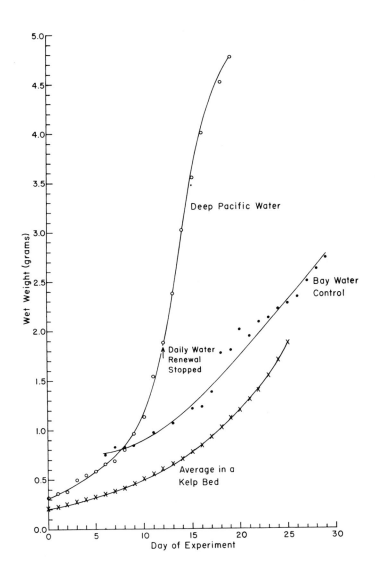

Figure 3. Wet weight changes observed among juvenile
<u>Macrocystis</u> sporophytes growing in Pacific water from
about 230 meters depth and in water from Newport Bay,
California. A computed average curve for small plants
in nature is also shown.

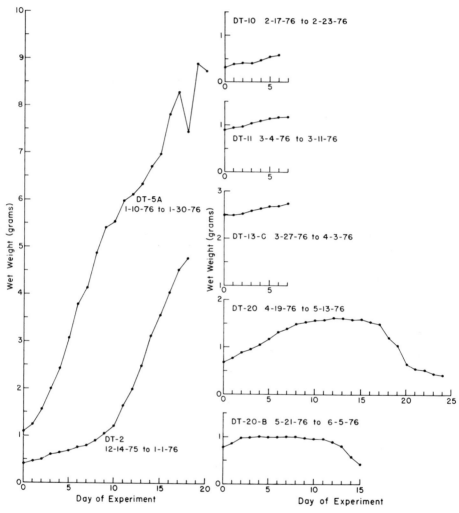

Figure 4. Wet weight changes observed among juvenile *Macrocystis* sporophytes growing in Pacific water pumped up from depths of 230 to 300 meters. Media collected in December and January supported much better growth than water recovered from February to June.

system whereas good growth would indicate that we had
overcome an insufficiency problem present in our batch
culturing technique.

We exposed juvenile plants to flowing deep water
pumped up from depths of 870 meters at the artificial
upwelling facility of the University of Texas on the
island of St. Croix. The test plants grew well in the
deep water (Table 1, Figure 5). Growth became

TABLE 1

Growth rates (percent daily weight increases) recorded
from 3 juvenile Macrocystis sporophytes utilized in
our study of effects of artificially upwelled water at
St. Croix and from a control plant held in flowing bay
water at our Corona del Mar headquarters. Rates (R)
were computed from the equation

$$R = \frac{\ln(W_t/W_o)}{t}$$

where initial and final weights are W_o and W_t and t is
time. Weights are in grams, time is in days.

PLANT NO.	W_o	W_t	W_t/W_o	t	R	REMARKS
1	1.548	12.788	8.261	23	9.2	in deep water from 8/12 to 9/3
2	0.868	6.450	7.431	23	8.7	in deep water from 8/13 to 9/4
3	1.379	12.609	9.143	23	9.6	in deep water from 8/13 to 9/4
11	1.095	9.950	9.087	28	7.9	control in tray at Corona del Mar, CA.

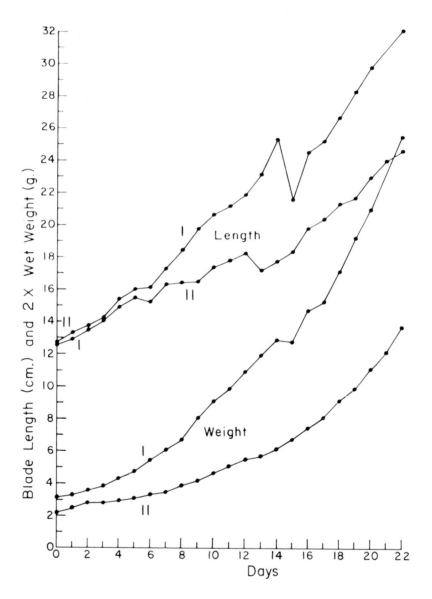

Figure 5. Daily changes in blade length and wet weight
among juvenile Macrocystis sporophytes grown in upwelled Atlantic water from 870 meters deep at St. Croix
(1) and in water from Newport Bay California (11).

outstanding when the deep water was diluted with equal
parts of surface water (the surface water was very low
in NO_3, PO_4, and NH_4) but declined to average values at
greater dilutions tested (Table 2). Trace metal

TABLE 2

Percent daily weight increases computed from 9/5/76 to
9/10/76 for <u>Macrocystis</u> sporophytes held in various
combinations of St. Croix artificially upwelled water
and inshore surface water as indicated. Growth rates
were computed as described in Table 1. Flow rate was
18 L/min. through the deep water tray and about
3 L/min. through the other trays.

Plant No.	Composition of Medium % deep	% surface	W_o gms	W_t gms	R x 100
21	100	0	0.413	0.919	16.0
22	"	"	0.480	1.040	15.4
23	"	"	0.720	1.445	14.0
24	50	50	1.047	2.754	19.3
25	"	"	0.333	0.884	19.5
8	"	"	1.721	4.505	19.2
26	10	90	0.489	0.625	4.5*
27	"	"	0.634	1.597	18.5
7	"	"	2.495	5.680	16.5
28	5	95	0.585	1.112	12.9
29	"	"	0.781	1.369	11.2
10	"	"	2.532	4.723	12.6
30	0	100	0.954	1.523	9.4
31	"	"	1.608	2.345	7.5
9	"	"	4.406	7.929	11.7

* Plant No. 26 was later identified as <u>Eisenia</u>.

composition of the St. Croix deep water, however, might differ significantly from Pacific deep water.

We recently constructed equipment near our California laboratory headquarters (Figure 1) to test kelp growth in flowing upwelled Pacific water and determine if differences existed between Pacific and Atlantic water. We used a wave-energized pump to force water up through a 300 meter vertical pipe permanently moored at a location about 400 meters deep (Figure 6). The

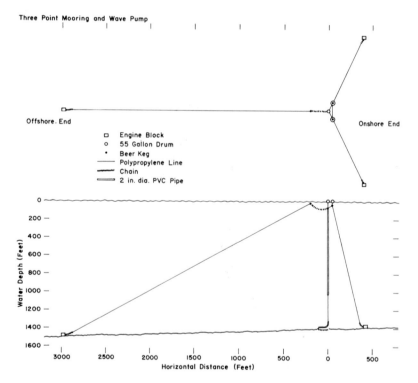

Figure 6. Diagram of three point mooring installed about six kilometers south of Newport Bay California (see Figure 1). The mooring holds a wave pump that consisted of a 300 meter long pvc pipe suspended vertically by an oil drum floating on the surface. A one-way valve in the pipe ensured unidirectional flow.

upwelled water was delivered to a plexiglass chamber
located 13 meters below the sea surface and containing
juvenile <u>Macrocystis</u> sporophytes. These plants displayed excellent growth (Figure 7). Thus flowing deep

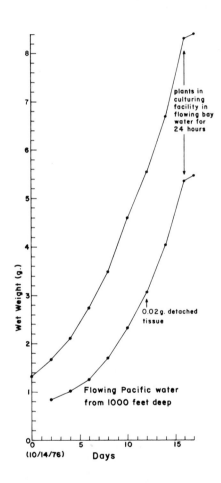

Figure 7. Changes in wet weights observed in two
juvenile <u>Macrocystis</u> sporophytes maintained in the open
sea off southern California at a depth of about 13
meters and exposed to water pumped up from a 300 meter
depth.

water from both the Atlantic and the Pacific Oceans
supported growth that was better than obtained in
certain of our batch cultures. The evidence thus
supports the hypothesis that inadequacy of one or more
trace constituents affected success of our batch
cultures using deep water media.

  We have also obtained some indications that deep
water may contain inhibitory factors. Thus addition of
EDTA to deep water nearly always improves capacity of
the water to support good growth (Figure 8). As noted

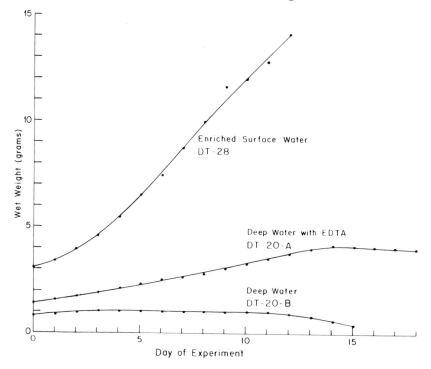

Figure 8. Comparisons of growth records among three
juvenile Macrocystis sporophytes held in different
media as indicated. The deep water came from depths of
about 230 meters. The surface water medium was enriched with NO_3, PO_4, I, Br, Fe, Mn, B, Mo, and Zn.

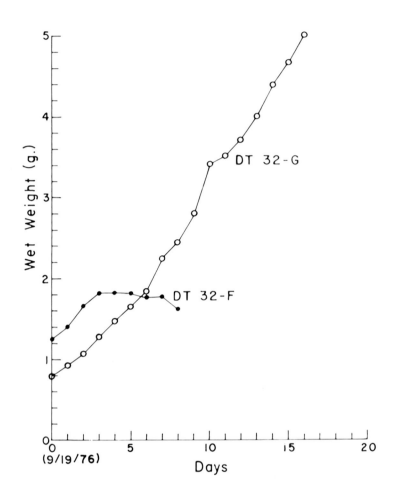

Figure 9. Comparison of growth records from two juvenile *Macrocystis* sporophytes grown in media prepared from water pumped up from a depth of about 45 meters with the following additives: Experiment DT-32G contained NO_3, PO_4, I, Mn, Fe, Zn, Mo, Co, and Li: Experiment DT-32F had all of these additives except Li and concentration of Co was tenfold greater than the level in DT-32G. Subsequent experimentation showed that failure of the plant in DT-32F was due to excess Co and not from a deficiency in Li.

above, mixtures of equal parts of surface and deep
water yield a medium better for growth than pure deep
water alone (see Table 2). The EDTA might bind some
inhibitory substance occurring in deep water. Mixing
deep with surface water might dilute the inhibitory
substance. Thus effects of deep water on kelp growth
as yet are not completely understood. Nonetheless
there is considerable evidence that young sporophytes
do grow very well in deep water media.

Before we can identify specific deficiencies in
samples of deep water, nutritional requirements of
Macrocystis must be defined. We have studied uptake
rates for phosphate, nitrate, and ammonium in Macrocystis. We are currently examining needs for trace
metals. It appears that deficiencies or excessive
amounts of iron, manganese, molybdenum, and cobalt in
the culture medium can result in poor growth relative
to controls (Figure 9). We have not as yet been able
to grow Macrocystis in a completely defined culturing
medium. All our successful media have been prepared
using seawater.

ACKNOWLEDGEMENTS

Grant support from the Energy Research and Development Administration and the National Science Foundation is gratefully acknowledged. Our thanks are due
to Prof. Oswald A. Roels of the University of Texas and
to the staff of this University's facility at St. Croix
for making possible our experiments with Atlantic upwelled water. The author is indebted to J. F. Devinny,
H. C. Fastenau, J. S. Hsiao, G. A. Jackson, L. G.
Jones, P. D. Kirkwood, V. A. Kuga, A. L. Long, J. J.
Morgan, V. A. Roberts, T. G. Stephan, L. M. Taylor,

M. Wheeler, and P. A. Wheeler for advice and assistance in all aspects of these studies. Certain other aspects of the Marine Farm Project have been funded by the American Gas Association.

MASS CULTURE OF *CHLORELLA* IN ASIAN COUNTRIES

OSAMU TSUKADA AND TAKAYOSHI KAWAHARA

Marine Development Division, Miyako Co. Ltd., Kushincho 1-6, Kita-ku, Osaka 530, Japan

AND

SHIGETOH MIYACHI

Institute of Applied Microbiology, University of Tokyo, Bunkyo-ku, Tokyo 113, Japan

Table 1. *Estimated production of* Chlorella *in Japan (1976)*

Company	Culture method	Annual production (tons/year)	Area of pond (m^2)	Productivity ($kg/m^2 \cdot year$)
A	Open culture	90	11,500	7.83
	C-source, acetate	40	6,100	6.56
B	Open culture C-source, acetate	35	5,150	6.80
C	Closed circulation C-source, CO_2 + acetate	5	1,620	3.09
D	Closed circulation C-source, CO_2	3	855	3.51
E	Tank culture C-source, glucose	60		
F	Tank culture C-source, glucose	60		
G	Tank culture C-source, glucose	48		
Other 2 or 3		?		
10 ?		*ca.* 350		

It is estimated that about 400 tons (d.w.) of *Chlorella* was consumed in Japan in 1975. About half of the consumed *Chlorella* was

Table 2. *Estimated production of* Chlorella *in other Asian countries (1976)*

Country	Company	Culture method	Annual production (tons/year)	Area of pond (m^2)	Productivity (kg/m^2·year)
Republic of China	A	Open culture	130	16,200	8.02
	B	"	100	14,500	6.89
	C	"	130	17,100	7.60
	D	"	90	12,700	7.09
	E	"	30	4,200	7.14
	F	"	18	2,400	7.50
	G	"	80	11,000	7.27
	H	"	18	2,300	7.83
	I	"	90	12,000	7.50
	J	"	25	3,600	6.94
	K	Tank culture	85		
	11		796		
Philippines		Open culture	70		
Malaysia		"	70		
		Total *ca.* 940			

produced in Japan and the rest was imported from Asian countries, mostly from Republic of China. Table 1 shows that about 350 tons of *Chlorella* will be produced in Japan in 1976. The table also shows that most of the factories are growing *Chlorella* by mixotrophic culture with the addition of acetate or glucose. Annual productivity ranges from 3 to 8 kg/m^2.

The amount of *Chlorella* to be produced in 1976 in other Asian countries is shown in Table 2. About 800 tons of *Chlorella* will be produced in Republic of China this year and the figure will be significantly larger in next year since more than 10 factories are under construction in this year. There is also a plan to start mass culture of *Chlorella* in Hong Kong. Many of the factories are adopting either one of the culture systems developed in Tokugawa Institute for Biological Research, Tokyo; closed circulation method or open culture system (1).

In the factories which will produce 100 tons of *Chlorella* annually in Japan it is estimated the cost to produce 1 kg of *Chlorella* is $ 5.4-10.8. Assuming that the protein content in *Chlorella* is 55%, the cost per kg of *Chlorella* protein is estimated to be $ 9.8-19.57, while the same cost for soybean protein is $ 0.5-0.7. Therefore *Chlorella* protein is more than 10 times as expensive as soybean protein. However, the whole sale price of *Chlorella* is $ 17-27/kg and the retail price as health food $ 100/kg. This is the reason why so many factories are producing *Chlorella* and the number is still increasing.

In the total amount of *Chlorella* produced only about 60 tons are being used as the feed for fish such as carb and gold fish. Most of the others is being used as health food.

Thus far only fresh water *Chlorella* were grown in large scale. From 1977 the mass culture of sea *Chlorella* to provide the feed for marine protozoa (the feed for various sea fish, crab and lobster) will start in the Marine Development Division of Miyako Co. Ltd.

Literature cited

1. Tamiya,H. 1955. Growing Chlorella for Food and Feed. Proc. World Symposium on Applied Solar Energy, Phenix, Arizona.

MASS PRODUCTION OF ALGAE: BIOENGINEERING ASPECTS.

JOEL C. GOLDMAN and JOHN H. RYTHER

Woods Hole Oceanographic Institution
Woods Hole, Massachusetts 02543

INTRODUCTION

Until recently, the major goal of researchers working on the
mass culturing of unicellular algae was to produce a protein product for animal and/or human consumption. In agriculturally advanced nations such as the United States the approach has been a
dismal failure to date for several reasons: 1) the ready availability of more conventional protein sources, 2) underdeveloped
technology for growing, separating, and processing the unicellular organisms, mainly due to insufficient research funding, and
3) consumer non-acceptance.

Now, because of the "energy crisis", there is renewed interest in mass algal culturing systems (both microscopic and macroscopic plants), but with the new goal of converting the grown organic matter to a highly combustable fuel such as methane. The
basic objectives involved in growing algae are still the same -
yield optimization of desired species - but the overriding consideration now is to produce more energy than is consumed during
process operation.

The major unanswered question thus is "what engineered process design will lead to maximum areal yields with minimum energy
input"? The task then of the bioengineer is to translate the
basic concepts of the photosynthetic process into a functional design in which sunlight conversion efficiencies are maximized with
minimum energy expenditures.

The main energy inputs in mass culture systems consist of
capital energy expenditures involved in initial plant development
(both materials and construction) spread over the life of the system, power for pumping, mixing and chemical addition, if required,
and energy consumption in the production and/or recovery of
nutrients for plant growth.

By considering individually the controllable and uncontrollable variables affecting culture growth, we can gain some insight into the problems facing the bioengineer in developing "energy plantations." The main controllable variables (or operating parameters) include type of culture system (batch, continuous, semi-continuous), nutrient source (external or recycled), culture configuration (surface area, depth), degree and type of mixing, and residence period (control of growth rate). The proper combination of these operating parameters clearly must be fitted to match the uncontrollable conditions of sunlight (intensity and duration) and temperature, which, in turn, are functions or geographical location.

ENERGY BALANCES

The production of useful energy through algal growth involves three crucial steps: 1) photosynthetic production of organic matter, 2) collection and processing of the plant material, and 3) fermentation of the organic matter, leading to methane formation and storage. Each step must be carried out in a sophisticated manner involving detailed engineering design and control; hence, there is a significant requirement for mechanical energy input.

The candidate algal systems that might be considered for solar energy conversion can be grouped into three basic schemes (Fig. 1): 1. *Open*-nutrients and water derived from external sources and then wasted, 2. *Semi-Open* - either nutrients or water recycled, and 3. *Closed* - both nutrients and water recycled.

ALGAL BIOMASS-ENERGY SYSTEMS

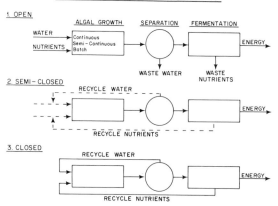

Fig. 1. Flow diagrams of candidate algal growth-energy production systems.

A simple energy balance for analyzing the efficiency of the different schemes can be described as:

$$f = \frac{E_O}{E_I} = \frac{AJY}{E_I} \qquad (1)$$

in which f is the ratio of gross energy out E_O (the caloric content of the plant material produced) to E_I, the energy input required (including losses) to develop and maintain the entire process, A is the surface area of the algae culture, J is the heat of combustion of algae (usually taken as about 5.5 Kcal/gr ash-free dry wt), and Y is the areal yield of ash-free dry wt per unit time. Equation 1 can be rearranged to:

$$\frac{E_I}{A} = \frac{J}{f} Y = CY \qquad (2)$$

in which C is a constant (J/f) for a given f-factor. The term E_I/A is then the maximum allowable total energy input per unit growth area for a given f-factor and yield (Fig. 2).

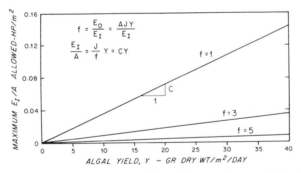

Fig. 2. *Relationship between E_I/A (the maximum allowed total energy input, including losses, per unit area) and algal yield for varying f-factors.*

Aside from the energy input requirements for growth there would be significant energy demands for algal separation and fermentation. Thus, in order to realize a net production of energy the f-factor must be >1. The exact magnitude of f depends on the summation of energy inputs from the various components. Clearly, after all the energy inputs and losses are tallied there should still be a net energy production for the system to be viable.

LAND AND MATERIALS REQUIREMENTS

To better analyze the energy demands on the system, it is instructive to first appreciate the magnitude of what a photosynthetic energy plantation might entail. The U. S. Energy Research and Development Administration (ERDA) has projected that fuels from biomass would have to account for about 5-10% of the total U.S. energy needs in the future for the technique to have potential (Ward, 1976). Summarized in Table 1 are the materials and land requirements necessary for satisfying 5% of the U.S. energy needs in 1990 through algal growth and energy conversion, assuming both a 100% and 10% conversion efficiency.

TABLE 1.

Materials and land required to meet 5% of 1990 U.S. energy needs with algal growth-energy production systems at 100% and 10% conversion efficiencies.[a]

Annual Quantities of Materials Required	Percentage of Energy Conversion Efficiency	
	100%	10%
Energy, BTU	6.1×10^{15}	6.1×10^{15}
Algal Biomass, gr dry wt.[b]	2.8×10^{14}	2.8×10^{15}
Land Area, m^2 [c]	7.6×10^{10}	7.6×10^{11}
Water, m^3 [d]	9.3×10^{11}	9.3×10^{12}
Major Nutrients, gr [e]		
Nitrogen	2.2×10^{13}	2.2×10^{14}
Phosphorus	2.5×10^{12}	2.5×10^{13}

[a] 1990 U.S. needs = 121.9×10^{15} BTU (Enzer *et al.*, 1975)

[b] based on a J value of 5.5 Kcal/gr dry wt

[c] based on a sustained yield of 10 gr dry wt/m^2/day

[d] based on a standing algal crop of 300 mg dry wt/ℓ

[e] based on 8% N content and N:P ratio (by atoms) = 20 in algae.

The land, water, and nutrient requirements necessary are clearly staggering. For example, if 100% conversion efficiency were possible, a land area exceeding that comprising the combined states of Massachusetts, Rhode Island, Connecticut, and Vermont, and a yearly water flow equal to almost twice the volume of Lake Erie (5.4×10^{11} m^3) or the annual discharge of the Mississippi River ($5.5 \times 10^{11} m^3$) would be needed. Nitrogen requirements, which, if derived from domestic waste water, would equal the daily wastewater discharge of about five billion people, which exceeds the current world population (Fig. 3). Similarly, nitrogen obtained from commercial fertilizers would greatly exceed the current annual U.S. production ($\sim 8 \times 10^{12}$ gr).

Considering a more realistic 10% conversion efficiency, land area equal to the state of Texas (6.9×10^{11} m^2) and an annual water flow greater than the combined volumes of Lakes Michigan, Ontario, and Erie (8.0×10^{12} m^3) or the combined yearly flows

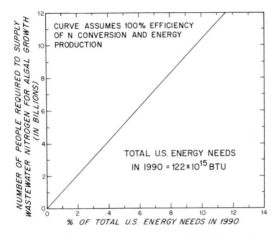

Fig. 3. Relationship between percent of 1990 U.S. energy needs (122 x 10^{15} BTU) and the number of people required to supply nitrogen from domestic wastewater for algal growth for 100% conversion to useful energy. Other assumptions: 1) 100% conversion of inorganic nitrogen to algal biomass; 2) wastewater nitrogen concentration = 30 mg/ℓ; 3) percent nitrogen in algal dry weight = 8; 4) per capita daily wastewater flow = 100 gal.

of the Amazon, Congo, Yangtze, and Mississippi Rivers (8.0 x 10^{12} m^3) would be necessary. The outlook for obtaining the needed phosphorus is even more dismal. The known U.S. reserves of phosphorus are 12.7 x 10^{14} gr (Brobst and Pratt, 1973). Using the 10% conversion efficiency requirement for phosphorus from Table 1, it would take slightly over 50 years to deplete the U.S. supply, assuming there were no other uses for this mineral.

ENERGY INPUTS

An energy balance sheet, including itemized energy inputs to an algal energy plantation is detailed in Table 2. This list includes the major items, but is not meant to be all inclusive because experiences with very large algal growth systems are non-existent-the largest intense algal culture systems in use are many orders of magnitude smaller in size than those required for the proposed energy plantations (Oswald, 1970); hence, there is little engineering experience upon which to develop rational estimates of energy inputs.

On the other hand, good data is becoming available on maximum attainable algal yields (Vendlova, 1969; Goldman *et al.*, 1975; Stengel and Soeder, 1975; Lapointe *et al.*, 1976), and short-term yields of up to ∿30 gr dry wt/m^2/day of both microscopic algae and seaweeds have been reported. This upper limit is in accord with Ryther's (1959) earlier estimate that, based on sunlight

TABLE 2.

Energy balance for plant growth - energy production system per gram ash-free dry weight of plant produced.

Energy Parameter	Energy Kcal
ENERGY OUT (Gross)	5.5
ENERGY IN (including losses)	
Nutrients	
Wastewater - not feasible	—
Fertilizers (N & P)	1.5
Recycle (Efficiency?)	???*
Water (Extraction & transport system)	
Freshwater	??
Marine	?
Pond Construction & Operation	?
Pumping (Water transport and/or recycle)	?
Pond Mixing (pumping/aeration)	0.2
Chemical Addition (CO_2)-pH control	??
Algal Separation (Efficiency?)	1.2*
Fermentation losses (∿50%)	2.7
Fermentation Unit Construction	???*
& Operation (Heat input)	
ENERGY OUT (Net)	
With Fertilizers	0
With Recycle	1.4-?

*Major uncertainty.

limitations only, yields of ∿27 gr dry wt/m^2/day can be achieved in aqueous systems. Sustained yields in the best of geographical locations would, because of seasonal variations in climatic conditions, inherent culture instabilities, and other uncontrollable factors, be considerably lower, but once again cannot be reliably estimated because of the unavailability of long-term growth data.

Considering the energy input items in Table 2, it becomes evident that any type of growth system, whether it be microscopic freshwater algae, marine seaweeds, or even terrestrial plants, will have to employ either partial or complete recycle in order to show a net energy production. External sources of nutrients such as wastewater are not available in the quantities required. Commercial fertilizers, even if production could be significantly expanded to meet the needs of an energy plantation, require considerable fossil fuel energy expenditures during production (about 1.5 Kcal/gr dry wt of raw plant material produced according to Pimental *et al.*, 1973 and assuming plant nitrogen = 8%), thereby reducing the net energy available from the grown plant by >25% (assuming J = 5.5 Kcal/gr dry wt).

Hence, recycling of nutrients appears to be the only parti-

cal alternative. Yet, this approach is beset with several major
obstacles. First, even if recycling were to be employed, to
initially develop the cultures for producing 5% of the 1990 U.S.
energy needs external sources of nutrients would be required for
stocking the cultures. Assuming that the retention time in the
cultures was two days (Goldman *et al.*, 1975), then, depending on
the efficiency of energy conversion, on the order of 4.4×10^{13}-
10^{14} gr of nitrogen and 5×10^{12} - 10^{13} gr of phosphorus (0.4 - 4%
of the total U.S. supply) would have to be initially supplied to
and stored in the system from an external source. Needless to
say, these values represent enormous quantities of fertilizers.

A second major problem with nutrient recycling involves both
the efficiency of recycling and the energy input required to recover the materials. Uziel *et al.* (1975) reported that about 2/3
of the nitrogen in digested freshwater algal slurry was associated
with the suspended solids and would not be readily available for
reassimilation by algae unless further oxidized by aerobic bacteria. Golueke and Oswald (1959) observed similar problems in
trying to recycle nitrogen in an integrated algal growth - fermentation system. Thus, if it is necessary to add an intermediate
bacterial oxidation step between the fermentation unit and the
algal growth system to achieve high nutrient recovery in recycled
systems then considerable and, as yet unknown, energy inputs
would be required.

The enormous requirements for water, as with nutrients, virtually rule out supply from an external source for freshwater systems. In addition, recycled systems would be subject to severe
evaporation losses and subsequent salt buildup since they more
than likely would be located in geographical areas favoring strong
sunlight and associated high temperatures. Make-up water from external sources would still be required.

Marine systems employing seaweeds on the other hand, could
be operated in a semi-open system with recycled nutrients, but
using unlimited external sources of seawater. Coastal land requirements, rather than water, would be the decisive limiting factor in this case. There are 40,595 km of open ocean coastline in
the U.S. To meet the land requirements outlined in Table 1, a
strip of land circumventing the entire U.S. coastline with a width
of 2-20 km would be needed. Energy inputs for any type of extraction and/or transport system associated with the above system
would depend on the quantities of water required and the pressure
head and could easily be estimated.

Similarly, energy inputs for the pond materials, construction
and operation (including maintenance and internal pumping) could
be determined once a pond design was chosen.

Actual requirements for culture mixing, however, are not well
defined. Mixing is necessary for a number of reasons: to prevent
settling and subsequent decay of organic matter, to prevent thermal stratification, to break down diffusion gradients of essential
nutrients which could develop in intense cultures (Gavis, 1976),

to provide uniform cell exposure to light, and to allow carbon
dioxide transport into solution from the atmosphere both for pH
control and for algal assimilation. Unless carbon dioxide addition to the culture (either through mixing, aeration, or supple-
mentary input) can keep pace with carbon assimilation, the aqueous
bicarbonate reservoir will be drained and the pH will rise according to the chemical reaction:

$$HCO_3^- \rightarrow CO_2 + OH^- \qquad (3)$$

Inorganic carbon and pH limitations can be severe in fresh
water mass cultures (Goldman et al., 1972) and, as Jackson (1977)
shows, macrophyte growth in marine systems that are buffered with
large quantities of inorganic carbon can also be limited by pH increases due to bicarbonate losses through Eq. 3.

Hence, mixing and possibly carbon dioxide addition is vital
for sustaining high yields and preventing carbon and/or pH limitations. Oswald (in press) estimates that about 0.2 kcal/gr of
dry wt of mixing energy is required for a yield of 10 gr/m^2/day
in fresh water algal systems using waste water as a nutrient
source. For recycled systems in which inorganic carbon would
have to be added from external sources due to carbon losses
through fermentation, mixing requirements (pumping or aeration)
could be significantly higher and result in major energy inputs.

The final energy inputs in Table 2 involve the processes of
algal separation and fermentation. The difficulties in separating and concentrating algae are obviously related to cell (or
plant) size. The very small unicellular algae such as Chlorella,
Scenedesmus (species typically found in freshwater cultures) and
marine diatoms can only be separated by processes such as centrifugation or chemical coagulation, costly and energy consuming
techniques (Golueke and Oswald, 1965). Efforts to grow filimen-
tous blue green algae that can be more readily separated by mico-
straining are in progress (Benemann et al., 1976). Seaweeds, in
contrast, because of their large size certainly would be the
easiest plant material to separate.

Oswald (in press) estimates that the cheapest form of mechanical separation for freshwater algae would require ∿1.2 Kcal/gr
dry wt (yield = 10 gr/m^2/day) of energy input. No information is
currently available on energy requirements for seaweed separation,
although they would most likely be lower than for microalgae.

The efficiency of converting stored plant energy to methane
appears to be in the range of 50-70% for both microscopic algae
and seaweeds (Uziel et al., 1975; Trojano et al., 1976). Energy
requirements appear to be exceedingly high if the process is to
be carried out in conventional type steel or concrete digesters,
as are commonly used in the wastewater treatment industry (Oswald
1976). Deep earthen ponds with covers could significantly reduce
energy requirements, but have not been field-tested and could
pose a serious explosion hazard due to sealing difficulties that

might arise. Possible requirements for chemical pre-treatment of
the plant material (Trojano *et al.*, 1976), digester heating and
mixing all compound the uncertainties related to the fermentation
step. Estimates of energy requirements are therefore difficult to
make, but appear to be major compared to other items in Table 2.

Summing up the known energy inputs in Table 2, we see that
with fertilizer use, the system is already energy consuming after
only four of the nine items listed are tallied. With nutrient recycle only 1.4 Kcal/gr dry wt minus all of the remaining uncertain
energy inputs listed are left.

It must be emphasized that all of the above calculations were
made assuming a yield of 10 gr dry wt/m^2/day. If sustained yields
could be significantly increased to approach the high short-term
yields already achieved (20-30 gr/m^2/day), then the energy requirements could conceivably be decreased. Shown in Fig. 4 are
four possible ways in which yield could affect the net production
of energy.

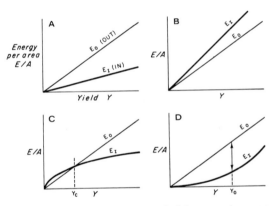

Fig. 4. *Possible effects of plant yield on net energy production*
A. $E_O > E_I$ *at all yields;* B. $E_I > E_O$ *at all yields;* C. $E_O > E_I$
when $Y > Y_C$; D. $E_O - E_I$ *a maximum at* Y_O.

In Fig. 4A is shown the most desired case: E_O is always
greater than E_I. Unfortunately, in practice this situation would
never occur. The opposite case (Fig. 4B) is when $E_I > E_O$ for all
yields, demonstrating that the entire concept is unfeasible. The
possibilities shown in Figs. 4C and 4D indicate the effects of
yield on net energy production. In the former case yields would
have to be greater than a critical yield Y_C if capital energy requirements such as construction inputs were the major items. The
situation in Fig. 4D indicates low energy at low yields, but
rapidly increasing inputs required to maximize yields (e.g. mixing, CO_2 additions, pumping, etc). Then there might be an optimum yield Y_O at which $E_O - E_I$ was a maximum. More realistically, if

energy plantations could produce energy then a combination of possibilities seen in Fig. 4C and 4D would result: fixed initial inputs resulting in a minimum required yield Y_c and rising energy inputs necessary to maximize yields, leading to a desired yield Y_o for maximum net energy production.

CONCLUSIONS

In this paper we have tried to demonstrate that the production of energy through algal growth and fermentation depends on our ability to maximize areal yields and fermentation efficiencies while minimizing the required mechanical and electrical energy inputs. The engineering problems we have outlined are not unique to algal cultures, but apply to all types of photosynthetic systems, both terrestrial and aquatic. The main difficulty, as outlined by Meyers (1977), is that solar energy, although vast in total input, strikes the earth at a low flux; hence, vast quantities of surface area and sophisticated collection systems are required to capture and convert radient energy in the quantities required to satisfy projected national needs. These problems are, by the way, common to all solar energy collection schemes (Res. & Educ. Assoc., 1975).

The vast land and materials requirements for energy plantations and the difficulties in minimizing energy inputs, as outlined in this paper, leads us to the pessimistic conclusion that photosynthetic processes such as algal systems may be difficult to justify solely as large-scale energy producers. Because modern agriculture is itself highly energy intensive (Pimental et al., 1973; 1975; Heichel, 1976), and because of the prospects for world-wide food shortages in the near future (Brown, 1975), it would seem prudent that the highest priority for photosynthetic processes be for food production. In this regard, the prime goal in promoting algal systems may be "energy conservation" rather than "energy production." As we outlined earlier (Goldman and Ryther 1976), algal systems, because of their versatility, can be employed in a variety of potentially energy conserving programs vital to maintaining a high standard of living: advanced waste-water treatment, the production of raw protein, fertilizers, drugs, and colloids, and waste recycling-aquaculture processes. The net result may be a savings in energy by reducing or replacing more energy intensive and conventional industrial and agricultural processes. Clearly, the scope of these operations would be on a much reduced scale than envisioned in Table 1, but the potential appears high for a noticeable net savings in energy. Perhaps by toning down the scale of requirements for photosynthetic systems and taking an integrated and multi-use approach, the long-sustained hope for developing mass algal cultures can become a reality.

ACKNOWLEDGEMENTS

Supported by contract No. EG-77-5-02-4151 from the U.S.
Energy Research Development Administration to Woods Hole Oceanographic Institution.

REFERENCES

Benemann, J. R., B. Koopman, J. Weissman, and W. J. Oswald (1976)
Biomass production and waste recycling with blue-green algae.
Presented at *Microbial Energy Conversion Seminar,* Gottingen,
West Germany (October 7, 1976).

Brobst, D. A. and W. P. Pratt (eds.) (1973) *United States
Mineral Resources,* Geological Survey Professional Paper 320,
U.S. Gov't. Printing Office, Washington, D.C.

Brown, L. R. (1975) The world food prospect. *Science 190:* 1053-
1059.

Gavis, J. (1976) Munk and Riley revisited: nutrient diffusion
transport and rates of phytoplankton growth. *J. Mar. Res.
34:* 161-179.

Enzer, H., W. Dupree, and S. Miller (1975) *Energy Perspectives*
U.S. Dept. of the Interior, U.S. Gov't. Printing Office,
Washington, D.C.

Goldman, J.C. and J. H. Ryther (1976) Waste reclamation in an
integrated food chain system. In J. Tourbier and R. W.
Pierson (eds.) *Biological Control of Water Pollution,* pp.
197-214. Univ. of Penn. Press, Phil., Pa.

Goldman, J. C., D. B. Porcella, E. J. Middlebrooks, and D. F.
Toerien (1972) The effects of carbon on algal growth-its relationship to eutrophication. *Water Research 6:* 637-679.

Goldman, J. C., J. H. Ryther, and L. D. Williams (1975) Mass Production of marine algae in outdoor cultures. *Nature 254:*
594-595.

Golueke, C. G. and W. J. Oswald (1959) Biological conversion of
light energy to the chemical energy of methane. *Appl. Microbiol. 7:* 219-227.

Golueke, C. G. and W. J. Oswald (1965) Harvesting and processing
sewage-grown planktonic algae. *J. Water Pollut. Control
Fed. 37:* 471-498.

Heichel, G. H. (1976) Agricultural production and energy resources. *Amer. Scient. 64:* 64-72.

Jackson, G. A. (1977) Biological constraints on seaweed culture.
This volume.

Lapointe, B. E., L. D. Williams, J. C. Goldman, and J. H. Ryther
(1976) The mass culture of macroscopic marine algae. *Aquaculture 8:* 9-20.

Meyers, J. (1977) Bioengineering approaches and constraints.
This volume.

Oswald, W. J. (1970) Growth characteristics of microalgae cultured in domestic sewage. In *Prediction and Measurement of*

Photosynthetic Productivity. Proc. of the IBP/PP Technical Meeting, Trebon, September 14-21, 1969. Center for Agricultural Publishing and Documentation, Wageningen, The Netherlands.

Oswald, W. J. (1976) Solar energy fixation with algal-bacterial systems. Presented at *Symposium on Energy Conversion*, Campinas, San Paulo, Brazil, July 8, 1976.

Oswald, W. J. (In press) Photosynthetic single-cell protein. In *Protein Resources and Technology: Status and Research Needs* U. S. Gov't. Printing Office, Washington, D.C.

Pimental, D., L. E. Hurd, A. C. Bellotti, M. J. Forster, I. N. Oka, O. D. Sholes, R. J. Whitman (1973) Food production and the energy crisis. *Science 182:* 443-449.

Pimental, D., W. Dritschila, J. Krummel, J. Kutzman (1975) Energy and land constraints in food protein production. *Science 190:* 754-761.

Research and Education Association (1975) *Modern Energy Technology* Vol. II. Res. and Educ. Assoc., New York, N.Y.

Ryther, J. (1959) Potential productivity of the sea. *Science 130:* 602-608.

Stengel, E. and C. J. Soeder (1975) Control of photosynthetic production in aquatic ecosystems. In J. P. Cooper (ed.) *Photosynthesis and Productivity in Different Environments*, pp. 645-660; Cambridge University Press. Cambridge, Great Britain.

Trojano, R. A., D. L. Wise, D. C. Augenstein, R. G. Kispert, and C. L. Cooney (1975) Fuel gas production based on a mariculture system. Dynatech R/D Company, Cambridge, Mass., unpublished report.

Uziel, M., W. J. Oswald, and C. G. Golueke (1975) Solar energy fixation and conversion with algal bacterial systems. Final Report to National Science Foundation. Sanitray Engineering Research Laboratory, University of California, Berkeley, Calif.

Vendlova, J. (1969) Les problemes de la technologie de la culture des algues sur une grande echelle dans les installations au dehors. *Annali Microbiol. 19:* 1-12.

Ward, R. (1976) U. S. Energy Research Administration, personal communication.

A CRITICAL ANALYSIS OF BIOCONVERSION
WITH MICROALGAE

WILLIAM J. OSWALD and JOHN R. BENEMANN

University of California, Berkeley, California

There is great promise to the proposition of growing algal
biomass using the nutrients in sewage and then harvesting and
digesting the biomass for methane production. Algae are already
being grown in large scale oxidation ponds as a method of
treating sewage and other liquid wastes. Harvesting of the algae
is presently the limiting technical and economic factor which
needs to be solved through applied research. The economics of the
proposed system would be based on high waste treatment credits,
leaving the methane as a "free" byproduct. The methane alone
might by some estimates pay for the operations of such systems,
allowing sufficient algal biomass production to meet natural gas
requirements. The algae could be supplied with additional
nutrients through carbon dioxide enrichment of wastes, nitrogen
fixation, recycling of nutrients from the algal fermentation
residues and even by mineral fertilization. Many additional
benefits can be cited, including water reclamation, fertilizer
production (several metric tons N/hectare/year), environmental
protection, and even aesthetic and social considerations. Algal
bioconversion thus seems an excellent solution to our energy
and many other problems. However, the more optimistic proposals
for bioconversion are often overrated and algal bioconversion
still faces many problems.

We wish to initially detail all the questions and problems
which have been raised from time to time in analyses of algal bioconversion. Then we will endeavor to give our analysis and
solutions to these problems. We hope that a clearer picture of
the potential and limitations of algal bioconversion will emerge
from this approach.

QUESTIONS AND PROBLEMS

1. PRODUCTIVITY:

The fundamental units in which algal biomass production is
measured are $gm/m^2/day$ and MT/HA/year. (These values should refer
to volatile solids only, to allow a reasonable comparison between
various algae). The key problem in discussing algal productivity
is that the little data available is in the first units while
MT/HA/year are known from extrapolations (or estimates) only;

usually from a dual extrapolation of both area and time. The
simple conversion factor 1 gm/m^2/day = 3.65 HA/year does not seem
appropriate. It is doubtful whether values obtained with systems
of less than 1000 m^2 are indicative of production rates obtainable with multihectare systems. Only a few algal production (as
opposed to oxidation) ponds actually exist. Many of these are
as small as a backyard wading pool, which they often are. All
but the largest would not produce enough algae if digested for
methane to supply a single American home. One exception is a
ten-hectare pilot plant near Mexico City which produces the
filamentous blue-green algae Spirulina, valued for its pigment
and protein content. This unit supposedly produces about one
metric ton per day. The product is sold for over $5/kilo (mainly
to Japan, where home-grown Chlorella is even more expensive).
This puts it out of the price range (by almost 500-fold) for
energy conversion.

The information on the Spirulina production plant of 36 MT/HA/
year is not published; no verifiable operational data exist. The
productivity value most often assumed is that of 50 MT/HA/year
(20 tons/acre/year), representing in the Southwest United States
about a 5% efficiency for visible sunlight energy conversion.
Although this is well above photosynthetic efficiencies for most
crop plants, the fundamental assumption in applied algology is
that algal systems can outproduce higher plants by several-fold,
perhaps even an order of magnitude. Only thus would it be possible to suggest algal systems as energy producers. However,
there seems to be neither data nor theory to support this view.
Algae have the same basic photosynthetic apparatus as higher
plants; they are limited by the same quantum requirement for
carbon dioxide fixation and also exhibit respiration and photorespiration. Although algae could grow year round, their pro-
ductivites would vary greatly during different seasons. Oxidation ponds, which are continuously pointed out as examples of
large-scale algal systems, are not algal production ponds. Most
oxidation ponds are of the "facultative" type with depths above
one meter, long detention times, and slow mixing; causing their
algal productivities to be low. Only shallow (less than 0.5
meters deep), well-mixed "high-rate" ponds are useful for algal
production. However, only a few operational high-rate ponds
exist and production data is incomplete. Thus, most estimates
of productivities are based on extrapolations from small-scale
systems. There are several reasons why large-scale systems would
be much less productive than small-scale systems: small-scale
systems are easily completely mixed, can be carefully controlled,
have no dead spots or short-circuits, and receive all the tender
loving care of which a concerned scientist is capable. The
mixing requirement appears to be an insoluble problem.

2. NET ENERGY:

A good measure of the effectiveness of algae-methane systems is whether they are net producers of energy and, if so, by what margin. A detailed analysis will be required for determining the energy costs of pond construction (grading, earthwork, lining, embankment protection, pumps, etc.). This might not be minor, particularly if present pilot designs for algal production ponds are used in large-scale systems. However, the main energy use will be during pond operations for pumping, mixing, harvesting, and digestion. The latter two will be dealt with later. Pumping is usually assumed minor; however, if sewage must be brought more than a few miles or over any grade, pumping could become a significant factor. It is mixing, however, which can be the real energy drain in algal pond operations.

Mixing is required for thermal destratification, suspension of the planktonic algae, avoidance of "dead spots" or "short-circuiting" of the ponds, and maintenance of a uniform environment in the ponds. Neglect of any of these requirements would result in a drastic reduction of productivity. The problems of achieving these tasks even under experimental conditions in small-scale ponds (above 100 m^2) indicates the difficulty of accomplishing adequate mixing in large-scale pond systems. In ponds, as in many agricultural systems, arithmetic increases in productivity are achieved only by geometric increases in energy inputs. This is only possible if the product has a much higher value than energy itself (such as Spirulina). Thus, it may be concluded that algal systems are only energy efficient at low productivities (as in oxidation ponds) and that net energy production is impossible above certain limits. Algal production is meaningless unless the algae are harvested and suitably concentrated for further conversion. For bioconversion, unharvested algae is like a bird in the bush.

3. HARVESTING:

This is the Achilles heel of algal ponds. There are at present no low-cost effective techniques for algal harvesting. Centrifugation is expensive, with energy requirements even greater than for mixing. Chemical flocculation (the method of choice for sanitation districts with EPA construction grants) is only a marginal improvement. Algal settling has often been described, but seldom reproduced. As the years continue without a satisfactory solution, new ideas for algal harvesting become more sophisticated--magnetic and electric separations, ultrafiltration, flocculation with polyelectrolytes, air flotation, etc. None appear promising economically on a large-scale. It seems that algal harvesting is only inexpensive when colonial or filamentous species such as Spirulina are present which can be removed by settling, filtration or screens. This has led to the proposal that such types of algae be selectively cultivated.

4. SPECIES CONTROL:

Ecologists, after one-hundred years of observation and experimentation, still have little knowledge of what makes certain algae dominate, how to predict algal blooms, or how to control algal populations. Oxidation ponds exhibit many of the phenomena found in eutrophic bodies of water. However, the intensity and concentration of algal growth, rapid shifts of populations, establishment of dominant species almost overnight, increase and decrease of algal predators, etc., gives these systems very unstable, rapidly changing populations which respond more readily to variations of climate or nutrients than to any manipulation the pond operator might be able to undertake in large-scale systems.

Any effects on algal species which might be achieved in small-scale systems (through controlled mixing, temperature, recycling, etc.) are likely to be either impossible to apply to, or without effect in, large-scale ponds. Even in small-scale systems, species control is difficult to achieve. About the only effective method is to develop specific media (high bicarbonate media for Spirulina, secondary effluents with seawater for diatoms) and select for the species which has the highest specific growth rate at any particular temperature using short detention times. This allows cultivation of certain algal species which are not necessarily the most desirable ones. (Or, in the case of Spirulina, can be cultivated only in alkaline lakes or at great cost for selective chemical media). Continuous cultivation of one species throughout a year or even a season appears to be unfeasible. The rapid growth rate of algae also assures a rapid response to varying conditions. Even two days of changing weather are sufficient to alter pond populations drastically. This has occasioned Chlorella manufacturers in Japan to abandon outdoor cultivation for indoor heterotrophic controlled growth (at a cost of over $10/kg). Although population dynamics are an interesting subject, the development of algal control techniques is not assured by its study. Methods based on continuous culture theory are not applicable to algal production systems which are a mixture of plug-flow and continuous systems. Thus, any algal harvesting methods which depend on algal population control are highly speculative and, at present, still to be demonstrated. Even if successful, they would be effective only part of the time and subject to continuous upset.

5. THE LIMITS OF SEWAGE:

Even assuming that algal productivity is adequate, mixing is minimized and algae can be harvested, the question arises of how much algae can be produced from available wastes. Of the various limitations, nutrients are the most important. Light, carbon, nitrogen and phosphorus are the principal nutrients of interest. Any algal system should always be operated with light

as the limiting nutrient to maximize productivity per unit area.
This is, however, counter to the requirements of waste treatment
which is maximized by making carbon (or other nutrients) limiting.
Since waste treatment is the main economic basis for operation
of algal ponds, this would have a significant effect on algal
productivity.

The total algal biomass produced on liquid wastes is determined by the limiting nutrient, usually carbon. The amount of
algae that can be produced on each person's sewage are only sufficient to allow production of at most 0.2% of the U.S. per capita
energy consumption. Indeed, the sewage produced by the entire
world's population would not be able to produce enough algae to
supply the United States with its natural gas requirements. The
amount of algae might be increased by injecting carbon dioxide
into the ponds to give the algae additional carbon. Of course,
the source of the carbon dioxide is a problem; air contains only
0.03%, natural sources are rare, and conventional power plants
are too few and seldom located in the right place. The latter
would make algae-methane production dependent on the fossil fuels
they are supposed to replace. Even if an adequate source of
carbon dioxide is located, the problem of how to distribute it
into the pond does not appear to be amenable to an inexpensive
solution. Also, carbonating sewage will increase the size of the
algal system beyond that which would be required for waste treatment, thereby removing its most important economic basis.
Finally, the amount of methane that might result as an effort of
carbonation is not sufficient to make a significant impact even
on the local demand. The only method by which significant amounts
of algae could be cultivated is to continuously recycle nutrients
from the residues of methane digestion. In one way this would
be good; it would be the best way to dispose of these residues
whose use as "fertilizers" is limited (see below). However,
only limited recycling appears feasible; the digester residues
might block light (or create an anoxic pond bottom) and it would
cause a build-up of refractory algal compounds. There might thus
only be limited opportunities (even should technical and economic
factors allow it) to expand algal production for bioconversion.

6. NITROGEN FIXATION AND FERTILIZER:

A common thought is that by adding carbon dioxide to ponds,
nitrogen would become the limiting nutrient, thus selecting for
nitrogen-fixing blue-green algae. The filamentous nature of
these algae allows filtration harvesting, this being the major
attraction of this scheme. As nitrogen-fixing algae are not
observed in oxidation ponds, it is speculative that such organisms
can be purposely grown in them. Anyway, the constant addition
of nitrogen in the sewage will allow enough non-nitrogen-fixing
algae to grow and severely reduce the efficiency of filtration
harvesting. Even the basic assumption of this idea is in doubt;
phosphorus, not nitrogen, is the limiting nutrient (after carbon)

of sewage for some of our largest cities that forbid P detergents
(New York and Miami, for example). This trend will continue in
this country. Anyway, it is not clear that the purpose of the
exercise--fertilizer production--is really worth the effort. The
sludge for the algal digestion reactor will only contain an (unknown) part of the nutrients; the supernatant will contain the
rest. Use of these fractions as fertilizer is limited by transportation distances and storage difficulties. Indeed, without
dewatering, it would not be feasible to utilize these. Thus, the
residue from methane fermentation of algae is similar to digested
sludge--an expensive disposal problem rather than a lucrative
fertilizer.

7. METHANE FERMENTATION:

Although it has not yet been carried out on a large-scale, it
is usually assumed from laboratory extrapolations that large-scale
methane fermentation of algal biomass can be effective and stable.
Methane fermentation of algae for energy production suffers from
the high cost of fermentations. In sewage treatment plants,
methane digestion of primary sludge is carried out only because it
is the cheapest and fastest method. The cost of the digestion
units is up to $10/ft^3$, making it impossible to pay for the units
from the value of the methane produced. Heating of the digesters,
another requirement, consumes a large share of the methane output.
Mixing, recirculation, pumping, foam control, and monitoring
consume further energy and personnel effort so that it is unlikely
that, even if the starting material (algae) and the digesters were
for free, that it would be possible to show a profit from such
operations. Finally, hydrogen sulfide and carbon dioxide removal,
although feasible, is not free. Thus, the arguments up to this
stage may be summarized into the statement that methane production
from algae would be an exclusively sanitary engineering process,
comparable to primary sludge digestion, and the methane produced
would be barely sufficient to even operate the ponds and digesters.
Any possible surplus would be insignificant and comparable to that
produced by conventional waste treatment systems. Little chance
exists of expanding algal biomass production beyond that of waste
treatment requirements.

8. POND COSTS AND LAND USE:

One of the assumptions often made is that pond construction is
inexpensive. This can be so for oxidation ponds where a hole in
the ground and a few embankments are sufficient. However, algal
production ponds, with their mixing requirements and shallow
depth, must be lined to prevent erosion of the pond floor. Lack
of depth requires more carefully grading and pond lining is
expensive. There are no current estimates of costs; $10,000/
hectare is a rough estimate for oxidation pond costs; high-rate
pond costs would be much higher.

It is not the pond construction costs but land costs which are
the most critical non-technical economic factor. Thus, all cost
estimates are made without consideration of land costs. The requirement for large areas of level land near population centers
will exclude the large urban populations from utilizing the potential (or purported) benefits of algal ponds. A further reduction must be made of that part of the population living in
climatologically unfavorable areas. Thus, it is doubtful if even
20% of the U.S. population could be served by oxidation ponds.
For sewage treatment alone, high rate ponds would require one
hectare for each 500 people while for bioconversion this might be
reduced to perhaps 50 people. A city of 1,000,000 would thus use
about 2,000 hectares for sewage treatment and over 20,000 hectares
(50,000 acres) for algal biomass production. This is comparable
to the urban land area! Despite such large systems, the energy
produced would barely provide 5% of the local energy needs.
Furthermore, with large biomass production systems, waste treatment credits will become a minor operating revenue. Bioconversion
would use needed agricultural land and reduce opportunities for
residential or industrial expansion and environmental preservation.
Land intensive systems such as algal ponds might have a role only
in the marginal areas far away from population centers. Thus,
development of large-scale algal bioconversion systems should
concentrate on systems which recycle their nutrients and water and
which are located in the sunny and empty Southwest where they
would feed the starving gas pipelines. The commercial feasibility
of such systems does not look promising.

9. COMMERCIAL APPLICATIONS:

The commercial opportunities of methane production as a byproduct of algal sewage treatment are small. Sewage treatment is
a function of local government who would, if they had any surplus
methane to dispose of, be in direct competition with private
enterprise. The unfair advantage of public agencies (no taxes,
eminent domain, non-profit, tax subsidies, etc.) would depress and
weaken the energy industry and hinder their development of alternative energy sources. The unreliability of algal bioconversion
would not allow any significant reduction in the physical plants
of the private energy corporations. However, they would increase
the differential between peak capacities and average demand,
thereby increasing capital costs and energy prices.
It is hard to imagine how anyone could make money with algal
systems. Perhaps, should technology be developed and simplified,
algal ponds could play a role in underdeveloped or communist
countries whose level of development and social organization is
more suitable for such technologies. For U.S. requirements,
only large-scale centralized private systems, capable of producing energy sufficient to run large power plants or supply
pipelines and cities are suitable.

10. PROGRESS TO DATE:

The proponents of algal bioconversion defend their "new"
technology by claiming it cannot be subjected too quickly to the
acid tests of practicality, feasibility, and economics. More
study is required; more research funds needed; more time essential.
A look at history reveals how little progress has been made.
Twenty-five years ago the first study of an algal pilot plant was
completed and the Carnegie Institute celebrated it shortly
thereafter with a Festschrift entitled "Algae Cultures from
Laboratory to Pilot Plant" (1). At that time, the emphasis was
on algal production for food (a sensible approach which should
continue). Nevertheless, the same problems were encountered then
as now--the high costs of the system, overheating, low productivity,
difficulties with nutrient and carbon dioxide supplies, harvesting
problems and costs, changes in algal populations, predation, etc.
It can truly be said that, aside from oxidation pond technology,
little has been accomplished in the past twenty-five years in
applied algology.

ANALYSIS AND SOLUTION

The foregoing dismal presentation of the potential problems of
large-scale algal cultures can be balanced by the following objective and factual analysis in which each major problem is discussed in reverse order of its presentation.

*10. APPLIED ALGOLOGY RESEARCH IS A LONG-TERM PROPOSITION; PRO-
GRESS IN THE PAST 25 YEARS DOES NOT ALLOW OPTIMISM ABOUT THE
NEXT QUARTER CENTURY*

Three significant developments have occurred in the last 25
years: 1) the development of engineering designs for large-scale
algal ponds for waste treatment (2, 3, 4); 2) the accumulation of
large amounts of data on the biochemistry, physiology and ecology
of algae (5, 6, 7); and 3) the recognition of an urgent need for
developing alternative sources of energy, fertilizer and protein.
It is this latter factor, relatively new, which is now providing
an impetus to a renewed study of algal bioconversion. The key
ideas and most problems have been around for a long time; their
development and solution is only now being seriously contemplated.
The reasons for the lack of apparent progress over the past
25 years in algal production and bioconversion is due to the low
level of interest in this research rather than any particular road-
block or insurmountable problem. However, it can be fairly stated
that the research and development required to make microalgal
bioconversion a large-scale, economically attractive, reality will
be difficult, costly, and long-term. It will require teams of
scientists, drawn from many disciplines, working together on

specific tasks. Only recently were the first applied research
projects funded which focused specifically on microalgal biomass
production and its conversion to energy. This is only a beginning;
however, it could result in novel harvesting technology for
oxidation pond algae within a few years. Although microalgal
production is still at the pilot stage, it has not remained where
it was 25 years ago. We now know that algal production, employing covered ponds and chemical fertilizers, is technically difficult and only economically feasible for high value products.
For low-cost algal production only open ponds which are fer-
tilized with wastes are suitable. This approach is based on the
considerable experience gained with algal systems in waste treatment. The available information on algal biology should find
application in the development of algae domestication and cultivation technology.

9. *ALGAL BIOCONVERSION SYSTEMS ARE INCOMPATIBLE WITH THE U.S.
   ENERGY MARKET*

These are political issues rather than technical or economic.
Algal bioconversion, as other bioconversion or solar energy
systems, cannot, on its own, provide all the energy required or
deemed essential. Of the over 70 "quads" (10^{15} BTU's) of energy
consumed in the U.S., it is doubtful whether even 5% could be
supplied through algal bioconversion. Thus, such systems cannot
be considered on their own but must be envisioned as part of an
energy supply system encompassing a variety of alternatives. The
search for a single, simple solution to the energy problem is
as futile as that conducted for Eldorado. The conflict between
alternative energy sources might be severe; however, each must
find its own applications and markets. Government involvement
would neither be new nor evil. Public competition to the private
energy interest would be welcomed by many. Algal bioconversion
systems would neither be unreliable nor marginal. Their productivity would be predictable (based on experience) and the
methane can be easily stored for daily peak demands. Algal bioconversion would be applicable in many countries and social orders;
development of practical size systems in the U.S. would assure this.
Very large, centralized systems, however, ould not be practical.
They would be single purpose (methane production) and could not
compete with the multipurpose systems (methane, fertilizer, sewage
treatment, water reclamation, etc.) made possible by integrating
algal bioconversion into a locality. There are, however, many
opportunities for private enterprises, both big and small, in the
development and exploitation of algal bioconversion systems. Any
new industry produces new fortunes.

8. *LAND COSTS FOR LARGE-SCALE ALGAL BIOCONVERSION SYSTEMS WILL
   RULE OUT THEIR USE FOR THE URBAN MAJORITY OF THIS COUNTRY*

Land costs are the key economic parameter dictating the

potential use of algal and other bioconversion systems. At a 5%
efficiency conversion of visible sunlight into energy, about one
acre of land in the lower half of the Continental United States
will be required to satisfy the total per capita energy needs.
The area required would be higher if biomass is fermented to
methane with only about two-thirds of the energy in algae being
converted to methane. Thus, about 10% of the land area in the
Continental United States or about 50% of the acreage used for
crop production would be needed in bioconversion. These numbers
are obviously too high and suggest that bioconversion would only
supply a fraction of total energy requirements. Algal ponds
would probably compete with agriculture for level acreage near
population centers. However, unlike other bioconversion systems,
algal ponds would not compete with agriculture for water and
fertilizer, but, indeed, could provide the water and fertilizer
required by intensive agriculture. Many land areas unsuited for
crop production might be used in algal systems--many millions of
acres of agricultural land have impregnable hardpan, while others
are being stripped of topsoil to the point that they are unsuited
for intensive crop production and, at best, are returned to
pasture or forest lands (8). Algal biomass production systems
could play a significant role in soil reclamation; the green
manure provided by the digestion residues should be a better soil
conditioner than the digested primary sludge. The ponds them-
selves could be rotated to provide rich new fields created by
deposition of algal sludge. Although these functions cannot be
assigned values (since the data on soil restoration is not available), soil degradation is a major national problem which needs
to be considered in a system analysis of bioconversion.

High land costs are a political problem; speculative manipulations in the absence of land use plans increases land costs near
urban areas. A more realistic economic comparison is the value
produced by algal bioconversion systems versus agricultural production. One-half of the 2.2 billion acres of the Continental
United States is contained in farms; gross annual production
amounted to less than $100 per acre nationwide. (However, only
400 million acres are actually in crop production). With bioconversion systems, one could probably expect (see below) about
200 million BTU's of methane per acre, worth about $500, and about
two tons of nitrogenous fertilizer, worth an equivalent amount.
Thus, gross revenue for algal ponds would be better than that of
the great majority of farm land in the United States, including 13%
of U.S. cropland that falls within the "Standard Metropolitan
Statistical Areas" (8). On the basis of such comparison, it
appears that land costs are not a necessary obstacle to bioconversion systems in the United States.

Algal production ponds do not need to be of significantly
higher costs than conventional oxidation ponds. Mixing can be
gentle enough to prevent bottom sediment suspension (see below)
and no special lining of the pond needs to be used. Indeed, high-
rate ponds can be more inexpensive than conventional "facultative"

oxidation ponds, as demonstrated by the example of Hollister,
California, where a recent comparative cost analysis showed that
a high-rate pond system costs 10% less than facultative ponds (9).

7. METHANE DIGESTION IS A CAPITAL AND ENERGY INTENSIVE OPERATION

This is a general problem of biological SNG (synthetic natural
gas) production. Current designs for digesters require high
capital investments in steel or concrete, and mixing and heating
of the units often utilizes much of the energy produced. Algal
biomass has been subjected to methane fermentations in laboratory
experiments but there have yet to be carried out large pilot
scale experiments. Laboratory experiments, however, allow certain
extrapolations: algal biomass is easily digested and efficiently
converted to methane in a stable fermentation system not subject
to upsets. About 50 to 60% of the energy in the algae can be
converted into methane (10, 11). Removing carbon dioxide from
the methane gas would only be necessary if it is transported over
a long distance (which is unlikely). Hydrogen sulfide removal
is a proven technology. High costs of methane digesters can be
overcome by simplified construction and operations. Anaerobic
ponds cost only about one percent of concrete digesters and they
could easily be fitted with floating plastic methane collectors.
Heating and mixing in such systems would be minimal, increasing
detention times. Since the ponds are inexpensive, this would
be allowable. Such low-cost fermentation technology still needs
to be developed. It can be expected that the present emphasis
in bioconversion research on methane fermentations of agricultural
wastes will accomplish this.

6. FERTILIZER PRODUCTION FROM ALGAL BIOMASS IS ACTUALLY A SLUDGE DISPOSAL PROBLEM

The nutrient content and soil conditioner values of algal
sludge residues from the methane fermentation process are roughly
similar to those of animal wastes. The ratio of ten N to one P
is generally valid on a dry weight basis for algae. After fermentations, volatile solid reductions are over 50% and N and P
(which are not lost) increase proportionally. The problem is
dewatering the residual algal sludge, as well as the liquid
fraction coming from the digester. The high moisture content
increases transportation and application costs. Storage of the
residue is another requirement. Sludge beds are a proven method
for sludge storage and dewatering. Transportation must be
minimized through proper siting of the ponds in agricultural
areas where the fertilizer can be used. If yields of 20 tons
algae/acre/year and 10% nitrogen content are assumed, it would
mean that algal ponds would supply all the fertilizer used by
10 to 30 acres of intensive agriculture. On the national average,
8 million tons of N fertilizer are used for 400 million acres
of crops, or one ton per 50 acres. This means that about 4

million acres of algal production ponds might be sufficient to
supply all agricultural nitrogenous (and most phosphate) fertilizer. Although this estimate might be high it is indicative
of the power of algal bioconversion systems. The fertilizer produced in excess of local agricultural requirements could be used
to increase the size of the system by recycling the digester
residues. Depending on the distribution of nutrients, the liquid
fraction alone could be recycled to the ponds.

Filamentous blue-green algae could play a significant role in
algal production systems because of their ready harvestability
and physiological properties (12). The widespread occurrence of
algae such as Oscillatoria suggests that they may be purposely
cultivated in algal production. Nitrogen fixation by blue-green
algae could be important in situations where algal biomass production is limited by available nitrogen in wastes or where phosphorus removal from low nitrogen wastes is desired. Whether
nitrogen fixation actually occurs in oxidation ponds is not known.
The reports of Anabaena and other heterocystous nitrogen-fixing
blue-green algae in industrial (13) and other oxidation ponds (14)
suggests that nitrogen fixation is already playing a role in some
of these systems. Nitrogen fixation would only be required
when algal biomass is increased above the N growth potential of
the wastes and sufficient phosphate is available. It is expected
that nitrogen fixation would play a secondary, though important,
role to nitrogen recycling in fertilizer production through algal
bioconversion (15). Cultivation of nitrogen-fixing blue-green
algae would be undertaken in ponds subsequent to the initial
receiving ponds. Selection for nitrogen-fixing organisms would
be determined not by the nitrogen content of the pond influent,
but by the concentration in the ponds.

5. *THERE IS NOT ENOUGH SEWAGE TO GROW ENOUGH ALGAE TO EVEN SUPPLY*
   1% OF U.S. ENERGY REQUIREMENTS

There is no typical sewage composition; the variations between
localities make the differences greater than the similarities.
However, from the C, N, and P content of sewage and the per capita
volume, the total algal biomass that may be grown under nutrient
limiting conditions can be calculated. For a sewage composition
of 100 mg TOC, 50 mg total N, and 10 mg P, an algal growth potential of 200 mg/1, 500 mg/1 and 1000 mg/1 for C, N, and P
respectively may be calculated (assuming a C:N:P ratio of 50:10:1
for dry algal biomass). An average daily per capita sewage flow
is about 500 1 which means that, at most, 182 kg (400 lbs) of
algae per capita per year may be cultivated using carbon dioxide
enrichment and nitrogen fixation to overcome C and N limitations.
Since about 5000 BTU can be recovered as methane from digestion
of one pound of algae, about 2.0 million BTU/capita/year may be
produced by stretching algal bioconversion to its limits. This
represents less than 1% of per capita energy consumption in the
U.S. (about 350 million BTU's). Thus, it is apparent that the

nutrient content of sewage is limited in terms of algal biomass
production potential. However, this ignores the much larger
amount of liquid wastes produced by agricultural and industrial
processes. It is difficult to compare their nutrient content
with that of municipal sewage. In many cases, particularly for
agricultural wastes, the algal biomass produced might have a higher
value as an animal feed or protein source rather than methane since
there is little contamination by metals, infectious agents, or
harmful chemicals to worry about. Nevertheless, if municipal,
industrial, and available agricultural liquid wastes are considered, up to 2% of the per capita energy consumption might be
generated by algal bioconversion. This, of course, assumes an
integrated system in which waste carbon dioxide from power plants
is used to feed the algal ponds the necessary carbon. This would
require a carbon dioxide injection system--a surface aerator
covered by a plastic bubble next to the pond mixing device might
be a promising approach. If such a system could double as a
power plant stack gas scrubber, a significant economic advantage
might be gained. Relatively small power plants would be required
to supply all the carbon dioxide requirements of a pond system.
These are actually economical in a "total energy system" where
waste heat is used by industry and, in this case, digester heating.
Although the power plant would require fossil fuels, coal is
expected to be available for several hundred years. Alternative
feedstocks might be used; for example, refuse and dry biomass
produced by energy plantations.

The economics of   carbon dioxide enriched waste treatment
processes are not easily determined; variable assumptions may be
made about the various costs and credits involved. In general,
as waste treatment processes are upgraded from primary to secondary, and then to tertiary, costs double and quadruple.   Since
tertiary treatment (N and P removal from effluents) can only be
accomplished in algal systems through carbon dioxide enrichment
of sewage, it appears likely that the whole process would be
operated for waste treatment. Compared to conventional tertiary
treatment systems (activated sludge, flocculation,denitrification),
algal systems are likely to retain the 2:1 cost advantage they
presently enjoy in secondary treatment by oxidation ponds. Thus,
waste treatment credits would support algal biomass production on
carbon dioxide enriched wastes.

Of course, algal bioconversion systems can only be considered
in climatologically favorable areas covering roughly one-half the
area of the Continental United States. This restricts the impact
of these systems in terms of the national energy supply. The
great variations in costs, available nutrients, and insolation
found in different localities restrict consideration of potential
energy supplies by bioconversion to local or regional demands.
Since the phosphorus content in algae can be well below 1% (under
P limiting conditions), its content in sewage is usually over 10
mg/1, and, since other liquid wastes are often available in large

amounts, it is possible that up to 5% of regional and even 10% of some local energy supplies might be met through methane fermentations of algal biomass. Other energy values must be included, e.g. fertilizer and waste treatment. The internal energy requirements in mixing of the pond and harvesting of the biomass would not significantly reduce their net energy production (see below). With favorable local conditions, nutrient recycling might be economically feasible, thereby increasing the size and output of the algal bioconversion systems. Thus, algal bioconversion can become a significant future energy resource whose impact could best be measured in terms of local or regional supplies rather than national requirements.

4. ALGAE SPECIES CONTROL CANNOT BE ACHIEVED IN LARGE-SCALE ALGAL PONDS

Three reasons exist why control of algal species, or at least types, would be desirable in algal production ponds: 1) to make pond operations and yields more predictable; 2) to avoid upsets in the digesters due to variable feeds; 3) to aid in algal harvesting. Since the third reason is the compelling one at the present time, it is emphasized. Algal population dynamics in ponds are complex and little understood. However, pond systems are relatively simple when compared to natural ecosystems. The possibility of algal species control through adjustment of mixing, depth, recirculation, algal and predator recycling, hydraulic detention time, nutrients, pond designs, chemical and biological agents, and inoculations have never been rationally studied. It is difficult to predict to what extent algal species control might be possible. Small-scale systems, even laboratory ones, would be sufficient to develop the necessary data and theory. Of course, larger-scale systems are not simple extrapolations of laboratory cultures, but they can be operated according to general principles learned in smaller-scale experiments. The actual cultivation of domesticated algal strains is a more difficult and longer range proposition. For the purposes of algal bioconversion, only a partial control over algal types would be sufficient. It is not expected that one strain of algae might be maintained throughout a whole year; occasional changes in algal types and even upsets would be expected.

Besides species control through nitrogen limitations (which would favor the harvestable filamentous blue-green algae), another algal species control method has been suggested which would involve a selective recycle of harvestable algae (12). The mathematical and experimental basis for this method is described elsewhere (16). In brief, it is based on the competitive advantage given to algae whose effective detention time in the ponds is longer than the hydraulic detention time. On its own, such a method might have only limited applications; its effectiveness would be decreased to the extent that ponds are not completely mixed systems. However, in combination with other operational

controls, selective biomass recycle could become a practical
method to help maintain desirable (harvestable) species in ponds.
Only through long-term experience will algal species control
techniques be perfected.   Until then, a variety of algal harvesting methods must be available to allow biomass production
even when algal populations change.

3.   *ALGAL HARVESTING METHODS ARE EXPENSIVE AND LITTLE CHANCE FOR
     IMPROVEMENT EXISTS*

    The relatively small size and low concentration of algae in
oxidation ponds has made their harvesting difficult.   Two
approaches exist: development of techniques specific for some
species or improvement of current non-specific methods which
are designed for the spectrum of algae normally found in ponds.
Among the former are size selective methods; if algae are large
enough, they can be inexpensively filtered.   Thus, it is possible
to concentrate filamentous blue-green algae with rotary screens
equipped with a backwash (microstrainers).   Harvesting algae with
such devices would cost less than ½¢ per pound of algae.   Even
more inexpensive methods would be settling or flotation; however,
at present, these are only feasible with additions of chemicals
(lime, alum, polyelectrolytes, etc.) which seriously complicate
the process and make it almost as expensive as centrifugation.
The use of specific environmental effects on algal settling
(darkness, nutrient limitations, carbon dioxide enrichment, etc.)
must be investigated to improve on algal settling processes.
Use of physiological capabilities of algae such as phototaxis
and flotation can also be considered.   If all else fails, in-
pond settling can be used to prevent unharvested algae from
escaping in the pond effluents and polluting receiving waters.
The twin problems of algal harvesting and species control are
the key technical difficulties which must be resolved before
algal bioconversion can be realized.

2.   *TOO MUCH POWER IS REQUIRED FOR MIXING AND OTHER POND
     OPERATIONS TO ALLOW SIGNIFICANT NET ENERGY PRODUCTION BY
     PONDS*

    Beyond an ill-defined minimum, there is no evidence that in-
creased mixing increases pond productivity.   Most oxidation pond
algae can be kept suspended by low mixing velocities, about 2
in/sec.   Many planktonic algae stay suspended on their own.   The
critical aspect for mixing is thermal destratification which may
also be accomplished through mixing velocities of 2 in/sec during
daylight hours.   From the Manning equation, it can be calculated
that by using paddle-wheels (at 25% efficiency), in a 1 foot
deep pond with a 50-ft wide channel, and two turns per acre,
about 900 KW HRS (3 M BTU's) would be used for a mixing speed
of 1 ft/sec for one hour a day and 2 in/sec for 23 hours a day.
Even considering a power conversion efficiency of 30%, this would

only represent about 5% of the energy output from such ponds.
Of course, other energy consuming operations will be required,
including harvesting, but none appear to be of a magnitude
higher than for mixing. There is, thus, good reason to expect
that a net energy analysis will be very favorable for algal
bioconversion. Present pessimision regarding this problem is due
to the overdesign of experimental facilities. However, even with
these, it is possible to calculate a positive net energy output
for algal production (17). Indeed, the net energy values calculated is greater than that achieved in most food or, for that
matter, energy production processes.

2. ALGAE BIOCONVERSION CANNOT BE PRODUCTIVE ENOUGH TO COMPETE WITH OTHER SOLAR ENERGY SYSTEMS

In solar energy conversion, the efficiency of sunlight energy
conversion is an important factor. The assumption made in this
paper of 200 M BTU's methane/acre/year produced by algal bioconversion corresponds to roughly 20 tons/acre/year of dry algae
or a 4% efficiency of conversion of visible sunlight at 30°
latitude. Whether such yields (equivalent to about 14 $gm/m^2/day$)
can be sustained year round in large-scale systems is yet to be
determined. Indeed, this is an optimistic prognosis; however,
several reasons can be suggested why such yields might be possible. Maximal theoretical photosynthetic conversion efficiencies
are about 10%, leaving a comfortable margin. Many higher plants
are reported to give yields above 20 tons/acre/year (19). Although algal biomass production might not be inherently more
efficient than higher plant production, there is no reason to
suppose it would be much less efficient. One key advantage of
algae in biomass production is that continuous production is
possible. Indeed, production rates from both small-scale and
large-scale algal systems are encouraging when the climatic and
other limitations under which the experiments were performed are
considered. Very high production rates, above 50 tons/acre/year
reported in some cases, need not be dismissed out of hand. The
organic constituents of wastes can be assimilated heterotrophically,   resulting in apparent photosynthetic efficiencies above
theoretical. If such anaerobic photosynthesis could be encouraged
in ponds, improved productivity would result. Small-scale cultures usually are of higher productivities than large-scale cultures; the difference must be ascribed to lack of engineering
designs for large systems rather than any specific factors which
increase productivity as an inverse function of size. Indeed,
even the conventional oxidation ponds do not produce negligible
amounts of algae--about 5 tons/acre/year is a reasonable estimate. It should be possible to easily double, probably quadruple,
these yields with high-rate ponds designed for maximizing algal
productivities.

To some extent, the arguments over algal production are pre-
mature. Only harvested algae are of interest in algal bio-

conversion. Just as a bird in the hand is worth two in the bush, ten tons/acre/year of harvested algal biomass is much better than twice as much in the pond effluents. The harvesting technology will set a lower limit to the allowable productivity-- the lower the harvesting costs, the lower the productivity that can be tolerated. Since the basic economics of algal biomass production are in waste treatment, lower productivities can be tolerated if they result in good waste decomposition and nutrient removal. Relatively low productivities of ten tons/acre/year might be economically feasible, even if not desirable, in algal bio-conversion.

CONCLUSIONS

No resolution of the key arguments pro and con algal bio-conversion is possible until the technical and scientific problems of algal harvesting, population control, and productivity are resolved. The great variability between regions and locali-ties of the basic parameters which determine algal bioconversion (insolation, waste flows and compositions, land costs, fertilizer needs, waste treatment credits, water reclamation potential, etc.) make it difficult to generalize the potential of algal biocon-version. In the absence of a detailed study, our estimates range from 0.1% of national to 10% of total local energy supplies in the form of methane. A cautious estimate would be that without nutrient integration, algal bioconversion could supply 0.1% of predicted national energy needs by the target date 2000 and 1% by 2020. A considerably shorter time scale might well be possible. The methane produced by algal ponds could be greatly increased by nutrient integration, i.e. recycling of nutrients from the methane digester to the ponds. In terms of energy supplies, the energy value of photosynthetic oxygenation (1 KW HR/kg of O_2) and the fertilizer production of (one ton of nitrogen is equivalent to about 50 M BTU's) would add to the net energy balance of such systems. Algal biophotolysis (19, 20)--the production of hydrogen from water and sunlight--an idea of as yet uncertain potential, might become a significant energy source in the future. It is likely that bioconversion with microalgae could supply 10% of the U.S. energy useage in the next century.

ACKNOWLEDGMENT

This work was supported by ERDA Contract E(04-3)-34 (239) and NSF Grant No. AER76-10809.

REFERENCES

1. Burlew, J.S. (ed.) (1953) *Algae Culture from Laboratory to Pilot Plant*, Carnegie Institute, Washington. *Publ. 600.*
2. Oswald, W.J. and C.G. Golueke (1960), *Adv. Appl. Microbiol.* *2*, 223.
3. Oswald, W.J. (1976) in *CRC Handbook of Applied Microbiology*, in press.
4. Gloyna, E.F., J.F. Malina, and E.M. Davis (eds.) (1976) *Ponds as a Waste Treatment Alternative*, Center for Research in Water Resources, University of Texas, Austin.
5. Steward, W.D.P. (ed.) (1974) *Algal Physiology and Biochemistry*, University of California Press, Berkeley.
6. Fogg, G.E. (1963) *Algal Cultures and Phytoplankton Ecology*, University of California Press, Berkeley.
7. Carr, N.G. and Whitton, P. (eds.) (1975) *The Blue Green Algae*, University of California Press, Berkeley.
8. Pimentel, D., E.C. Terhune, R. Dyson-Hudson, S. Rochereau, R. Samis, E. Smith, D. Denman, D. Reifschneider, and M. Shepart, (Oct. 8, 1976) "Land Degradation: Effects on Food and Energy Resources," *Science, Vol. 194*, No. 4261.
9. CSO International (1975) *City of Hollister Project Report.*
10. Golueke, C.G., W.J. Oswald, and H.B. Gotaas (1957) *Appl. Microbiology, 547*, 551.
11. Uziel, M., W.J. Oswald and C.G. Golueke (1975) *Solar Energy Fixation and Conversion with Algal Bacterial Systems*, Final Report, SERL, University of California, Berkeley.
12. Benemann, J.R., B. Koopman, J. Weissman, and W.J. Oswald (1976) *Proc. Conf. Microb. Eng. Conversion, Gottingen*, in press.
13. Martin, J.D., V.D. Dutcher, T.R. Frieze, M. Tapp, and E.M. Davis (1976) in *Ponds as a Waste Treatment Alternative*, (Gloyna et al. ed) p. 191.
14. Palmer, M.C. (1962) *J. Phycol. 5* 78.
15. Oswald, W.J. and J.R. Benemann (1976) *Proc. NSF-RANN Conf. II* (in press).
16. Weissman, J., J.R. Benemann, B. Koopman, and W.J. Oswald (1976) in preparation.
17. Soeder, J.C. (1976) *Proc. Conf. Microbial Energy Conversion*, Gottingen (in press).
18. Alich, J. and R. Inman, 1976 *Proc. Conf. Clean Fuels from Biomass Sewage, Urban Refuse and Agricultural Wastes*, Orlando, Florida.
19. Weissman, J. and J. Benemann, *Appl. Environmental Microbiology*, (in press).
20. Benemann, J. and J. Weissman, *Proc. Microbiol. Energy Conversion*, Gottingen (in press).

Using Sugar Crops to Capture Solar Energy

E. S. Lipinsky and T. A. McClure

Battelle Columbus Laboratories

The Energy Research and Development Administration (ERDA) is evaluating and developing alternative energy resources for future use. Solar energy methods are particularly attractive because they make use of free sunlight and have low pollution potential. Furthermore, safety and environmental problems frequently associated with other energy sources generally are absent in solar energy ventures.

Energy systems utilizing plant biomass as sources of liquid fuels (such as ethanol and methanol) and gaseous fuels (such as substitute natural gas and synthesis gas) are being investigated, in addition to electric power generation via steam. Sugar crops, trees, grains and grasses, and various aquatic fuel sources are under consideration. Sugar crop systems have some potential advantages over trees or aquatic sources in that much of the biomass is in the form of directly fermentable simple sugars. Sugar crops are also noted for high yields per unit of land area. On the other hand, trees can store biomass until needed. Kelp or water hyacinths can be grown on territory that has little or no alternative economic value. Therefore, there exists a high level of uncertainty as to which specific biomass is the most promising for development.

RELATIVE SIGNIFICANCE OF VARIOUS SUGAR CROPS

Three sugar crops capable of producing sizeable quantities of biomass per unit of area are sugarcane, sugar beets, and sweet sorghum. The potential availability of sugar-containing biomass is a function of the potential yield and the potentially available area for cultivation of the sugar crop. Sugarcane has the highest yield but the smallest cultivatable area of the three crops. Sugar beets have lower yields, but can be grown over a wide geographic area. However, pest control necessitates sugar beet crop rotations which limit the useful area to about one quarter of that which is suitable on climate and soil criteria. Sweet sorghum is primarily an experimental crop at this time. Experimental sweet sorghum yields vary widely, but

the annual land area that is potentially suitable for sweet
sorghum far exceeds that of sugar beets because frequent crop
rotation does not appear necessary.

The above generalizations are the authors' primary conclusions from more detailed research on the three crops. The yield/
geographical area tradeoffs described above appear to give
sugarcane the greatest near-term potential as an energy source,
with sweet sorghum having greater potential over the long run
of the next two to three decades. Sugar beets appear to occupy
a middle position. Much more is known concerning the problems
encountered in large scale, long-term production of sugar beets
compared with very limited information regarding sweet sorghum.
When comparable information is available concerning sweet
sorghum, its attractiveness compared with sugar beets may change.

The remainder of this paper is limited to a more detailed
discussion of the prospects for utilizing sugarcane biomass as
an energy source. Sugarcane, as previously indicated, produces
the highest biomass levels per acre of the three sugar crops.
Cultural practices and processing technology are well known.

SUGARCANE YIELDS

Average commercial sugarcane yields are compared with current
high yields and experimental high yields in Table 1. Commercial
results are considerably below the experimental highs for many
reasons that concern crop management decisions, rather than biological potential. In Louisiana for example, the harvesters work
best on 6-foot rows, but the optimum productivity for biomass is
at much closer spacings of the plants. Also, Louisiana is subject to hurricanes. Texas sugarcane has experienced frost damage
in most years since the reinitiation of production in this state.
In Hawaii, some producers effectively break up the "hard pan"
which is the subsoil into which the sugarcane roots must grow,
while others do a more superficial job. Average yields for the
state reflect these differing crop management practices.

Physiological Factors

The high photosynthetic capacity of the sugarcane plant was
thought to be the source of the high yields attained for this
crop. The C-4 photosynthetic pathway initially appeared to be
the key to high yields but recent research (Loomis and Gerakis,
1975), Bull (1975), and Glover (1973) indicate that other
factors are operating. It is becoming increasingly clear that
the length of the growing season after canopy closure has
occurred is the most important single factor in achievement of
high yields. The growing season length determines exposure to
high levels of solar insolation and favorable temperatures.

Table 1.   Sugarcane Yields and Growing Seasons

Location	Yields in Metric Tons of Dry Matter Per Hectare Including Tops and Leaves			Length of Growing Season of (Months)
	Commercial		Experimental High	
	Recent Average	Recent High		
Florida	24	60	88	10-12
Louisiana	17	36	60	7
Texas	26	63	68	12
Hawaii	71	158	180 [a]	24
Puerto Rico	23	52		12
South Africa	--	51 [b]		12

Sources: (a)   Reference 3 under List of References
           (b)   Reference 4 under List of References
           Others are the authors' estimates, based on field
           interviews with sugarcane specialists in the
           respective geographical areas.

Either deficient or excessive soil moisture levels are a
critically important brake on vegetative growth (Slavik, 1975).

The availability of appropriate quantities of macronutrients
and micronutrients are necessary for fulfillment of the potential
inherent in the sugarcane plant.   Crop physiologists have found
that crops other than sugarcane will equal or exceed sugarcane
yields on a daily basis during the growing season (Bull and
Glasziou, 1975).   However, the determinate growth of many plants
caused yields to be low, compared with long-season plants.

In forecasting sugarcane yield increases, the researchers
emphasized the importance of close row spacing.   Without this
recourse, future yield increases are expected to be rather
small.   The potential for close spacing as a means of yield
improvement is demonstrated by the work of Bull (1975).   Similar
research has been conducted in Louisiana, Hawaii, Texas, and
Florida.   However, only Bull's work is described here because
it is well documented in the literature.

Bull (1975) has found that canopy closure occurs in 14 weeks
when an experimental row spacing of 0.45 meters is employed but
24 weeks are required to achieve canopy closure at commercial
spacings at 1.4 meters.   In these Australian trials, close row
spacings have a very favorable impact on yield both of the total
dry matter and sucrose.   However, increasing the number of
plants per row as well as decreasing row width has a generally
adverse effect on yield.   Bull found that best results were
obtained with unselected populations.   Apparently, the present

10-year breeding and selection process at a specific, wide row spacing leads to cultivars that are not well adapted for close spacing.   Bull recommends that special cultivars be developed using close spacing throughout the breeding and selection process.

The authors' estimates of future sugarcane yields in several geographical areas are compared with Bull's results and recent commercial high yields in Table 2.   These estimates reflect the concept that yield improvements of 50 to 100 percent and more in dry matter are possible with close spacing and with more than 20 years to develop the new varieties that can tolerate close spacing.   Because yield is quite sensitive to land quality, separate estimates are made for high quality and adequate quality land.   It is uncertain at this time whether close spacing regimens will provide only more biomass or also more sucrose at the same or lower unit costs.   Bull (1975) and Irvine (1975) independently obtained increases in sucrose but other unpublished experiments yielded mostly additional fiber.

Table 2. Sugarcane Yield Forecasts for the Year 2000 Using Close Spacing, Compared with Recently Observed Yields (Metric Tons of Dry Matter Per Hectare)

Geographical Area	High Quality Land		Adequate Quality Land	
	Yield[a] in Year 2000	Recent Commercial High Yields	Yield in Year 2000	Current Average Yields
Florida	81	60	47	24
Louisiana	47	36	34	17
Texas	85	63	54	26
Bull's Experi-mental Australian Results[b]		70	--	

(a)   All yields include tops and leaves.
(b)   Bull (1975), adjusted to include tops and leaves.

YIELDS BASED ON ENERGY

The ultimate resources in classical economics are land, labor, and capital.   Recently, energy has joined the ranks of the ultimate resources.   The methodology of energy input/output analysis is in a state of turmoil, in part because both the energy outputs and inputs vary in both quality and scarcity. There are many alternative accounting techniques for measuring both the input and the output and for expressing the profits and losses.

Starting from the premise that no fully satisfactory energy-based yields can be derived at this time due to a combination of methodological problems and data availability/cost problems, Battelle has chosen to make use of data recently developed by the Economic Research Service (ERS) of the U.S. Department of Agriculture under a cooperative agreement with the U.S. Federal Energy Administration (USDA, ERS, 1976)*. These data are based upon information developed by ERS commodity specialists and subjected to review and update by agricultural extension economists, engineers, and others in each production location. Although this approach omits factors that may merit inclusion (the energy used to clear the land for farming and the energy required to manufacture farm machinery, for example), the ERS approach has the outstanding compensation that it involves an operationally meaningful procedure that can be repeated annually to create a useful time series.

Yields on the basis of unit area and unit energy consumption are compared in Table 3. The yields on both bases are distinctly better in Florida and in Texas than they are in Louisiana. Apparently, the high energy cost of irrigation in Texas renders the yield on the basis of energy consumed virtually equivalent to that of Florida, even though the Texas yield per acre is much higher.

The ratio of the potential energy content contained in sugar-cane production in these three states to energy directly consumed in its production also is displayed in Table 3. The differences between Florida and Texas are negligible but Louisiana product-ivity on this basis also is low.

Table 3. Comparison of Energy Contained with Energy Utilized Per Hectare, Selected U.S. Areas

	Florida	Louisiana	Texas
Dry Matter Yield, Metric Tons Millable Cane Per Hectare	18.3	12.8	22.5
Energy Contained, Megajoules Per Metric Ton (1000)	18.0	18.0	18.0
Energy Contained, Megajoules Per Hectare (1000)	329.4	230.4	405.0
Energy Used, Megajoules Per Hectare (1000)	37.0	35.0	43.0
Ratio of Energy Content Contained to Energy Directly Consumed	8.9	6.6	9.4

Sources: Battelle and the Economic Research Service of the U.S. Department of Agriculture.

*The authors added an estimated quantity of fuel utilized in trans-porting sugarcane from the field to the mill to USDA on-farm data.

These data are for millable cane only, and do not include
utilization of tops and leaves. These numbers need to be evaluated carefully for the following reasons:

(1)   The Florida data are a composite of production on muck
       lands that require no fertilization at the present time
       and Florida sandy/peat soils that require considerable
       fertilization.
(2)   Florida sugarcane production benefits from extensive
       flood control and water table control investments made
       by the United States Corps of Engineers and the full
       energy cost of this program is not borne by these
       beneficiaries of this irrigation and drainage technique.
(3)   Louisiana energy consumption data need to be averaged
       over a considerable period of time because of the impact
       of unusual weather considerations (e.g., hurricanes) that
       have a major impact on yields in this geographical area.
(4)   The Texas data represent a very small area that has just
       recently been devoted to sugarcane production.

Nevertheless, energy contained in sugarcane grown where it grows
well is so much more than the direct energy inputs that this crop
appears to have some real potential as a source of fuel.

AREA FORECASTS

United States' mainland production of sugarcane occurs on a
very small area of land in a few states. If sugarcane is to play
an important role in the future production of industrial chemicals and fuels, increases in yield on current sugarcane lands
will not be sufficient. Therefore, the additional land with the
appropriate attributes must be sought. The two major factors
that are required are an appropriate temperature profile (Figure
1) and appropriate water resources. The temperature profile is
especially important since a frost to the depth of 4 inches
during the winter will kill the sugarcane plant as it over-
winters. At 32 F the seed cane germinates. When root
temperatures reach 80 F, the sugarcane plant is in fine
photosynthetic operation. The productivity of the plant
increases as temperature increases so that the rate of growth is
virtually doubling for every 10 degrees increase in average
temperature in some areas.

Producers have been quite cautious about moving onto new
lands where freezes might be expected to be more prevalent than
on present lands. The temperature profiles for air and soil
temperatures in three major sugarcane states were determined and
they indicate that much more land could be put into sugarcane
production, as Table 4 indicates. These areas have appropriate
temperature profiles and are located in areas where water is
potentially available, although investment in irrigation or
drainage facilities may be needed.

Figure 1. Geographical Distribution of Mean Annual Freeze-Free Periods for Texas.

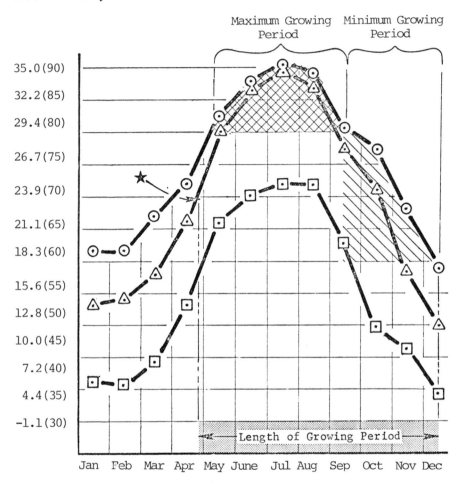

Temperature in Degrees Centigrade (Fahrenheit)

⊙ Maximum Air Temperature

▣ Minimum Air Temperature

△ Average Soil Temperature
4-inch depth

Table 4.   Estimated Area Capable[a] of Supporting Sugarcane Growth, Contiguous United States, 1969 (Hectares in thousands)

	Florida	Louisiana	Texas
Candidate Farmland	5,679	1,342	2,622
Harvested Cropland	904	522	1,194
Sugarcane [b]	116	125	14
Other	788	397	1,180
Cropland Used for Pasture	405	248	595
Woodland	1,544	339	438
Other Farmland	2,826	234	394
Irrigated Cropland	546	62	456

(a)   Based on frost-free period and water availability.
(b)   1975 area.

It should be recognized that much of the new land would be lower quality than that currently devoted to sugarcane agriculture. Average yields associated with such land were estimated to be lower than on present sugarcane land.

SUGARCANE PRODUCTION COSTS

The basic information for estimating 1976 sugarcane production costs have been supplied by the U.S. Department of Agriculture's (USDA) Firm Enterprise Data System (FEDS). Impetus for the FEDS resulted from a clause in the Agricultural and Consumer Protection (ACP) Act of 1973 which requires estimates by USDA of the production costs associated with certain commodities. The Enterprise budget system utilizes the Oklahoma State University budget generator for building and maintaining budgets. This method utilizes a computer to process input data into completed enterprise budgets and facilitates storage, modification, and updating of the budgets. This system, which has just become operational within the past year, has resulted in budgets being prepared for different sugarcane producing regions in the mainland United States. Detailed data for two regions are shown in this section--Texas Rio Grande Valley and Louisiana. A comparison of present and projected costs for Florida's peat soils is shown in Table 5. Comparable cost breakdowns are not available for Hawaii or Puerto Rico. Therefore, they are not included in this analysis.

The production costs represent current average technology, since the major anticipated use of the USDA budget is in dealing with aggregate supply questions in response to government supply management policies and programs. Data to develop the

production costs come from a variety of sources, including USDA's
Statistical Reporting Service, Economic Research Service, and
various state experiment station and extension service staff
knowledgeable in sugarcane production.

Table 5. Estimated 1976 Sugarcane Production Costs in

Texas and Louisiana

	Texas		Louisiana .	
	$/Hectare	$/Ton[a]	$/Hectare	$/Ton[a]
Preharvest costs				
Seed	29.65	0.35	40.72	0.73
Fertilizer	102.18	1.20	89.08	1.59
Chemicals	101.93	1.20	64.99	1.16
Labor	95.41	1.12	131.61	2.35
Fuel and Lubricants	15.91	0.19	25.70	0.46
Repairs	24.61	0.29	44.06	0.78
Interest on Operating				
Capital	29.28	0.34	17.12	0.31
Miscellaneous	42.01	0.50	17.99	0.32
Subtotal	440.98	5.19	431.27	7.70
Harvesting Costs				
Labor	85.13	1.00	85.30	1.52
Hauling	186.80	2.20	74.13	1.32
Fuel and Lubricants	50.90	0.60	23.08	0.41
Repairs	95.18	1.11	76.60	1.37
Interest on Operating				
Capital	8.75	0.10	2.10	0.03
Subtotal	427.76	5.01	261.21	4.65
Machinery Ownership Costs	142.18	1.66	133.95	2.39
Land Charge	222.39	2.61	185.33	3.31
Management Charge at 7%				
of Gross Receipts	97.04	1.14	71.26	1.27
TOTAL COSTS	1,330.35	15.61	1,083.02	19.32

(a) Tons = Metric Tons
Source: Authors' estimates based on unofficial U.S. Department
of Agriculture crop budgets and Texas A&M University crop
budgets (adjusted to 1976 price levels where necessary).

The cost estimates in Table 5 are total costs of production,
including variable costs for purchased inputs and fixed costs
for machinery, overhead, management, and land. Also, an estimated hauling cost was added to account for transportation of
sugarcane from the field to the processing facility.

Estimated 1976 sugarcane production costs for the two
regions are shown in Table 5. Yields of millable cane per unit

of area are differentiated according to expected productivity for
each region employing current average technology. This breakdown
shows the decrease in productivity and increase in production
costs per ton in regions having various soil and climatic
conditions. For example, total production costs per hectare are
lower in Louisiana than in Texas--$1,083 per hectare versus
$1,330 per hectare. However, since anticipated Texas yields
of millable cane are 85 tons (fresh weight) compared to only
56 tons per hectare in Louisiana, production costs per hectare
are about 20 percent less in Texas.

Dry Matter Costs

In producing sugarcane for fuel and/or chemicals production,
it is assumed that the entire aerial portion of the plant would
be utilized. Currently, tops and leaves are left on the field.
Table 6 shows a summary of estimated current sugarcane and
production costs, assuming that tops and leaves are included.

*Table 6. Summary of Estimated 1976 Sugarcane and Biomass
Production Costs, Including Tops and Leaves*

	Texas	Louisiana	Florida, Peat Soil
Millable cane ⎰ Per hectare,	85	56	78
Tops & leaves ⎱ in metric tons	25	18	25
Delivered Costs Per Metric Ton			
Millable cane, dollars	15.61	19.32	18.71
Millable cane + tops & leaves, dollars	13.28	15.69	15.75
Delivered Costs Per Metric Ton Biomass [a]			
Millable cane, dollars	57	72	69
Millable cane + tops & leaves, dollars	50	58	58
Costs Per Kilogram Biomass			
Millable cane, cents	5.7	7.3	7.1
Millable cane + tops & leaves, cents	5.1	6.0	6.0

(a) Assumes dry weight content of aerial part of mature
sugarcane = 27 percent.
Source: Calculated by authors from selected sources.

The quantity of tops and leaves per hectare was obtained by multiplying the millable cane production by 1.3. It had been reported that tops and leaves ranged from 20-40 percent on handstripped cane, depending upon the variety (Hipp, 1976). Actually, reliable statistics do not exist on the quantity of sugarcane tops and leaves relative to millable cane, because heretofore, tops and leaves have not had any real economic value.

Another assumption noted in Table 6 is that the dry weight content of the aerial part of mature sugarcane is 27 percent (Atlas of Nutritional Data on United States and Canadian Feeds, 1971). The dry weight content will naturally vary according to the time of harvest; that is, more mature cane will have a higher dry weight percentage than green cane. Also, the present dry matter of different components of sugarcane varies; for example, the dry matter content of fresh sugarcane leaves is indicated to be 35 percent, while the dry matter percentage of fresh leaves and tops combined is 25 percent (Atlas of Nutritional Data on United States and Canadian Feeds, 1971) or less (Warner, 1976).

In estimating delivered costs, including tops and leaves, it was assumed that total harvesting costs would also increase by 30 percent over current costs. This estimate may be somewhat low in that tops and leaves are less dense than cane stalks, thus making transportation costs higher.

Under these assumptions, the delivered costs per metric ton of dry matter, including tops and leaves, range from $50 per metric ton in Texas to $58 per metric ton in Louisiana.

Fermentable Solids and Combustible Organics Costs

Since the total biomass of sugarcane consists of fermentable solids plus combustible organic material (fiber), it is desirable to make some allocation of production costs among these two components due to their different values. That is, the fermentable solids portion has a higher value than does combustible organic material. A ratio of 4.62 to 1 was used in allocating total costs between fermentable solids and combustible organic material. This ratio reflects the smaller economic value associated with the fiber. Only the costs of harvesting and transportation are shared in proportion to their weight, other costs are borne mostly by the fermentable solids.

Table 7 indicates estimated 1976 raw materials costs of fermentable solids and combustible organic material from sugarcane based on the above methods of calculations. Yields per hectare are based upon average reported sugarcane yields for each growing region, plus 30 percent additional biomass for tops and leaves. Based on these calculations, the raw material cost per ton of fermentable solids ranges from $102 per ton in Texas to $119 per ton in Louisiana. The cost of combustible organic material ranges from $21 per ton in Texas to $32 per ton in the Florida peat soils.

Table 7. Estimated 1976 Raw Materials Costs of Fermentable Solids and Combustible Organics from Sugarcane

	Texas	Louisiana	Florida, Peat Soils
Production costs per hectare, dollars			
Fermentable solids	1,200	950	1,330
Combustible organics	260	210	290
Yield per hectare, metric ton			
Fermentable solids	11.7	8.0	13
Combustible organics	12.3	7.8	9.2
Cost per metric ton			
Fermentable solids	102	119	103
Combustible organics	21	27	32

Sources: Calculated by the authors.

Speculative Cost Forecasts

Closer spacing of sugarcane rows could increase yield by achieving canopy closure sooner. A speculative forecast of possible cost changes associated with this change in technology is shown in Table 8 for the Florida peat soil region. If the costs and the yield increase as shown, total production costs per metric ton might decline from $18.10 to $14.61 or approximately 22 percent. Verification of these changes are naturally subject to results of actual field experiments.

The effect of new cultural practices on the composition of the sugarcane plant is another unknown factor at this time. Therefore, it is difficult to speculate on projected costs of fermentable solids and combustible organic material. Quite possibly, close spacing would lead to a higher proportion of combustible organic material relative to fermentable solids in the plant. This is another topic for future research. Assuming no compositional changes, the fermentable solids costs based on the projected costs shown in Table 8 would be approximately $80 per metric ton, while the cost of combustible organic material would be approximately $17 per metric ton based on 1976 price levels.

Table 8. Sugarcane Production Costs on Florida Peat Soil,
Current and Projected(a)

Florida Peat Soil	Current Production Costs Dollars/ Metric Ton[b]	Cost Increase Factor	Projected Production Costs Dollars/ Metric Ton[c]
Preharvest Costs			
Seed	0.53	1.50	0.41
Fertilizer	1.09	2.00	1.12
Chemicals	0.89	1.10	0.51
Labor	1.69	1.50	1.30
Fuel and Lubricants	0.37	1.50	0.25
Repairs	0.44	1.50	0.34
Interest on Operating Capital	0.24	1.50	0.19
Miscellaneous	0.43	1.50	0.33
Subtotal	5.68		4.04
Harvest Costs			
Labor	3.18	1.25	2.05
Hauling	1.32	1.95	1.32
Fuel and Lubricants	0.54	1.90	0.53
Repairs	0.83	1.90	0.80
Interest on Operating Capital	0.29	1.50	0.22
Miscellaneous	0.24	1.50	0.35
Subtotal	6.61		4.80
Machinery Ownership Costs	1.50	2.00	1.54
Land Charge	3.63	1.00	1.86
Management Charge	1.29	2.25	1.49
TOTAL COSTS	18.71	1.43	14.61

(a) All costs at 1976 prices.
(b) Yield = 78 metric tons per hectare.
(c) Yield = 150 metric tons per hectare which represents 30
percent over millable cane due to addition of tops and leaves,
plus another 50 percent increase for close spacing of cane plants.
Source: Authors' estimates.

Acknowledgment

The research providing the basis for this paper was conducted by Battelle's Columbus Laboratories under contract to the U.S. Energy Research and Development Administration. The authors also express their appreciation to Dr. W. T. Lawhon of Battelle for his assistance on sugarcane agronomic topics.

LIST OF REFERENCES

(1) Atlas of Nutritional Data on United States and Canadian Feeds, National Academy of Sciences, Washington, D.C. (1971).
(2) Bull, T. A., "Row Spacing and Potential Productivity in Sugarcane", *Agronomy Journal*, 67, 421-423 (May-June, 1975).
(3) Bull, T. A. and Glasziou, K. T., "Sugarcane" in Crop Physiology, edited by L. T. Evans, Cambridge University Press.
(4) Glover, J., "The Rate of Apparent Photosynthesis of Whole Sugarcane Plants", *Annals of Botany*, 38, 909-920 (1974).
(5) Hipp, B. 1976, personal communication. Soil nutritionist, Texas A&M University Agricultural Research & Extension Center, Weslaco, Texas.
(6) Irvine, J. E., 1975, personal communication. USDA, ARS Sugarcane Research Laboratory, Houma, Louisiana.
(7) Loomis, R. S. and Gerakis, P. A., "Productivity of Agricultural Ecosystems", in Photosynthesis and Productivity in Different Environments, edited by J. P. Cooper, Cambridge University Press, Cambridge (1975) pp 145-172.
(8) Slavik, B., "Water Stress, Photosynthesis and the Use of Photosynthates", in Photosynthesis and Productivity in Different Environments, edited by J. P. Cooper, Cambridge University Press, Cambridge (1975) pp 511-536.
(9) Warner, J., 1976, personal communication. Hawaiian Agronomics, Inc., Honolulu, Hawaii.

FUEL GAS PRODUCTION FROM SELECTED
BIOMASS VIA ANAEROBIC FERMENTATION

D.L. WISE*, R.L. WENTWORTH & R.G. KISPERT
Dynatech R/D Co.
Cambridge, Massachusetts 02139

Presented at the U.S.-Japan Joint Cooperative Seminar
Entitled "Biological Solar Energy Conversion"
Miami, Florida, November 15-18, 1976

*Speaker

ABSTRACT
    There is substantial potential for production of meaningful
supplies of fuel gas from selected organic materials via anaerobic fermentation. Fuel gas production from animal waste appears
to have the greatest potential for short term alleviation of the
natural gas supply problem in the United States. Plans for development of a mariculture system appear to be merited due to the
very large quantities of organic material that may be produced for
the subsequent bioconversion to fuel gas. Employment of select
bioconversion concepts also has practical application in the area
of coal gasification; here, the gases CO, CO_2 and H_2 are anaerobically fermented to CH_4. Fuel gas production from solid waste is
determined at being at an advanced development stage. It is concluded that anaerobic digestion has progressed rapidly and clearly
has practical merit beyond the traditional sewage sludge digestion
application. Further development must stress larger scale processing units to obtain substantial supplies of fuel gas. Supportive engineering control and improvements associated with larger scale process development will be required.

INTRODUCTION
    The conversion of certain organic materials, primarily
wastes, to fuel gas utilizing biological fermentation processes
has been practiced and investigated for many years. Processes in
Europe, Japan and the United States for the reduction and stabilization of sewage sludge produce sufficient energy to provide mechanical and electrical power as well as process heat in many
treatment plants. In other parts of the world, including India
and Taiwan, simple, small-scale anaerobic processing units operate on animal and crop wastes. More recently, attention in the
United States has focussed on the utilization of anaerobic conversion for a number of more plentiful organic wastes. Due to
twin incentives of fuel gas production and waste disposal, anaerobic digestion of the organic fraction of municipal solid waste has

411

been investigated in some detail (1, 2, 3). Experimental studies
have shown that such processes are technically feasible at both
mesophilic and thermophilic temperatures (1). A detailed engineering study and economic analysis of the process has led to the
construction of a large-scale facility for demonstration of the
concept (2, 3). Other experimental programs have investigated
anaerobic conversion of animal and agricultural wastes (4). As
shown in Table 1, these materials can potentially contribute even
greater quantities of organic wastes than municipal refuse. Over-
all, it appears that fuel gas production from waste materials by
the bioconversion route has substantial potential.

TABLE 1

Sources of Waste

Municipal Solid Waste	230×10^6 ton/year
Industrial	140
Mineral	1700
Animal	1740
Crop	640

Whereas bioconversion has most often been considered with respect to the utilization of organic waste materials, the nature of
the process also makes it readily adaptable to other energy production concepts. Waste materials by their nature are limited in
supply. Bioconversion is also applicable, however, to renewable
energy sources such as sea-based mariculture systems or land-based
energy plantations. Recent studies are aimed at showing its application to coal conversion technology. Such concepts which may
produce significant quantities of energy at a single site will require application of bioconversion technology to ever larger scale
operations.

While many bench scale studies continue to be carried out
utilizing bioconversion techniques, it is becoming increasingly
clear that the experimental feasibility of converting organic materials to fuel gas has been established. Future programs utilizing waste materials must concentrate on the demonstration of this
technology at pilot scale and eventually in commercial plants.
Additional laboratory scale experiments with these materials must
focus on methods of improving process efficiency. Because of the
limited sources of waste materials, however, in order for bioconversion to provide a major contribution to fuel gas supplies in
the United States, processes utilizing primary sources of organic
materials, such as renewable mariculture and coal, must be developed.

In the following sections, an overview of the current technical background available for utilization of select organic substrates having potential for fuel gas production by anaerobic digestion is presented. Fuel gas production from municipal solid
waste and animal waste are examined in detail. A renewable source

of energy, mariculture, is discussed in the light of the seemingly unlimited potential biomass available. Finally, a process by which bioconversion may be applied to coal conversion and utilization is described.

FUEL GAS PRODUCTION FROM MUNICIPAL SOLID WASTE

Because of the twin problems of need for supplemental sources of fuel gas and concern about municipal waste disposal, considerable interest has recently been shown in applying anaerobic digestion to municipal solid wastes. The digestion of the organic matter in municipal refuse can be carried out much as the digestion of sewage sludge is done. The organic matter in typical solid waste is observed to be predominantly cellulose; therefore, the chemical conversion and stoichiometry of concern may be represented by:

$$C_6H_{10}O_5 + H_2O \rightarrow 3CO_2 + 3CH_4 \qquad [1]$$

In the process for preparing methane from municipal waste, the organic material is slurried with water and innoculated with the proper microorganisms. This innoculation is spontaneous in an operating digester since organisms are already present. The organic matter is partially solubilized or digested and then fermented, forming methane gas, carbon dioxide and a residue of undigested material. Under these circumstances one kilogram of chemically convertible waste will yield 0.41 cubic meters (1 lb → 6.65 cubic feet) of methane at standard conditions of temperature and pressure (5). The methane will be accompanied by an equal volume of carbon dioxide. The methane is scrubbed free of carbon dioxide and traces of hydrogen sulfide, then dried. The undigested residue is disposed of by incineration or landfill. Ferrous and non-ferrous metals and glass may be recovered prior to digestion, if that is economical.

The potential for a process which converts refuse to methane is significant. Residential and commercial refuse is produced at a rate of 1.36 - 2.26 kg (3-5 lbs) per person per day. At 4 lb/person-day, a 1000 ton per day (tpd) facility would service a population of approximately 500,000 people. According to the 1970 U.S. Census, there were 26 cities in the country with populations in excess of 500,000. More significant, there are 65 Standard Metropolitan Statistical Areas (SMSA's) in the U.S. with populations in excess of 500,000. The aggregate population of these SMSA's is in excess of 100,000,000, half of the nation's population. In terms of a 907,000 kg/day (1000 tpd) solid waste to methane facility, there is a potential market for over 200 plants in the urban areas of the United States.

A 907,000 kg/day (1000 tpd) bioconversion facility will produce approximately 105,000 cubic meters (3.7 million cubic feet) of methane per day, or 0.21 cubic meter (7.4 cubic feet) of methane per person. The 65 SMSA's with populations in excess of 500,000 have a potential for gas production from waste in excess of 21,225,000 cubic meters (750,000,000 cubic feet) per day. Based on published figures (6), this process, if implemented in these

65 SMSA's could produce approximately 1.5% of the total natural gas consumed in the United States. In order to further identify those regions of the country in which fuel gas production from solid waste could have significant impact, this study has been extended to each of these major SMSA's. The population of each was determined from the latest U.S. Census (1970). Potential gas production from each SMSA was determined based on an average daily production of 0.21 cubic meter of methane per person. Actual gas consumption in each SMSA was estimated based upon figures published in Brown's Directory of North American Gas Companies (7). Because gas consumption figures were not usually broken down by SMSA, estimates were developed based upon the ratio of customers within the SMSA to the total number of customers served by each gas company within the SMSA. On average, approximately 9% of the gas consumed in these SMSA's could potentially be produced by the municipal solid waste generated by their residents. The median ratio of potential production to estimated consumption was 4%. A geographic evaluation of these results shows that these plants would have the greatest impact along the Boston-Washington corridor, an area highly dependent upon imported energy (5).

Fuel gas production from municipal solid waste shows significant promise to provide additional sources of natural gas while reducing the magnitude of the solid waste disposal problem. The process is at an advanced state of development, as shown by ERDA's commitment to the construction of a 50-100 tpd demonstration facility in Pompano Beach, Florida. However, the limited amount of waste material available implies that this bioconversion process will be unable to provide the large quantities of natural gas necessary to meet supply requirements.

FUEL GAS PRODUCTION FROM ANIMAL WASTE

Market forces and weather conditions lead to the presence on U.S. farms and ranches of varying numbers of animals, but the order of magnitude may be appreciated by taking the animal population statistics of the U.S. Department of Agriculture for 1973. Using these figures, Lauer (8) estimated that the following quantities of manure are produced annually in the United States.

TABLE 2

Manure Production in the United States

Animal	Mass of Dry Manure, million metric tons
Cattle	195.0
Swine	22.4
Horses	12.6
Poultry	4.6
Sheep	2.9
Total	237.0

These statistics of manure production, obtained from animal populations, depend on estimates of typical rates of production by individual animals under different conditions of raising as well as on typical manure solids figures. For example, Ifeadi and Brown (9) cite the following characteristcis of waste derived from two classifications of cattle.

TABLE 3

Characteristics of Manure

	Beef Feeder	Dairy Cow
Animal weight, live, lb	1200	900
Manure production, wet		
lb/day · 1000 lb beast	82	60
Solids content, %	12.7	11.6
Volatile solids, % of solids	82.5	85

These figures for the characteristics of manure are supported by other publications (10, 11).

The digestion of manure will yield a gas mixture of carbon dioxide and methane, nominally at the proportion of 13.3 ft^3 (STP) per pound of digestible manure solids converted to gas (12), or 14 ft^3 at 60°F. The methane content of the gas obtained by digestion of manure has been observed to vary according to digester operating conditions and animal characteristics, a range of 60-80% by volume being observed (9).

These data enable a calculation of the gross amount of fuel gas theoretically available from all of the manure produced in the U.S. annually. It is clear that only a part of the manure can be collected and processed economically, but it is of interest to calculate the gross figure in order to comprehend the potential impact on national energy supplies. For this purpose it will be assumed that 83% of the waste solids (Table 2) may be converted to a gas mixture containing 70% methane. Using the conversion 2205 lb/metric ton:

$$237 \times 10 \times 0.83 \times 2205 \times 14 \times 0.7 =$$

$$4.2 \times 10^{12} ft^3 CH_4 \qquad [2]$$

Four trillion cubic feet of methane is equivalent to about 18% of our present annual consumption of natural gas, i.e., 23 $\times 10^{12} ft^3$ /year.

As pointed out above, beef cattle feedlots are appropriate locations for operation of gas generation plants. As a guide then to the magnitude of practical manure utilization an estimate may be made of the gas potentially available from feedlot operations.

Of the 112 million cattle in the United States, 35 million are slaughtered each year (Census of Agriculture figures); about

half of these slaughtered cattle are processed through feedlots.
The cattle feeding industry is characterized by relatively
large scale commercial operations rather than by many small operations. In Kansas, for example, virtually every feedlot has greater than 10,000 head of cattle and these operations have been initiated within the last ten years (personal communication with Mr.
Virgil Huseman, Director of Cattle Feeder Service, Kansas Livestock and Feeder Association). There are feedlots in Texas and
Colorado, however, which have 80,000 - 100,000 head capacity (13).
The total cattle on feedlots according to 1973 census figures was
17 million and it is assumed that this estimate is valid for purposes of calculations. Since cattle are on feedlots for 6 months
prior to slaughter, then the number of cattle on feedlots at any
one time is one-half this estimated number. Further, the numbers
of cattle on feedlots on July 2, 1972 (14), in the seven states
having the largest number of cattle on feed are as follows.

TABLE 4

Number of Cattle on Feedlots Ranked by States

State	Cattle on Feedlots July 1, 1972 (USDA Figures, millions of head)
Texas	2.12
Iowa	1.7
Nebraska	1.4
California	1.17
Kansas	1.13
Colorado	1.02
Arizona	0.5
	9.04 million head on feedlots

Using the data cited above it may be shown that the manure
available on the feedlots of these major feeder states could produce $3.7 \times 10^{11} ft^3$ methane (60°F) per year. This is nearly 10% of
the total theoretically available. These orientation calculations
make it clear that the quantity of gas available through manure
digestion will make recovery, where economical, well worthwhile.
The way in which large, progressive feedlots are managed from
the standpoint of manure handling makes an operation to collect
and convert the manure to gas feasible. Closed system waste management for livestock such as the use of high pressure wash systems to clean specially designed floors or concrete pads is being
developed (15, 16), and one system is specifically designed to
collect and digest the cattle waste (11, 17). Furthermore, because of increasing pressure to adopt waste management techniques
(18, 19, 20), more and more feedlot operators are installing efficient collection systems. For example, earlier the General Electric Company developed a process for the thermophilic aerobic

conversion of collected waste to protein for use as animal feed
(21) thus indicating industrial interest in the potential for collected animal waste. Over-all, it appears practical to consider
a process for the production of fuel gas from the solid waste of
large cattle feedlots.
    The composition of animal solid waste is dependent upon the
content of protein, fiber, and other items in the feed ration.
Animals in confinement are fed a composition of materials to
achieve the greatest economical weight gain in the shortest period
of time. Highly efficient consumption of feed by the animal is
subordinate to continuous and rapid weight gain. Animals that are
fed concentrated diet excrete more of the nutritive material present in the feed. Feed additives such as antibiotics, copper, and
grit will affect the biochemical properties and the physical characteristics of the waste. Quantitative bioassays of fresh feedlot
manure, for example, revealed that when the diet of cattle was
supplemented with chlortetracycline, approximately 75% of the di-
etary chlortetracycline was excreted (21). After the waste has
been excreted, it can be further altered by bedding and wasted
feed. Despite these compositional variations, the figures presented in Table 2 may be taken as representative of the manure
produced on typical feedlots.
    If a representative feedlot is taken to be one processing
30,000 cattle at a time, then calculations like those presented
above will show that such a feedlot is capable of producing 1.1 x
$10^9 ft^3$ methane per year. This may be compared with production of
12.0 x $10^8 ft^3$ (60°F) CH/year from 1000 ton/day of municipal solid
waste from a population of 400,000 people (2).
    Thus, a nominal sized modern cattle feedlot may produce me-
thane equivalent to the size of plant considered practical for
conversion of municipal solid waste. Costs for municipal solid
waste preparation are known (1) to contribute significantly to
over-all gas production costs for large sized gas-from-municipal-
solid waste plants. It is anticipated that very little such preparation of cattle feedlot waste will be required. It may be,
therefore, that smaller sized gas production plants may be operated on animal waste at a reasonable cost per unit of methane. Further, while land costs have been excluded from present cost estimates, for a real system such costs may be less for animal solid
waste processing than for municipal solid waste processing.

FUEL GAS PRODUCTION FROM MARICULTURE
    A path to follow towards finding renewable resources is the
sun. Rather than consider devices for direct conversion of the
sun's energy, stored energy in plants may be converted to fuel.
Specifically, anaerobic digestion can be carried out on almost
any organic material under the proper conditions. Recent attention has been directed at fuel gas production from municipal solid waste by anaerobic digestion (1,22). Fuel gas production from
solid waste clearly has limited potential for supplying substan-

tial amounts of energy(2). It has been suggested that special
crops may be raised for the purpose of bioconversion to fuel (23).
Suitable land in the U.S. and indeed, throughout the world, may
not be available to use for production. This leads back to the
consideration of the sea for growing of plants to be digested to
methane.

A first step in the evaluation of the proposed bioconversion
process is to find a suitable primary fuel. The seaweed chosen
must be in abundant supply, accessible, and capable of being biologically converted to methane. There is one species of seaweed
that seems most appropriate for this purpose. It is the so-called
giant kelp, Macrocystic pyrifera. This colossus of the marine
world commonly grows up to 200 feet in length and has the phenome-
nal growth rate of two feet or more per day (24). The fronds are
kept afloat by tiny gas bladders, so the collection of this plant
is in some respects as simple as harvesting wheat. M. pyrifera
is indigenous to waters whose average temperature is less than
20°C. Japan or the west coast of America would appear to be well
suited for a sea farm of this species.

One proposed ocean food and energy farm has been by Wilcox
(25). In this concept, a network of polypropylene kelp supporting
lines extend out from a central processing station. The kelp
plants are manually attached to a submerged raft and, when they
mature, are harvested by special ships for processing. Since the
nutrient concentration in the surface waters is low, a further
idea (25) is to use a wave actuated pump which will take cold, nu-
trient-laden water from a thousand feet below the floating station
and distribute it among the plants. If the proposed process is a
success, it has been predicted that future farms may be 100,000
acres or over 14 miles in diameter and, some day, up to 100 miles
in diameter (25).

In the proposed processes the harvested kelp is drained of
excess seawater while still on the harvesting barge, perhaps by a
colander-type device. The kelp is then conveyed from the harvesting ship to the processing plant and shredded. The shredded kelp
is stored, continuously fed to the digesters, and converted to
fuel gas. The effluent from the digesters may be further processed to yield fertilizers ahd animal fodder.

Early in this century an attempt was made to estimate the total kelp available in the kelp bads along the American Pacific
coast. A complete account was published in 1915 by Cameron et al
(26) and the survey resulted in the mapping of about 390 square
miles of kelp beds between Alaska and lower California. When
these plants are cut near the ocean surface, they are not damaged
and regenerate; currently in the California operations a mower
cuts the kelp to a depth of 1.25 meters (about 4 feet). Under
these conditions, two harvests may be collected from a kelp bed
per year. Therefore, according to the Cameron et al survey, as
well as others, a total estimated annual natural kelp harvest
could reach almost 60 million tons (26-29). It is of interest to

note that bed densities vary greatly in different parts of the
world. Biomasses up to 22 kg m^{-2} (4.5 lb ft^{-2}) fresh weight have
been reported in the relatively thick California beds while, for
example, Indian Ocean biomasses range from 95 to 606 kg m^{-2} with
an average of 140 kg m^{-2} (28.7 lb ft^{-2}). (30) The net annual bed
productivity in California was found to be 400-820 g carbon/m^2 ·
year (31,32).

Using these figures an estimate can be made of the possible
methane production from harvesting the natural kelp beds. However, the 390 square miles of natural kelp beds are spread from
Alaska to Mexico. The collection and delivery to a few central
locations appears to be too expensive to be useful. On the other
hand, a large kelp mariculture system or "farm" that is 23 miles
in diameter will contain as much kelp as the total estimated west
coast beds and it will be confined to one small area.

To arrive at the methane production from a mariculture
system, certain facts about the system must be known or assumed.
As a first estimate, the assumptions for the annual fuel gas production from a 100,000 acre mariculture system will be presented.

- A sea farm will support 436 Macrocystis pyrifera plants
  per acre (25).

- Each acre is expected to yield about 300 to 500 wet tons
  of harvested organic material per year (25, 33, 34).

- Wet (fresh) kelp contains approximately 85 percent water.

- The yearly average of volatile matter in M. pyrifera is
  approximately 70 percent.

- Assuming a simple digestion system, conversion of 40 percent of the volatiles is achieved (based upon experiments
  conducted by the authors).

- Kelp digestion is assumed to follow the conversion

$$C_6H_{10}O_5 + H_2O = 3CO_2 + 3CH_4$$

(Here 1 kilogram of digestible organic matter yields
0.41 m^3 CH$_4$ @ STP.)

The total volume of methane produced annually from a 100,000
acre mariculture system will be 2.6 x 10^{10}ft^{3}CH$_4$(2.78 x 10^{13}Btu).
Since the total U.S. energy consumption in 1970 was 69 x 10^{15}Btu,
this one mariculture system would contribute only 0.04 percent to
the total energy supply. However, the 1970 natural gas consumption was 22 x 10^{15}Btu; therefore this one mariculture system
would contribute 0.13 percent to the total natural gas supply.

On the other hand, if a larger 100 mile diameter system were
operating, this system could satisfy about 2 percent of the total
U.S. energy consumption or 6 percent of the natural gas consumption. To satisfy 100 percent of the nation's energy needs, a system 700 miles in diameter would be needed; for total U.S. natural
gas needs, a system 400 miles in diameter would be required.

BIOMETHANATION OF COAL GASIFIER PRODUCTS

The manufacture of methane from coal has become recognized as having substantial potential for meeting the needs in the U.S. for a clean energy fuel from abundant coal resources secure within national borders (35). The projected 3.5% annual increase in U.S. energy needs (36), the 6%/year increase in natural gas use (37), and the declining gas reserve/production ratio (37) make the manufacture of synthetic natural gas from coal attractive. As many as 36 gasification plants are projected to be constructed by 1985-1990 (38, 39).

In judging the state of technology with respect to our readiness to adopt coal gasification, attention soon becomes focussed on methanation. The catalytic process for this essential processing step has not yet been demonstrated adequately on a commercial scale. Methanation requires further development. Methanation catalysts are sensitive to sulfur poisoning. The exothermic nature of methanation introduces design complexities. Carbon deposition on catalysts is a further problem. The unresolved development problems inherent in methanation, plus the need to operate a shift conversion in conjunction with it, make it well worthwhile to investigate alternative methods for methanation. The process of biomethanation offers such an alternative.

It has been established through laboratory experiments that both CO_2 and CO can be combined readily with H_2 under conditions of anaerobic fermentation to produce methane. Biomethanation offers a number of advantages: accepts gasifier product directly, insensitive to sulfur, heat of reaction utilized in part to form cell mass, minimal water use, simplified gas clean-up, valuable single cell protein by-product formed, nitrogen contaminants may be utilized, and the process operates at pressures of the coal gasifier.

The basic chemistry of methanation can be described using only a few reactions. When the hydrogen to carbon monoxide ratio in synthesis gas is equal to or greater than 3, the conversion of CO and H_2 to methane can be described by the reaction:

$$3H_2 + CO \rightarrow CH_4 + H_2O + 49.3 \text{ k cal/mol} \qquad [3]$$

Two other reactions also act to produce methane from carbon oxides:

$$2H_2 + 2CO \rightarrow CH_4 + CO_2 + 59.1 \text{ k cal/mol} \qquad [4]$$

$$4H_2 + CO_2 \rightarrow CH_4 + 2H_2O + 39.4 \text{ k cal/mol} \qquad [5]$$

However, hydrogenation of carbon dioxide, Equation 5, does not occur in the presence of carbon monoxide. It may be pointed out that Equation 4 can be considered to be a combination of Equation 3 and the so-called shift reaction:

$$CO + H_2O \rightarrow CO_2 + H_2 + 9.9 \text{ k cal/mol} \qquad [6]$$

Although the shift reaction does not produce methane, it plays an important role in the catalytic conversion process by altering the

H_2/CO ratio to increase utilization of CO.

Based on a thorough chemical thermodynamic analysis of these reactions by Mills and Steffgen (37), it is concluded that

a) Catalytic methanation must be operated at the lowest temperature consistent with acceptable catalyst activity and with H_2/CO ratios at or above the limiting carbon deposition boundary ratios. Exacting demands on temperature control are introduced through the fact that the reactions of concern are highly exothermic. Furthermore, methanation of gas from coal will be practiced at high reactant concentrations.

b) Catalytic methanation at high pressure may enable use of lower H_2/CO ratios without encountering deposition of carbon on the catalyst - however, such operation at high pressure entails the release of large quantities of heat/ unit volume which results in increased catalyst-bed temperatures and decreased methane yield.

c) The formation of alcohols or other higher hydrocarbons is thermodynamically possible and will be pronounced if adequate control of the reactor cannot be achieved. The catalyst life is thereby reduced.

Methanation in the technology of coal gasification has been practiced only on the pilot scale. Present industrial applications of catalytic methanation are limited to the conversion of relatively small amounts of harmful carbon monoxide to methane in circumstances such as ammonia synthesis where carbon monoxide would interfere with catalytic utilization of gas mixtures. Thus present industrial application of catalytic methanation is far from the scale and complexity that will be imposed on large scale processes for the manufacture of "high Btu" or pipeline gas.

Over-all, while a substantial amount of work has been done, fundamental problems exist and are inherent in the entire catalytic conversion process. It appears that alternative processing concepts should be given careful review. One alternative processing concept is that of carrying out the entire conversion of coal gasifier product gases to methane by anaerobic fermentation.

Biomethanation involves anaerobic fermentation of CO, CO_2, and H_2 to methane. Anaerobic fermentation is most often considered as the splitting of a substrate into two or more fragments, part oxidized and part reduced - relative to the composition of the original substrate compound. Over-all, the anaerobic decomposition of animals and plants is a complex association of many successive fermentations by an array of different microorganisms. At the end of these complex metabolic pathways a comparatively small number of intermediate fermentation products (for example, acetic acid) are converted to CO_2 and CH_4. Indeed, Volta in 1776 (40), is given credit for first identifying these gases which he found to be generated in the neighborhood of

decomposing vegetation in bodies of water and in the soil.
Modern chemistry and biology have confirmed that CH_4 and CO_2, in
some ratio, are the final end products of anaerobic fermentation.

With the above definition of fermentation, CO may be viewed
as intermediate between carbon dioxide (totally oxidized carbon)
and methane (reduced carbon). With this insight, Fischer, Lieske
and Winzer (41), were the first workers to demonstrate experimentally that microorganisms (derived from an anaerobic sewage
sludge digester) were able to bring about an anaerobic conversion
of CO into CO_2 and CH_4.

The starting point of these investigations is to be found in
the classical observations of Söhngen (42) in 1906 that select
microorganisms are able to convert a mixture of CO_2 and H_2 into
CH_4. The reaction is:

$$CO_2 + 4H_4 \rightarrow CH_4 + 2H_2O \qquad [7]$$

In fact, it has been shown only recently (43) that this reaction
is common to all methane bacteria, i.e., those microorganisms
that produce methane. It is therefore seen that the above bioconversion of CO_2 and H_2 to methane is a long established fact.

The work of Fischer, et al, with (a) CO and (b) CO and H_2
pointed at the conclusion that the primary reaction in which CO
takes part is always:

$$CO + H_2O \rightarrow CO_2 + H_2 \qquad [8]$$

In the presence of a sufficient amount of hydrogen, these primary
products are converted according to the secondary reaction of
Equation 7. These proposed reactions and experiments of Fisher,
et al, have since been fully experimentally established by
Kluyver and Schnellen (44). Two major experiments of these workers are of telling value. In the first experiment CO and H_2 were
converted to methane by Methanosarcina Barkerii. Hydrogen was
added in excess and in all cases the CO was completely consumed.
The conclusion is that the reaction proceeds as:

$$CO + 3H_2 \rightarrow CH_4 + H_2O \qquad [9]$$

This, indeed, is an over-all summation of the above reactions
proposed by Fisher, et al (41). The second key experiment by
Klyuver and Schnellen was the conversion of CO by Ms. Barkerri
without the addition of hydrogen. This conversion proceeded to
completion with the restriction, however, that the hydrogen
produced in situ is sufficient to reduce one-quarter of the
amount of CO_2 produced, in keeping with the chemistry involved.
Over-all, it was therefore established that the anaerobic fermentation of carbon monoxide will proceed according to the equations:

$$4CO + 4H_2O \rightarrow 4CO_2 + 4H_2 \qquad [10]$$

$$CO_2 + 4H_2 \rightarrow CH_4 + 2H_2O \qquad [11]$$

$$4CO + 2H_2O \rightarrow 3CO_2 + CH_4 \qquad [12]$$

Since it is an established scientific fact that CO, CO_2, and H_2 are converted to CH_4 it is of interest fo consider conversion rates. Recently Zeikus (45, 46) isolated a new anaerobic auto-trophic, extreme thermophile from sewage sludge that has doubling times of less than 5 hours at temperatures above 70°C when con-verting only CO_2 and H_2 to pure CH_4. These short generation times, relative to methane bacteria, distinguish this organism as one of the fastest growing methane bacteria known. This unique bacterium, converting CO_2 and H_2 to CH_4, is named Methanobacteri-um thermoautotrophicum. Recent experiments have demonstrated that this microorganism can be grown at up to 80°C when convert-ing CO_2 and H_2 to CH_4 (47).

Also rather recently, Bryant, McBride, and Wolfe (48) car-ried out rate determining experiments using the bacteria Methano-bacillius omelianskii which oxidized H_2 and reduced CO_2 to CH_4. These workers established that this conversion of CO_2 and H_2 to CH_4 proceeds at very high rates. Methane production rates of 6 m moles/minute in a 12 liter reactor were obtained. The bioconver-sion was carried out at 40°C or mesophilic conditions. Converted to engineering units, the methane production is 17.1 ft^3 CH_4 (@ 60°F)/day · ft^3 reactor volume at 1 atmosphere pressure. This is a very high methane production rate per cubic foot of reactor volume in view of the fact that methane production from anaerobic sewage sludge digesters is generally observed to be only approxi-mately 1 ft^3 CH_4/day · ft^3 reactor volume at these conditions. If it is assumed that the chemical activity of the dissolved gas-es remains constant as the pressure is increased, then at 30 at-mospheres pressure (the nominal pressure of product gases from a coal gasifier) the methane production will be 513 ft^3 CH_4 (@ 60° F)/day · ft^3 reactor volume. The use of thermophilic cultures reported by Zeikus (47) should further raise the potential meth-ane production per unit volume of reactor. For example, if thermophilic bioconversion of CO_2 and H_2 to CH_4 at 80°C, follow-ing Zeikus (47), doubles the conversion rate found at mesophilic conditions by Bryant, et al (48), then methane production in the order of 1,000 ft^3 CH_4 (@ 60°F)/day · ft^3 reactor volume appears possible. These experiments establish a sound basis for evalua-ting the biomethanation process for conversion of coal gasifier product gases to methane. Such a program is presently being undertaken by the authors under the sponsorship of the Energy Research and Development Administration (49).

SUMMARY AND CONCLUSIONS

Selected organic materials having potential for fuel gas production by anaerobic digestion have been discussed. It is concluded that animal waste has the greatest potential in being useful for the short term alleviation of the gas supply problem in the United States. On the other hand, due to the very large

424   D. L. WISE *ET AL.*

fuel gas consumption in the United States, plans for development of a mariculture system appears merited. Employment of certain bioconversion concepts was seen to have practical application in the area of coal gasification. It was pointed out that fuel gas production from solid waste is now at an advanced development stage and at this time appears to clearly offer an alternative to incineration or other energy intensive solid waste processes. In conclusion, it is to be noted that anaerobic digestion of organic materials other than sewage sludge has advanced rapidly. Feasibility of bioconversion of selected organic matter in most cases has been established, or may be inferred with appropriate engineering care. Development, therefore, should stress larger scale processing units and the supportive engineering requirements associated with such process development.

REFERENCES

1. Cooney, C.L. and D.L. Wise, Biotech. & Bioengr. 17, 1119 (1975).

2. Kispert, R.G., S.E. Sadek and D.L. Wise, Resource Recovery & Conservation 1, 95 (1975).

3. Wise, D.L., S.E. Sadek, R.G. Kispert, L.C. Anderson and D.H. Walker, Proceedings of the Berkeley Conf. on Cellulose as a Chemical & Energy Resource, 1974 (published in Biotech. & Bioengr. 5, 285-301, 1975).

4. "Energy, Agriculture and Waste Management", W.J. Jewell, editor, Ann Arbor, Mich., Ann Arbor Science Publishers, Inc., 1975.

5. Wise, D.L., S.E. Sadek and R.G. Kispert, 1974, Fuel Gas Production from Solid Waste, Progress Report NSF/RANN/SE/C-827/PR/73/4, Dynatech R/D Company, Cambridge, Massachusetts, Report No. 1151.

6. Hedin, D.G., Ed., Brown's Directory of North American Gas Companies, 87 Ed., Moore Publishing Co., Inc., Deluth, Minnesota, 1973.

7. Hedin, D.G., Ed., Brown's Directory of North American Gas Companies, 85 Ed., Moore Publishing Co., Inc., Deluth, Minnesota, 1971.

8. Laur, D.A., "Energy, Agriculture and Waste Management", W.J. Jewell, editor, Ann Arbor, Michigan, Ann Arbor Science Publishers, Inc., 1975, p.409.

9. Ifeadi, C.N. and J.B. Brown, Jr., ibid., p.373.

10. Anon., The Farm Quarterly 27 (4) 52 (August, 1972).

11. Wells, D.M., W. Grub, R.C. Albin, G.F. Meenaghan and E. Coleman, "Advances in Water Pollution Research", Volume II, S.H. Jenkins, editor, Pergamon Press, 1971.

12. Morris, G.R., W.J. Jewell and G.L. Casler, "Energy, Agriculture and Waste Management, S.J. Jewell, editor, Ann Arbor, Mich., Ann Arbor Science Publishers, Inc., 1975, p.317.

13. Loehr, R.C., Inc. Water Engr. 7, 14, 1970.

14. Anon., "Cattle on Feed", U.S. Dept. of Agr., Statistical Report, Crop Reporting Board, August 1972 (a monthly).

15. Ngoddy, P.O. et al, report entitled "Closed System Waste Management for Livestock" on Project #13040DKP for EPA, June, 1971.

16. Butchbaker, A.F. et al, report entitled "Evaluation of Beef Cattle Feedlot Waste Management Alternatives: on Project #13040FXG for EPA, November, 1971.

17. Meenaghan, C.F, D.M. Wells, R.C. Albin and W. Grub, Paper No. 70-907 entitled "Gas Production from Beef Cattle Wastes" presented at the 1970 Winter Meeting of the Am. Soc. of Agr. Engrs.

18. The Economics of Clean Water Vol. II, Animal Wastes Profile, U.S. Dept. of the Int., March, 1970.

19. Agricultural Pollution of the Great Lakes, U.S. Gov't Printing Office #5501-0134, July 1, 1971.

20. Gilbertson, C.B. et al, J.W.P.C.F. 43, (3) Park 1, 483, 1971.

21. U.S. Patent #3,462,275.

22. Pfeffer, J.T., "Reclamation of Energy from Organic Refuse", University of Illinois, Urbana, Dept. of Civil Engineering, Report to U.S. Environmental Protection Agency, Grant EPA-R-800776, April 1973.

23. Kemp, C.C., and G.C. Szebo, "The Energy Plantation", Proceedings of the 34th Annual Conf. of the Chemurgic Council, 1973.

24. North, Wheeler J., "Giant Kelp: Sequoias of the Sea", National Geographic 142, 250-269, August, 1972.

25. Wilcox, Howard A., "The Ocean Food and Energy Farm Project", a paper presented at the 141st Annual Meeting of the American Association for the Advancement of Science, Jan. 29, 1975.

26. Cameron, F.K. et al, (1915), U.S. Dept. Agric. Rep. 100, Washington.

27. Mann, K.H., "Seaweeds: Their Productivity and Strategy for Growth", Science, 182, 975-981, December 7, 1973.

28. Proceedings of the Fourth International Seaweed Symposium, Pergamon Press Ltd., 1964.

29. Silverthorne, Wesley and Philip E. Sorensen, "Marine Algae as an Economic Resource", a paper presented at the 7th Annual Conference of the Marine Technology Society, Aug. 16-18, 1971.

30. North, W.J., Ed. Nova Hedwigia, 32 (Suppl.) 1 (1971); P. Grua, Terre Vie, 2, 215 (1964).

31. Clendenning, K.A., Nova Hedwigia, 32, (Suppl.) 259 (1971).

32. North, W.J. (principal investigator) (1971): Kelp habitat improvement project. Annual Report 1 July, 1970 - 30 June, 1971. 150 pp.

33. Idyll, C.P., "The Harvest of Seaweed", Sea Frontiers, 17, 342-348, November, 1971.

34. Iversen, E.S., Farming the Edge of the Sea, The Garden City Press Ltd., 1968.

35. Mills, G.A., Environ. Sci. Technol. 5, 1178, 1971.

36. Dupree, W.G., and J.A. West, United States Energy, U.S. Dept. of Interior, December, 1972.

37. Mills, G.A., and F.W. Steffgen, Cat. Reviews 8, 1973.

38. National Petroleum Council, U.S. Energy Outlook, Vols. I and II, 1971, Summary Report, December. 1972.

39. Federal Power Commission, National Gas Supply and Demand 1971-90, Staff Report No. 2, February, 1972, p.96.

40. Pine, M.J., "The Methane Fermentations" in Adv. in Chem. Ser. 105, Am. Chem. Soc., 1971.

41. Fischer, F., R. Lieske and K. Winzer, Biochem. Z., 236, 1931.

42. Söhngen, M.N.L., Rec. Trav. Chim. 29, 238, 1906.

43. McBride, B.C. and R.J. Wolfe, "Biochemistry of Methane Fermentation" in Adv. in Chem. Ser. 105, Am. Chem. Soc., 1971.

44. Kluyver, A.J., and C.G.T.P. Schnellen, Arch. Biochem. 57, 1947.

45. Zeikus, J.B., and R.S. Wolfe, J. of Bact. 109, 707, 1972.

46. Ibid., Int. J. of Syst. Bact. 22, 395, 1972.

47. Zeikus, J.G., personal communication, 1974.

48. Bryant, M.P., B.C. McBride and R.S. Wolfe, J. of Bact. 1118, 1968.

49. Augenstein, D.C., D.L. Wise and C.L. Cooney, "Biomethanation :Anaerobic Fermentation of CO_2/H_2 and CO to Methane", preprint of presentation at the 69th Annual Meeting of the AlChE, Nov. 28-Dec. 2, 1976, Chicago, Ill. (work carried out under ERDA Contract E(49-18)-2203).

APPLICATION OF SOLAR ENERGY BIOCONVERSION
IN DEVELOPING COUNTRIES

ULRICH HORSTMANN

Institut fuer Meereskunde, Universitaet Kiel, Germany

(Marine Station of the University of San Carlos,
Cebu City, Philippines)

The term "developing country", which was applied to various
nations of the world in the early 1950's, is not uniformly defined.
Wherever it has been used, it helps to identify countries which
have a number of features in common as, among others: low living
standards for a majority of their populations, extremely uneven
distribution of capital and property, low productivity, high un-
employment, little developed health services, and an insufficient
food supply.   As of 1971, 118 states and dependent areas were
internationally acknowledged as Developing Countries (DAC).   The
United Nations, following the so-called UNCTAD list, distinguishes
between the Least Developed Countries (LLDC), comprising 28
nations with 215 million people in 1973, and the Most Severely
Affected Countries (MSAC), a category to which in the same year 49
states with one billion inhabitants (including some of the LLDC
countries) were assigned.   The developing countries are also often
spoken of as Third World Nations, in contrast to those of the First
and Second worlds, or as the Southern Nations opposite those in the
West (First World) and the East (Second World).   However, "south"
here does not refer so much to the southern half of the globe but
to the tropical belt around the equator.   For the approximate
location of the Least Developed Countries, see Fig. 1.

People living in the tropics have originally been favored by
nature, which, in the past, has provided about everything needed
for daily life and not demanded the same amount of ingenuity and
technology required for survival from the inhabitants of the more
temperate climate zones.   It was only more recently that the
situation in the tropical countries began to change as a consequence of the importation of western progress.

One result of this importation is the rapid population
growth, caused primarily by improvements in hygiene, nutrition,
and health services; another the on-going process of population
concentration in emerging industrializing urban centers, where
life is progressively being pressed into western forms and becoming increasingly dependent on western technology.

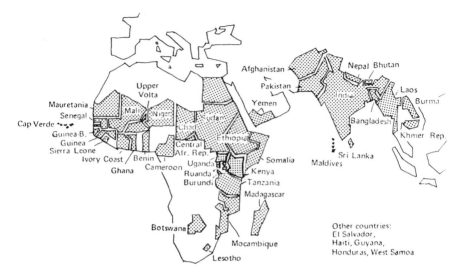

Figure 1. The world's least developed countries.

Two of the larger problems which the developing tropical countries have to solve in their present struggle for modernization are (1) the production of adequate food supplies for their growing populations, and (2) the securing of sufficient fuel for their multiplying energy demands. The need of the Third World Countries for more energy is growing by leaps and bounds. Only a few of them, belonging to the "privileged" group of the OPEC countries, have at least no short-term problems of meeting these needs; the vast majority, however, will have to overcome tremendous difficulties to satisfy actual and anticipated demands.

Table 1, by way of example, presents some pertinent information on anticipated demands for one of the countries, the Philippines. This example cannot be generalized because situations differ from one developing country to another, but it serves to illustrate the vast gap between presently available energy resources and future requirements. The data indicate, among other things, that by the year 2000 the Philippines will have to secure 10 percent of its needed energy from non-conventional sources of whatever kind.

To what extent developing countries will have to increase their energy supply if they wish to do nothing more than draw even with the present energy consumption of the developed world is made obvious by Table 2. This task of "just catching up" appears even more difficult when we consider the high rate of population

TABLE 1.   PROJECTED ENERGY REQUIREMENTS BY SOURCE AND USE
Philippines 1975 - 2000

		1975		1985		2000	
		MMB*	%	MMB	%	MMB	%
A.	Electrical Use						
	Hydro	3.6	4.8	14.6	8.8	37	8.0
	Geothermal	-	-	14.0	8.5	46	10.0
	Nuclear	-	-	6.0	3.6	41	9.0
	Oil	13.6	18.2	10.7	6.5	5	1.0
	Coal	-	-	3.7	2.2	9	2.0
	Non-conventional	-	-	-	-	38	8.0
	TOTAL A	17.2	23.0	49.0	29.6	177	38.0
B.	Non-Electric Use						
	Oil	56.9	76.5	108.5	65.6	240	52.0
	Coal	0.3	0.5	7.0	4.2	37	8.0
	Non-conventional	-	-	1.0	0.6	9	2.0
	TOTAL B	57.2	77.0	116.5	70.4	286	62.0
	TOTAL ENERGY	74.4	100.0	165.5	100.0	462	100.0

*MMB - million barrels
Source: Sectoral Reports of the Energy Committee Presented
at the International Conference on The Survival of Humankind.
Manila, Philippines, September 6-10, 1976.

TABLE 2.   ENERGY CONSUMPTION IN SELECTED COUNTRIES, BY
DEVELOPMENTAL STAGE OF COUNTRY: 1973

DEVELOPING COUNTRIES		DEVELOPED COUNTRIES	
Country	Energy Consumption*	Country	Energy Consumption*
Sri Lanka	0.7	United States	55
Indonesia	0.8	Canada	26
India	0.9	Japan	20
Pakistan	0.9		
Thailand	1.6		
Philippines	1.7		
Malaysia	2.7		
South Korea	4.5		

*In barrels of oil per capita

TABLE 3.   RELEVANT PROJECTS

PROJECTS	INTENDED APPLICATIONS	PRINCIPLE/DESIGN
A. DIRECT SOLAR		
1. Water Heaters	Buildings requiring hot-water	Flat-plate collectors; Retrofit conventional heaters.
2. Process Steam	Process Industries	Special Flat-plates or Focusing Collectors.
3. Space-Cooling	Small Office Bldgs. Supplement for Large Bldgs.	Flat-plates w/Reflectors; LiBr-H_2O system w/energy storage.
4. Refrigeration	Food Preservation in Powerless Areas	Flat-plates w/auxiliary power; NF_3-H_2O or NH_3-NaSCN System
5. Crop Drying	Controlled Drying of Grain & Cash Crops	Flat-plates/forced convention
6. Stills	Identified sites w/lack of potable water.	Roff-type with Rain-water collector
7. Heat Engine	Small-scale irrigation or stock water low-power for rural needs.	Flat-plates or Focusing Closed cycle system w/ Organic Fluid.
8. Photovoltaic	To power small TV sets, communication systems in powerless areas.	Array of silicon coupled to storage batteries.
B. WIND ENERGY		
1. Windmills for Water pumping	Small-scale irrigation, stock water	Multiblade; Comm'l Type.
2. Windmills for Lower-Power Electricity	To power small TV sets, communication systems in powerless areas.	Fewer Blades; Coupled to Alternators & Storage Batteries
3. Windmills for Med-Power Electricity	Cottage industries refrigeration, etc.	-do-
C. BIOCONVERSION		
1. Family-Size Biogas Generators	Rural cooking and lighting	Anaerobic Fermentation of Animal Manure.
2. Large-Scale Biogas Generators	Fuel supplement; Waste Disposal; Fertilizer prodn.	Fermentation of Animal Manure, Agricultural Wastes.

*NOT SHOWN BUT INCLUDED IN PROGRAM:

Basic Researches, Establishment of Monitoring Network, Info/Promotion Program, Surveys, OTEC Exploratory Study.

Source:Sectoral Reports of the Energy Committee Presented at the International Conference on The Survival of Humankind. Manila, Philippines, September 6-10, 1976.

FOR SOLAR ENERGY PROGRAM*

DESIRED CAPACITY	TECH-NOLOGY STATUS	R & D EMPHASIS	OTHER REMARKS
1 gal hot water/sq.ft. collector	95	Materials substitution; Lower cost.	Expected to replace about 30% of fuel use of conventional systems.
100 tons steam/day	75	Development of special but low cost collectors	Expected to significantly reduce industrial oil use.
100 tons cooling capacity.	70	Pilot systems design & operation; Materials substitution.	May avail of some commercially available components.
100 kg ice/day	60	Systems design of ice-maker; All-solar or assisted by bio-gas or wind-mills.	Should result in higher income level in power-deprived areas.
5 cavans/day	90	Optimal design for local conditions; Materials substitution.	Should result in large savings on crop wastage
20 gal fresh water/day	90	-do-	Suitable sites shall depend on rainfall data, other surveys
1-10 KW	40	Developmental work based on most advanced existing systems.	SOFRETES Co. of France has commercial units but presently very expensive.
36 W/unit	100	Assembly of commercially available components.	Expensive but successfully used elsewhere for information & population control programs.
2000-3000 gal/day	95	Optimization of components; Adaptation to Phil. Conditions.	Widely used in other countries of the region.
0.2-1 KW	80	-do-	For information & population control programs.
10-100 KW	50	-do-	Will need special wind map data for siting.
150 cf/day	95	Optimization of capacity, efficiency; Lower cost	
up to 50,000 cf/day	90	Optimal Design. Solar Heat Assist.	

Scale of 100:
10 - Highly Experimental
100 - Immediately Adaptable.

increase which the developing countries are experiencing at present.

Among the non-conventional energy sources, at the present stage of science and technology, solar energy is considered the most important; it is at least one of the most readily and abundantly available. In the countries around the equator, sun radiation is a theoretically unlimited source of energy. In the area between 30° north and 30° south, for example, the average daily intensity of sun radiation is 500 calories per cm^2. With an only 10 percent efficiency of conversion to usable power, an area of 300 m^2 will yield an energy equivalent of 150 KW.

There are several fields in which solar energy conversion could be directly applied: (1) operation of water heaters, (2) operation of steam generators, (3) space cooling, (4) crop drying, (5) refrigeration, (6) operation of heat engines for low-power generation, and (7) photovoltaic systems. Some developing countries have already begun to draw up extensive plans for the utilization of solar energy and have set up a series of experimental or immediately usable projects. Table 3 summarizes such projects which either are in progress or in the planning stage in the Philippines.

However, when it comes to applying solar energy through direct (conventional) methods, the task of efficient energy collection and storage has not yet been solved satisfactorily. That developing countries have not applied the methods more fully is primarily due to the fact that materials and equipment for solar collectors are still very expensive, and that there exists a lack of information on new developments in the field. To overcome the need for sophisticated and expensive collector materials and to make storage of solar energy possible, solar energy by bioconversion is proposed as the solution.

In its classical form, solar energy bio-conversion takes the path of mass production of organic substances through photosynthesis. These substances can then be used as fuel for the production of energy. However, more efficiently, they can be converted into a mixture of methane and carbon dioxide through the process of anaerobic bacterial fermentation, a process which has the additional advantage of producing fertilizer as a by-product.

The production of methane from plants or organic waste materials through anaerobic fermentation has been started in a number of developing countries. In various places one can find simple but effective devices which produce methane gas for cooking purposes from animal manure. An example for the commercial exploitation of this process may be seen in an integrated agro-industrial complex near Manila in the Philippines. It demonstrates how bio-gas from pig manure can be a very profitable source of fuel for an attached meat cannery (Maramba 1975). Datta (personal communication) estimates that there are 28,000 bio-gas producing units in India utilizing mainly cow manure. Many more methane-producing machineries could be set up in developing countries using not only animal manure but also the abundant waste material which, most of the time, is either disposed of anywhere on the land, or simply dumped into the sea, a doubly wasteful thing to do since it is also one of the main causes of pollution (Horstmann 1976). Furthermore, use could be made of the masses of seaweed that drift ashore and rot on the beaches. In India, for example, Datta (personal communication) has engaged in successful experiments of adding seaweed to cow manure in order to increase methane production. Marshall (1969) and Chan (1970) have described simple and inexpensive sewage treatment units that can be used for methane production on tropical islands. As a similarly profitable energy source, one can use the water hyacinths (Eichhornia crassipes) which float on tropical rivers in great quantities and, at times, cover entire lakes. Instead of being hazards to fishermen and aquaculturists, these plants, which are known to be among the most efficient primary producers, could be put to good use. Corresponding experiments are presently being conducted in the United States (Eggers 1976). A SMF-NASA study claims that through the utilization of organic wastes in the United States two to three per cent of the country's gas needs could be met.

Aside from the just described classical form of solar bioconversion, which is yet far from being fully exploited in developing countries, a new and more promising method has been developed in the last few years: solar hydrogen production from algae and photosynthetic bacteria. Some 25 years ago, Gaffron and Robins (1942) discovered that, under certain environmental conditions, blue-green algae are capable of producing hydrogen. This method offers promising possibilities for the generation of energy and for food production. While anaerobic fermentation burns down organic substances, the hydrogen method preserves valuable protein-rich plant materials, which can be further utilized for human as well as animal consumption.

The application of photosynthetic hydrogen production in marine environments of tropical countries is advantageous in many respects; in developing countries it is highly called for. It is advantageous because of factors like (1) high temperature throughout the year, (2) regular and intensive solar radiation, (3) relatively stable and, therefore, predictable weather and wind conditions, and (4) the ubiquity of marine environments with extensive coral reefs which provide protective spaces for marine cultures whose species show rather promising qualities for conversion efficiency. Bio-energy conversion is called for in developing countries because of (1) the absence of, or the difficulties related to, the installation of centralized power supply systems, (2) the increasing demands for energy even in remote areas or isolated islands, (3) the need for larger amounts of protein for food and feeding purpose and (4) because of the inexpensive methods available for collecting and storing of energy.

The generation and distribution of electrical power is often restricted to industrially developed or urban areas. Rural areas are largely neglected, especially those that are geographically isolated and sparsely settled. In such areas, solar bio-conversion would permit power generation which does not have to rely on imported fuel. The amount of energy to be generated can easily be adapted to the size of the area to be served, which may be only one small settlement or a larger village. A further consideration as far as the size of the plant is concerned is the amount of needed by-products which can be derived from the process of bio-energy conversion. Tropical blue-green algae and photosynthetic bacteria contain between 25 and 60 per cent of protein (Mitsui 1975) which can serve as food source for humans as well as for the feeding of animals. Seafarming, because of its per-hectare yield, which is much higher than that obtainable through soil farming, is increasingly suggested to help meet rising protein demands. One of the large problems encountered in developing countries which attempt to augment their protein supply through aquaculture is the production of sufficient phytoplankton. The latter is needed to feed the larvae of desired species of fish, shellfish, or mollusks. This difficulty could at least partly be solved by the use of the hydrogen method. Furthermore, carbohydrates can be converted into alcohol, and various kinds of metabolic substances produced by micro-algae can be utilized for medicinal purposes. An additional advantage is gained through the related production of nitrogen for fertilizing. This can be obtained through the culture of nitrogen-fixing blue-green algae. While the process of nitrogen fixation through blue-green algae is considered a harmful one in a natural aquatic environment, since it is a source of organic pollution (Horstmann 1977), the latter can be avoided when the process takes place under controlled conditions.

Once the system for solar hydrogen conversion has been developed to the point where it becomes economically feasible, third-world nations have abundant opportunities for its use. The warm climate and the favorable light conditions which prevail in tropical countries throughout the year guarantee the rapid and continuous production of algae. Most tropical countries are situated in the meteorological region of the doldrums, the trade winds, and the horse latitudes; aside from the monsoon-influenced regions, all of them are areas with steady and comparatively predictable weather conditions. The extensive shallow-water coral reefs along their shores provide excellent protective locations, which permit

the building of large algae culture tanks. Such conditions are
needed for large-scale hydrogen production when extended areas
have to be prepared for the culture of hydrogen-producing plants.

Lien and San Pietro (1975) have estimated that, at an efficiency of solar energy conversion of 12 per cent, a theoretical
maximum of 0.90×10^8 Kcal per acre and year (average) can be
obtained in the United States. The observed agricultural efficiency, in contrast to aquacultural efficiency, using sugar cane
as an example, is estimated at 2.5 per cent. One word of caution
seems to be called for at this point: In the early stages of
development, sunlight-energized hydrogen production appears to be
suitable not for large-scale power generation but only for application at the domestic level, e.g., cooking, lighting, and refrigeration.

The large-scale application of photosynthetic hydrogen production seems to belong to the more distant future. Present
research work is following two main directions: (1) the development of the cell-free approach in which the essential hydrogen-
producing phases are extracted from the cell (Mitsui and Arnon
1962), and (2) the selection and cultivation of more efficient
hydrogen-producing species and strains. How fast progress in this
field and immediate application in developing countries can be
achieved, may be illustrated by an example of seaweed cultivation
in the Philippines.

As recently as 1972, the seaweed industry (Euchema striatum)
of the Philippines was almost entirely dependent upon naturally-
grown resources which did not meet the market demands and were in
danger of becoming depleted. The introduction of a culturable and
fast-growing strain of E. striatum, the Tambalang strain, in 1974,
resulted in a flooding of the world market with carrageenin-
producing seaweeds. Using this strain, growth rates of up to
twelve per cent per day can be obtained at the present time. Lest
the foregoing should have created the impression that in the near
future all energy needs of the developing countries or any country,
for that matter, can be satisfied with these new energy sources, I
hasten to add that for now they will only help to reduce the
dependence on presently used energy sources imported at great
expense from abroad. Nevertheless, all efforts should certainly
be bent to achieve this goal as speedily as possible.

LITERATURE

Chan, G.L. 1970. Use of potential lagoon pollutants to produce
    protein in the South Pacific. Paper presented to FAO Techni-
    cal Conference on Marine Pollution.   Rome, Italy,
    FIR: MP/70/E-10:4 pp.

Datta, R.L. (personal communication). Chemical Research Institute. Bhaynagar, India.

Egger, A.J. (personal communication). National Science Foundation. Washington, D.C., U.S.A.

Gaffron, H. and Rubin. 1942. Fermentative and photochemical production of hydrogen in algae. J. gen. Physiol., 26:219.

Horstmann, U. 1974. Eutrophication and mass production of blue-green algae in the Baltic. Merentutkimuslait. Havforskningsinst. Skr. No. 239 (1974).

Horstmann, U. 1976. Solar Energy Bioconversion in the Philippines. Paper presented at the International Conference on the Survival of Humankind: The Philippine Experiment. Manila, Philippines.

Lien, S. and A. San Pietro. 1975. "An Inquiry into Bio-photolysis of Water to Produce Hydrogen". NSF-Indiana University.

Maramba, F.D., Sr., et al. (mimeographed). Bio-gas Plant in an Integrated Agro-Industrial Complex at the Maya Farms, Manila.

Mitsui, A. and D.I. Arnon. 1962. Photoproduction of hydrogen gas by isolated chloroplasts in relation to cyclic and non-cyclic electron flow. Plant Physiol., 37S: IV.

Mitsui, A. 1974. Utilization of solar energy for hydrogen production by cell-free system of photosynthetic organisms. Proceedings of the Hydrogen Economy Miami Energy Conference, Miami Beach. (Ed.), T.N. Veziroglu. University of Miami. pp. S5-41.

Mitsui, A. 1975. Multiple utilization of tropical and subtropical marine photosynthetic organisms. The 3rd International Ocean Development Conference, Volume 3, pp. 13-29. Seino Printing Co., Tokyo.

BIOLOGICAL CONSTRAINTS ON SEAWEED CULTURE

GEORGE A. JACKSON
Woods Hole Oceanographic Institution
Woods Hole, Massachusetts   02543

Use of seaweeds for conversion of solar energy to a convenient form will require an extensive knowledge of seaweeds, their growth and their destruction. Unfortunately, seaweed study is a recent field, without the centuries of experience agriculture has provided terrestrial botany. Practical experience has resulted from small scale, labor-intensive cultivation by oriental farmers. Such information is of little use in designing the large algal farms envisioned for capturing solar energy. These farms must have an energy output greater than the energy needed to build and maintain them. The diffuse nature of sunlight will make this difficult.

Jackson and North (1973) reviewed seaweed properties important to their culture. W. J. North and J. Ryther, their collaborators and others have since added to the biological knowledge. I will use their results to develop some properties of large mariculture systems.

There are two seaweed systems proposed for energy harvest. The first envisions giant kelp *Macrocystis pyrifera* growing attached to large rafts far from land. Nutrients would be supplied to these beds by pumping nutrient-rich water from oceanic depths. Because this cold water is denser than surface water, such a farm would probably have a physical barrier to contain the water. The second system would grow the smaller red algae on land in seawater-containing channels. One nutrient source would be sewage effluent.

NUTRIENTS

Plants need more than photons to grow. To make tissue they need carbon, nitrogen, phosphorus, other elements and vitamins. Nutrient content and growth rates of seaweeds vary with the stage of the seaweed and nutrient concentrations of the water (e.g.,

Topinka and Robbins 1976, DeBoer, Lapointe and D'Elia 1976).
Algal C:N ratios range from 10:1 to 40:1 in *Macrocystis pyrifera*
and from 5:1 to 20:1 in the red alga *Neoagardhiella baileyi*.

Algae take up most of their nitrogen in the form of nitrate
or ammonia. Nitrogen concentrations in most surface waters are
low enough to limit seaweed production for at least part of the
year (Dawes et al. 1974, Topinka and Robbins 1976, Jackson 1975).
Nutrients will be supplied to seaweed culture facilities by adding sewage effluent (for the land system) or by pumping high
nutrient water to the surface. Seawater at 1000 m has about
30 µM-NO_3.

The various CO_2 forms in solution (CO_2(aq), HCO_3^-, CO_3^{2-}) form
the carbon pool used by algae. Seawater concentration of this
inorganic carbon is about 2 mM. Algal uptake for photosynthesis
is equivalent to removal of CO_2(aq) and therefore raises solution
pH (Fig. 1). Shacklock et al. (1973) investigated the pH effect

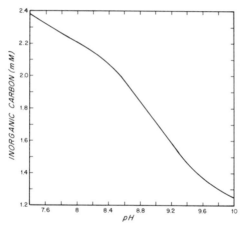

Fig. 1.   Inorganic carbon in seawater as a function of pH.   Car-
bonate buffering only is assumed.   Alkalinity = 2.3 meq. liter⁻¹,
pK_1 = 6.08, pK_2 = 9.28.

on *Chondrus crispus* growth. *Chondrus* growth rates in seawater
vats regulated to pHs of 7.3 and 7.8 were greater than in a vat
regulated at pH 6.7 and in an unregulated vat which reached pH
9.4. For seawater of pH 8 and alkalinity 2.3 meq liter⁻¹, with-
drawal of 700 µM CO_2(aq) for photosynthesis would raise pH to 9.4
(Fig. 1). Plants with a C:N ratio of 40:1 would also lower the
nitrate concentration by 18 µM. This is less than the nitrate in
the 1000 m water. The implication is that more water would have
to be supplied per unit of photosynthesis if pH control is important than if only nitrate supply is. The deleterious pH effect
has not been closely examined for the seaweeds proposed for energy

harvest. The critical pH might be higher or lower than 9.4.

TRACE METALS

Algal growth can also be harmed by excessive concentrations
of trace metals. Huntsman (1974) investigated diatoms grown in
deep ocean water. This water had the nitrogen, phosphorus and
silicate needed for algal growth but needed to have chelators
added before it would support growth. Jackson and Morgan (1976)
analyzed their experimental results to show that the factor controlling growth was the free copper ion concentration. North
(1976) has tested the sensitivity of *Macrocystis* gametophytes to
unchelated deep water.

Most surface waters seem to be detoxified. Presumably, organic chelators excreted by organisms lower free metal ion concentrations below deleterious levels. Barber and Ryther (1965)
has found that upwelling areas, where deep oceanic waters come to
the surface, need chelators added before they will support phytoplankton growth. This is a potential problem for seaweeds grown
on artificially upwelled water. However, many trace metals
adsorb onto kelp tissue (Cowen et al. 1976). The high seaweed
concentrations may be able to lower trace metal concentrations in
a kelp farm sufficiently to detoxify the deep water.

TRANSPORT

Algae differ from terrestrial plants by taking all of their
nutrients from solution through photosynthesizing tissue. The
analogous process on land is uptake of CO_2 from the atmosphere
through the leaves. Terrestrial photosynthetic rates are limited
by CO_2 concentrations and transport rates. One would expect that
a similar transport limitation could be important for algae in
water, where diffusion rates are 10^{-5} those in air. Such transport limitation could result from a slow nutrient supply rate or
from a build up of metabolic waste products. Munk and Riley
(1952) calculated that transport rates to a hypothetical seaweed
surface varied as v^2 (where v is the velocity). Wheeler (1976)
measured oxygen production by blades of giant kelp, *Macrocystis
pyrifera*, in a water tunnel. Experimental results showed that
oxygen production varied linearly with water velocity. My extrapolation of his results to $v = 10$ cm/sec is:

$$P = P_o (1 + 0.5\ v)$$

where
v = velocity, cm s^{-1}
P_o = oxygen evolution at $v=0$.

This can be used to estimate organic matter produced if oxygen
and organic production are linearly related and if maximum organic

production occurs at 10 cm s^{-1}. Assuming a maximum net production of 30 g-dwt m^{-2} d^{-1} which was measured in a well mixed system (DeBoer and Lapointe 1976); and an energy content of 3 kcal g^{-1}dwt = 3.8 x 10^5 joule g^{-1} (Paine and Vadas 1969), net seaweed production is given by:

$$P = 6.3 \times 10^4 \ (1 + 0.5 \ \nu) \ J \ m^{-2} \ d^{-1} \qquad (2)$$

This result is not easily applied to aquatic environments because there are few situations where water flow is uniform. Presence of seaweeds where waves and currents meet the bottom puts them in high turbulence environments. Algal-turbulence relationships have not been analyzed but seaweeds have been frequently described as lusher in more turbulent areas (e.g., Conover 1968).

In the simple environment of a seaweed culture facility eq. 2 should describe one facet of seaweed growth. For example, seaweeds grown in a channel, where a sloping bottom keeps water flowing, should depend on water flow rates. The energy expended to keep water flowing can be calculated. I envision a flat-bottomed channel wide enough so that the sides offer negligible flow resistance. The slope, S, necessary to maintain a given water velocity can be found by using Manning's formula (Chow 1959) and a coefficient of roughness equal to 0.2 (p. 182, Chow 1959):

$$S = 3.7 \times 10^{-6} \ d^{-2/3} \ \nu^2 \qquad (3)$$

where

d = depth, m
ν = water velocity, cm s^{-1}.

The mass flux through a plane 1 m wide perpendicular to the flow is $\nu d \rho$. Energy lost by water flowing 1 m downstream through a 1 m wide cross section is:

$$
\begin{aligned}
L &= \text{mass flux x height difference} \ *g \qquad (4)\\
&= \nu d \rho S g \\
&= 3 \times 10^1 \ \nu^3 \ d^{\ 1/3}
\end{aligned}
$$

where

L = energy loss/m^2, J m^{-2} d^{-1}.

The cost of pumping water around will be higher than the loss to the plants because of pumping and energy conversion inefficiencies. If the pump is 30% efficient and the motor diving pump uses electricity at 30% efficiency, then pumping is only 10% efficient in converting electricity to water energy. Because this is an energy farm, the energy to drive the pump must come from seaweed, with an assumed fermentation efficiency of 50% and an electrical generation efficiency of 20%. Thus the number of seaweed calories needed to drive the system is 100 times the number dissipated in the stream:

$$\text{Seaweed energy cost} = 3 \times 10^3 \, v^3 \, d^{1/3}, \, J \, m^{-2} \, d^{-1}. \quad (5)$$

Thus the yield for such a system depends on the water velocity:

$$
\begin{aligned}
\text{Yield} &= \text{Production} - \text{cost} \\
&= 6.3 \times 10^4 \, (1 + 0.5 \, v) - 3 \times 10^3 \, v^3 \, d^{1/3} \quad (6) \\
&\quad J \, m^{-2} \, d^{-1}.
\end{aligned}
$$

When the depth is 1 m, maximum yield is $10^5 \, J \, m^{-2} \, d^{-1}$ at a water velocity of 1.9 cm s^{-1} (Fig. 2). This is 27% of the maximum

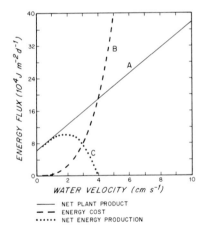

Fig. 2.   Energy fluxes as a function of water flow.
   A.   Net production by plants.
   B.   Energy costs of pumping water.
   C.   Net energy production, A-B.

growth. Thus, yield from channel cultivation is one quarter what maximum growth rates would indicate. It is important to know the effect of water motion on the species of interest before the value of a farm can be calculated.

The role water motion will play in a kelp farm is less clear. Energy in currents or waves could subsidize photosynthetic production. Because kelp beds dampen waves travelling through them, it may be necessary to limit bed sizes. However, the problem of keeping high nutrient, high density water near the surface may result in the water flowing in channels near the surface. Mixing would be provided by water flow, then the energy cost of pumping water will become the significant factor.

SELF-SHADING

Growth of seaweeds in cultures have reached as high as 20% d^{-1} (DeBoer and Lapointe 1976, North 1976). Growth in natural stands is rarely that fast. Even if there are requisite amounts of nutrients and turbulence, shading of one plant by another reduces the light available for growth. As a result, relative growth of seaweed beds is much less: *Macrocystis pyrifera* replaces itself 2-3 times annually off southern California (North 1971) and six times annually off central California (V. Gerard, personal communication). These are equivalent to growth rates of 0.6% d^{-1} and 1.1% d^{-1}. Low growth rates off southern California are also caused by low nutrient conditions during summer (Jackson 1975).

The important parameter in energy fixation is not the growth per plant but the growth per unit surface area. This is the net production, the difference between energy photosynthetically fixed and energy metabolically respired. Yield, the amount actually harvested, is net production less any other losses. Net production increases initially as photosynthesizing biomass increases. At some density, additional photosynthesis by added plants is balanced by additional respiration. Further increases in plant biomass increase respiration and decrease net production. DeBoer and Lapointe (1976) have investigated the effect of plant density on net photosynthesis with the red alga *Neoagardhiella baileyi* (Fig. 3). I have used their results to develop a simple

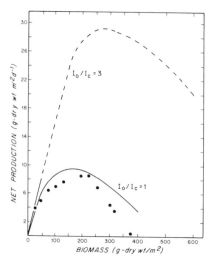

Fig. 3. Net production as a function plant density. Points represent measured production at 136 ly d^{-1} (DeBoer and Lapointe

1976). The two curves represent model predictions for light intensities of 141 and 423 ly d^{-1}.

growth model.

Algal photosynthetic rates are linearly related to light intensity at low light levels and constant at higher, saturating intensities. Photosynthesis in some phytoplankton is inhibited at very high intensities (Yentsch 1974) although Clendenning (1971) found that *Macrocystis pyrifera* is not. I have assumed that the photosynthetic rate is proportional to light intensity, I, below a critical intensity, I_c, and constant above that. For a given amount of tissue, b,

$$\text{Photosynthetic rate} = \gamma b \qquad \text{if } I_c < I \qquad (7)$$
$$= \gamma \frac{I}{I_c} b \text{ if } I_c \geq I$$

When light is absorbed only by uniformly distributed plant tissue,

$$I = I_0 e^{-kx} = I_0 e^{-\beta Bx/l} \qquad (8)$$

where

I_0 = surface light intensity
x = depth below the surface
l = depth of the bottom
B = biomass beneath the surface, g-dry wt m^{-2}
β = absorption constant, $m^2 g^{-1}$.

Lastly, respiration should be proportional to biomass:

$$R = \alpha B \qquad (9)$$

An expression for net photosynthesis, derived by combining eqs. 7 and 8 and then integrating, is:

$$N = \frac{\gamma}{\beta} \frac{I_0}{I_c} (1-e^{-\beta B}) - \alpha B \qquad \text{if } I_0 \leq I_c \qquad (10a)$$

$$= (\gamma-\alpha)B \qquad \text{if } I_0 > I_c \text{ and } B \leq \frac{1}{\beta} \ln(\frac{I_0}{I_c}) \qquad (10b)$$

$$= \frac{\gamma}{\beta} (\ln(\frac{I_0}{I_c}) + 1-\frac{I_0}{I_c} e^{-\beta B}) - \alpha B \qquad \text{if } I_0 > I_c \text{ and } B > \frac{1}{\beta} \ln(\frac{I_0}{I_c}) \qquad (10c)$$

where

N = net production, g-dry wt. $m^{-2} d^{-1}$.

Eq. 10a describes the low light case, where no plant is saturated; eq. 10b photosynthesis when all plants are light saturated; and

eq 10c describes the case where there are both saturated and un-
saturated plants. Assuming $\gamma = 0.20$ d^{-1} (about the highest
growth rate found), I used the net production curve of DeBoer and
Lapointe (1976) to derive the following parameter values: $\alpha =$
0.04 d^{-1}, $I_c = 141$ ly d^{-1}, $\beta = 0.01$ m^2 g^{-1}-dry wt (Fig. 3).
    This model can be used to predict the maximum amount of bio-
mass production in a given time period. More than 50% of the
potential algal growth at high and low light intensities takes
place within 30 days, 80% in 60 days (Figs. 4, 5). Harvesting of

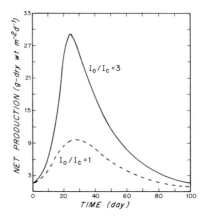

Fig. 4. Daily growth increment through time of plant with initial
biomass of 10 g-dry wt. Daily growth was modelled with eq. 10
Note that maximum photosynthesis occurs for a short period. Maximum production occurred when biomass was 280 g at high and 160 g
at low light intensities.

the alga grown at high light intensity every 30 days would yield
40% more than harvesting every 60 days, but would demand an expenditure of twice as much energy and other resources. Thus the
energy cost of harvesting is a crucial variable in determining
net energy output.
    This growth model is simplistic. It does not include sea-
sonal effects on photosynthesis, changes in light absorption or
non-respiratory losses. It may not be accurate for giant kelp.
However, it does incorporate the self-limiting aspect of algal
growth to show that short-term growth measurements cannot be
readily extrapolated to annual yields. Energy, capital and man-
power costs associated with seaweed harvesting must be considered
before the actual return can be calculated.

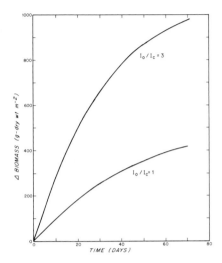

Fig. 5. *Maximum growth possible within a given time period.*
This was calculated from Fig. 4. Maximum growth is 1100 g at high
light intensity and 500 at low intensity. Note that half the
maximum yield is achieved in 23 days at high and 30 days at low
intensity.

OTHER LOSSES

   These processes--turbulence, self shading and nutrient concentration affect the net production rate. The amount of organic
matter actually harvested, the yield, will be less. Plant matter
will be lost to diseases, grazers and storms. For example,
Macrocystis pyrifera, beds are harvested off California. Yields
are approximately 0.2 kg-dry wt. $m^{-2} yr^{-1}$ while net production is
estimated to be 2 kg-dry wt. $m^{-2} yr^{-1}$ (Jackson and North 1973).
Most of the 90% net production not harvested is lost when fronds
decay or are ripped out in storms. Crude estimates suggest that
only about 10% of net production is consumed by herbivores. How
much yields can be improved in a managed seaweed farm is still
unknown.
   Diseases have caused significant losses to *Porphyra, Laminaria*
and *Macrocystis* (Andrews 1976). The seagrass *Zostera marina* was
virtually eliminated on the east coast of the United States by
the pathogen *Labrinthula.* Will selection of seaweeds for high
yielding strains result in increased disease susceptibility? It
has in terrestrial crops.
   The role of grazers in marine ecosystems is poorly understood.
Under normal circumstances they seem to consume little of the
algal production. Nicotri (1976) found that crustaceans consumed an insignificant fraction of the net production in a small
mariculture facility. However, sea urchins have been able to
decimate beds of *Macrocystis* (Leighton 1971) and *Laminaria* (Lang

and Mann 1976). These are believed to result from declines in
sea urchin predators. As a result, man has had to become the
predator controlling urchin numbers. Will man have to control
other herbivores in a mid-ocean farm? We cannot know until we
know what the herbivore controls in the nearshore area.

CONCLUSION

There will be no miraculous energy yields in seaweed farms.
Energy and capital will be devoted to propagating, nourishing,
protecting and harvesting seaweeds. Actual yields will be determined by the costs of such details as pumping water, harvesting tissue or controlling grazers. Modern agriculture is an
energy intensive process (Heichel 1976); aquaculture will be also.
It is attention to the details of mariculture that will determine
the success, not any 25% increases in photosynthetic breakthroughs.
Marine farm designs must be carefully optimized if they are to
produce energy rather than consume it. Further research on seaweed biology will be needed to define the details. Careful
engineering will be needed to define the systems. Only when
these are done, will we know if we can harvest energy from the
sea.

LITERATURE CITED

Andrews, J. H.   1976.   The pathology of marine algae. *Biol. Rev.*
        51: 211-253.
Barber, R. T., and J. H. Ryther.   1965.   Organic chelators: fac-
        tors affecting primary production in the Cromwell current
        upwelling. *J. exp. mar. Biol. Ecol. 3:* 191-199.
Chow, V. T.   1959.   Open channel hydraulics.   McGraw Hill.
Clendenning, K. A. 1917.   Photosynthesis and general development
        in *Macrocystis,* p. 169-190. *In* W. J. North (ed.), The
        Biology of giant kelp beds (*Macrocystis*) in California.
        Cramer.
Conover, J. T.   1968.   The importance of natural diffusion gra-
        dients and transport of substances related to benthic marine
        plant metabolism. *Bot. Mar. 11:* 1-9.
Cowen, J. P., V. F. Hodge and T. R. Folsom.   1976. *In vivo* accu-
        mulation of radioactive polonium by the giant kelp *Macro-
        cystis pyrifera. Mar. Biol. 37:* 239-248.
Dawes, C. J., J. M. Lawrence, D. P. Cheney and A. C. Mathieson.
        1974.   Ecological studies of Floridian *Eucheuma* (Rhodophyta,
        Gigartinales), III.   Seasonal variation of carrageenan,
        total carbohydrate, protein and lipid. *Bull. Mar. Sci.
        24:* 286-299.
DeBoer, J. A., and B. E. Lapointe.   1976.   Effects of culture
        density and temperature on the growth and yield of *Neoagard-*

hiella baileyi. Unpublished manuscript.

DeBoer, J. A., B. E. Lapointe and C. F. D'Elia. 1976. Effects of nitrogen concentration on growth rate and carrageenan production in *Neoagardhiella baileyi.* Unpublished manuscript.

Heichel, G. H. 1976. Agricultural production and energy resources. *Amer. Scient. 64:* 64-72.

Huntsman, S. 1974. Unpublished results.

Jackson, G. A. 1975. Nutrients and productivity of the giant kelp, *Macrocystis pyrifera,* in the nearshore. Ph.D. thesis, Calif. Inst. Technol.

Jackson, G. A., and J. J. Morgan. 1976. Trace metal-chelator interactions and phytoplankton growth in seawater media: theoretical analysis and comparison with reported observations. Limnol. Oceanogr. In press.

Jackson, G. A., and W. J. North. 1973. Concerning the selection of seaweeds suitable for mass cultivation in a number of larger open-ocean, solar energy facilities ("marine farms") in order to provide a source of organic matter for conversion to feed, synthetic fuels and electrical energy. Final rept. to U.S. Naval Weapons Center, China Lake. Keck Laboratory, Calif. Inst. Technol., Pasadena.

Lang, C. and K. H. Mann. 1976. Changes in sea urchin populations after the destruction of kelp beds. *Mar. Biol. 36:* 321-326.

Leighton, D. L. 1971. Grazing activities of benthic invertebrates in kelp beds, p. 421-453. *In* W. J. North (ed.), The biology of giant kelp beds (*Macrocystis*) in California. Cramer.

Munk, W. H., and G. A. Riley. 1952. Absorption of nutrients by aquatic plants. *J. Mar. Res. 11:* 215-240.

Nicotri, M. E. 1976. The impact of crustacean herbivores on cultured seaweed populations. Unpublished manuscript.

North, W. J. 1971. Introduction and background, p. 1-97. *In* W. J. North (ed.), The biology of giant kelp beds (*Macrocystis*) in California. Cramer.

North, W. J. 1976. Evaluating oceanic farming of seaweeds as sources of organics and energy. Ann. Prog. Rept. for 1975. Keck Lab., Calif. Inst. Technol.

Paine, R. T., and R. L. Vadas. 1969. Calorific values of benthic marine algae and their postulated relation to invertebrate food preference. *Mar. Biol. 4:* 79-86.

Shacklock, P. F., D. Robson, I. Forsyth and A. C. Neish. 1973. Further experiments on the vegetative propagation of *Chondrus crispus* T4. T.R. 18, Atl. Reg. Lab., Nat. Res. Counc. Can.

Topinka, J. A., and J. V. Robbins. 1976. Effects of nitrate and ammonium enrichment on growth and nitrogen physiology in *Fucus spiralis. Limnol. Oceanogr. 21:* 659-664.

Wheeler, W. N. 1976. Transport limitation of photosynthesis in

Macrocystis. Unpublished manuscript.
Yentsch, C. S. 1974. Some aspects of the environmental physio-
   logy of marine phytoplankton: a second look. *Oceanogr.*
   Mar. Biol. Ann. Rev. 12: 41-75.

BIOENGINEERING APPROACHES AND CONSTRAINTS

JACK MYERS

I have to presume that I am here under a grandfather clause.
I have concerned myself in the past with some problems that were
essentially a bioengineering of photosynthesis. In that sense
I am a grandfather. However, I am not now so engaged and I have
no new and hot ideas to contribute to the problem. Hence I serve
you best if I look at our problem broadly.

Though I shall speak of bioengineering, I do so with some
hesitancy. First, I am not an engineer. Secondly, American
bioengineering is not notably distinguished by its
accomplishments whereas in Japan Bioengineering is a recognized
discipline and highly productive. It is likely that my Japanese
friends could treat the subject better than I.

Bioengineering strategies. For biosolar conversions we have
available the basic photochemical process of photosynthesis. It
is useful to consider levels of strategy at which bioengineering
may be pursued. First level: we select a particular plant
material, chosen for production of a desired output, and then
search for ways to maximize production and minimize capital and
operational costs. At this first level we are using machinery
developed by long-time evolutionary processes which respond to
competitive selection in nature. This is the initial approach
of agriculture and was applied also in attempts toward mass
culture of algae. Second level: we attempt improvement in
selected plants, as by genetic manipulation, to further maximize
the desired product and minimize management costs. This has
given us great improvements in agricultural production and also
has been used successfully with microorganisms in the
fermentation industry. Third level: we dissect and understand
the *in vivo* machinery to the point at which we can reconstruct
it for *in vitro* use. The goal is then a biochemical improvement
as viewed by efficiency in making the desired product. A
practical achievement in this direction is the use of immobilized
enzymes as catalysts for desired reactions. Although the third
level is clearly an ultimate goal, it has not yet been achieved
for any very complex biological process. All of us who have
worked on photosynthesis have also looked forward to the day
when our knowledge of the process will allow *in vitro*
reconstruction. I think we still have some way to go. We deal
with machinery organized in membrane structures, likely dependent
on some degree of membrane fluidity. We do not yet know all the
membrane components and only a little about their topology.
Probably we should begin efforts toward reconstruction even now.
We should be learing how to go about it. But we cannot expect

449

early success.

The major constraint on solar energy conversion. Sunlight
provides us an option with two features so attractive that we
are compelled to explore all its possibilities as an energy
source. For all practical purposes it is an inexhaustible
source; and, if viewed in analogy to a raw fuel, its initial cost
is zero. All possible methods for solar energy conversion are
faced also by a serious constraint: a low energy flux.

Energy sources as viewed in terms of energy flux. As
pointed out by Kapitza (2), a critical parameter for practical
energy conversion is the energy flux, W (power/area). For
some analyses W can be broken down into an energy density (U) and
a velocity of propagation (v) such that $W = vU$. For present
purposes, W provides sufficient information. Power (P) output
is simply the product of W and area (A): $P = W \cdot A$. The only reason
to write this out is to focus attention on dependence of Area on
the attainable W of the system.

In Kapitza's analysis A is taken, for example, as the
actual area between rotor and stator of an electric generator.
However, if we are to think about total energy producing systems,
then two additional functions need to be added. In Fig. 1 I
label these Collection and Efficiency. Conventional fuel sources
need to be collected. In evaluating economics we need to know

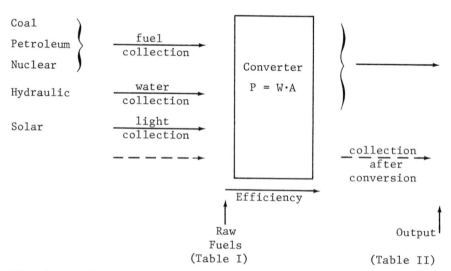

Fig. 1. Basis for comparison of energy producing systems.

efficiency and cost of input energy at the conversion site.   Then
Area is governed only by the energy converter.

It is instructive to consider the hydraulic (or
hydroelectric) conversion.   The initial W energy flux, rainfall
per square meter at higher elevations, is small.   However,
collection costs are zero to the point of a dammed reservoir.
Hence A becomes the Area of the reservoir and W is dependent
on its depth.

The four conventional cases have merit only in providing
analogs for thinking about solar energy.   Here there is a very
low W energy flux and attendant large Area requirements.   In
some conversion systems we can envision a light collection and
increase in W by optical means.   However, if we take the input
energy cost as zero, then A necessarily remains the original
surface area and all costs of collection must be included.

For photosynthetic conversion even the low W energy flux
may already be too high for maximum efficiency.   Green plants
have developed a machinery which is constructed and operated
by a low W.   For biosolar conversion we usually have not thought
of initial light collection.   Instead we necessarily are faced
with a low energy density in the product and face costs of
collection *after* energy conversion.   This is the procedure in
conventional agriculture even though value of the product
usually is not reckoned in energy units.

Comparisons with other energy conversions.   Although the low
value of W for solar energy flux is well known to all of us,
I have chosen to consider it explicitly.   My intent is to go
further and make some actual cost analyses.   Although these are
so simple and obvious that they must have been done by others,
I have not been able to find them reported.

Table I presents some simple calculations which seek to
evaluate solar energy as a *raw* fuel (no consideration of
efficiency) in current monetary terms.   Even in high insolation
areas (as the southwest United States) the value of sunlight as
raw energy is somewhere between 1 ¢ and 10 ¢ per square meter
per day.

It is more instructive but also more difficult to evaluate
solar energy in terms of useable output value.   For this one
needs an assumption about efficiency and data on cost breakdowns
for conventional energy producing systems.   Table II is designed
to ask an inverted question: to make a solar energy system
competitive with conventional energy systems, how much can we
afford to spend for construction and maintenance.   The answer
is simply that 2 ¢/KW-hr must cover construction and operation

Table I   *Present Value of Solar Energy as a Raw Fuel*

1. Take (for high insolation areas)

   260 W m^{-2} (1) $= 6.2 \text{ KW-hr m}^{-2} \text{ day}^{-1}$ (3)

   $= 21 \times 10^3 \text{ BTU m}^{-2} \text{ day}^{-1}$

   $= 5300 \text{ kcal m}^{-2} \text{ day}^{-1}$

2. Take coal @ \$9/ton and 17×10^6 BTU/ton (4)

   Raw energy cost = \$ $0.5/10^6$ BTU

3. Take Petroleum @ \$11/barrel and 5.8×10^6 BTU/barrel

   Raw energy cost = \$2/10^6 BTU

4. Value of Raw Solar Energy

   Compared to Coal: \$ $0.01 \text{ m}^{-2} \text{ day}^{-1}$

   Compared to Petroleum: \$ $0.04 \text{ m}^{-2} \text{ day}^{-1}$

Table II. *Solar vs Conventional Energy on Output Basis*

Costs	c/KW-hr			
	Petroleum[a]	Coal[a]	Nuclear[a]	Solar
Capital	0.4	0.8	1.0	2.0[c]
Operation	0.1	0.2	0.1	
Fuel	1.7	0.4	0.3	0
Total, production	2.2	1.4	1.4	2.0[b]
Efficiency, %	40	40	30	?

a Based on ref. (5); assumes \$11/barrel oil, \$9/ton coal
b Chosen arbitarily to be almost a curently competitive energy
  source
c Translation of this number is the designed end use of the Table.
  If efficiency is projected to be 16% and input flux is 6 KW-hr
  $\text{m}^{-2} \text{ day}^{-1}$, then output will be 1 KW-hr $\text{m}^{-2} \text{ day}^{-1}$. Then Capital
  and Operation costs must be not more than 2 ¢ $\text{m}^{-2} \text{ day}^{-1}$ or
  \$7.3 $\text{m}^{-2} \text{ year}^{-1}$.

expense. Translation to an Area basis requires knowledge of
input flux and efficiency. As shown by footnote c, even choice
of a relatively high input flux and unreasonably high projected
efficiency (16%) leaves only $7.3 per square meter per year for
construction and maintenance. I take this to be a sobering
figure, especially since it must include necessary costs of land
and costs of collection of a dilute product.

A number of objections may be raised to my treatment of the
W·A problem. (a) I have resorted to cost estimates which do not
have high precision. I doubt that more precise estimates will
substantially change the result. (b) I have resorted to *current*
costs for conventional systems, costs which necessarily will
escalate. Expendable fuel costs will increase because collection
from lower grade natural deposits will become more expensive.
However, you will note that fuel costs for coal and uranium are
not now major fractions of output costs; hence their escalation
will not proportionately increase output costs. (c) I have
compared solar energy to systems which achieve best efficiency
in large central installations. To this I find a real objection
which deserves further thought.

In developing ideas about energy conversion I think we have
fallen into the trap of bigness. Because conventional sources
favor large installations, must all future sources fit the same
mold? In looking at the W·A problem I do not see how solar energy
conversion is favored by large installations. As noted in Fig. 1,
there is a necessary collection problem whatever mode of
conversion is chosen. I find it difficult to see how collection
from larger and larger areas increases efficiency.

One further objection of a different kind may be made to my
treatment of Table II. Conventional sources are treated in terms
of output electrical energy. For a biosolar system, output
presumably is in the form of stored chemical energy (a fuel) and
subject to a further conversion efficiency if converted to
electrical energy.

In considering the consequences of what I have set down,
I have had to wonder how agriculture can be economically
successful. Of course the answer is that agricultural products
have value measured, not by energy content *per se*, but by
utility as a food or even more costly organic material. We live
in a world foreseeably almost as short on food as it will be on
energy. Photosynthesis accomplishes, not only a storage of
chemical energy, but also a concentration of carbon and a
reductive synthesis of organic compounds, even with the
biologically required stereoisomers. If we do engineer an
economically attractive facility for biosolar energy conversion,
and also have a choice of product, I wonder whether we will

choose a product serving best as a fuel or a product serving best as a food.

Having set this down, I feel as one who has carried a bomb into a peaceful gathering. I hope to be shown that I do not have a bomb at all, or maybe only a small one. At any rate my intent is entirely peaceful. There are constraints which we need to recognize. They are constraints which need to be kept in mind in choosing between the several levels of bioengineering strategies. It is not my intent to demonstrate that biosolar conversion is impossible; only that it is difficult. And that is something we all knew to begin with.

REFERENCES

1. Calvin, M. *Science 184*: 375-381 (1974)

2. Kapitza, P. *New Scientist 72* (No. 1021): 10-12 (1976)

3. Kok, B. *Proc. Workshop on Bio-Solar Conversion* (NSF-RANN) pp. 22-30 (1973)

4. Osborn, E. F. *Science 184*: 477-481 (1974)

5. Rose, D. J. *Science 184*: 351-359 (1974)